Dual Specificity Phosphatases

Dual Specificity Phosphatases

From Molecular Mechanisms to Biological Function

Special Issue Editors

Rafael Pulido
Roland Lang

MDPI • Basel • Beijing • Wuhan • Barcelona • Belgrade

MDPI

Special Issue Editors

Rafael Pulido
Biocruces Health Research Institute
Spain

Roland Lang
University Hospital Erlangen
Germany

Editorial Office
MDPI
St. Alban-Anlage 66
4052 Basel, Switzerland

This is a reprint of articles from the Special Issue published online in the open access journal *International Journal of Molecular Sciences* (ISSN 1422-0067) from 2018 to 2019 (available at: https://www.mdpi.com/journal/ijms/special_issues/DUSPs).

For citation purposes, cite each article independently as indicated on the article page online and as indicated below:

LastName, A.A.; LastName, B.B.; LastName, C.C. Article Title. *Journal Name* **Year**, *Article Number*, Page Range.

ISBN 978-3-03921-688-8 (Pbk)
ISBN 978-3-03921-689-5 (PDF)

Contents

About the Special Issue Editors

Rafael Pulido is an Ikerbasque Research Professor at Biocruces Bizkaia Health Research Institute (Barakaldo, Bilbao, Spain). His laboratory investigates the role of protein tyrosine phosphatases in human disease, and studies the involvement of phosphatases and kinases in human cancer and cancer-associated hereditary syndromes, with a focus on the PTEN tumor suppressor. His professional interest is the identification, characterization, and validation of new molecular biomarkers and targets for anti-cancer precision therapies.

Roland Lang is an MD specialized in Medical Microbiology and an Associate Professor of Innate Immunity at the Institute of Clinical Microbiology, Immunology and Hygiene at the University Hospital Erlangen, Germany. He studied medicine in Regensburg and at the TU Munich, where he was trained in Medical Microbiology. From 1999–2002 he was a postdoctoral fellow at St. Jude Children's Research Hospital in Memphis, TN. From 2002–2008 he was an established junior investigator at the Institute of Medical Microbiology at TUM. In 2008 he was recruited to Erlangen. His main research interests are innate immune receptors and pathways of pathogen recognition, how innate immune activation is regulated by cytokines and modulators of intracellular signaling, and the implications of these interactions on inflammation and infection, and during immunization.

International Journal of
Molecular Sciences

MDPI

Editorial

Dual Specificity Phosphatases: From Molecular Mechanisms to Biological Function

Rafael Pulido [1,2,*] and Roland Lang [3,*]

1   Biomarkers in Cancer Unit, Biocruces Bizkaia Health Research Institute, 48903 Barakaldo, Spain
2   IKERBASQUE, Basque Foundation for Science, 48011 Bilbao, Spain
3   Institute of Clinical Microbiology, Immunology and Hygiene, Universitätsklinikum Erlangen,
     Friedrich-Alexander-Universität Erlangen-Nürnberg, 91054 Erlangen, Germany
*   Correspondence: rpulidomurillo@gmail.com or rafael.pulidomurillo@osakidetza.eus (R.P.);
     roland.lang@uk-erlangen.de (R.L.)

Received: 3 September 2019; Accepted: 4 September 2019; Published: 6 September 2019

Dual specificity phosphatases (DUSPs) constitute a heterogeneous group of enzymes, relevant in human disease, which belong to the class I Cys-based group of protein tyrosine phosphatase (PTP) gene superfamily [1–4]. DUSPs possess the capability to dephosphorylate Ser/Thr and Tyr residues from proteins as well as to remove phosphates from other non-proteinaceous substrates, including signaling lipids [5]. Catalytically inactive pseudophosphatase DUSPs also exist which regulate phosphorylation-related cell signaling [6]. DUSPs include, among others, mitogen-activated protein kinase (MAPK) phosphatases (MKPs) and small-size atypical DUSPs. These proteins are non-transmembrane enzymes displaying variable substrate specificity and harboring a single PTP catalytic domain with a HCXXGXXR consensus catalytic motif [7,8]. MKPs are enzymes specialized in regulating the catalytic activity and subcellular location of MAPKs, whereas the functions of small-size atypical DUSPs are more diversified. DUSPs have emerged as key players in the regulation of cell growth, differentiation, stress responses and apoptosis. In physiology, DUSPs regulate essential processes, including immunity, neurobiology and metabolic homeostasis, and have been involved in tumorigenesis, pathological inflammation and metabolic disorders [9–11]. Accordingly, alterations in the expression or function of MKPs and small-size atypical DUSPs have important consequences in human disease, making these enzymes potential biological markers and therapeutic targets. Although major biochemical, structural, functional and physiological properties of many DUSPs are currently known, their developmental stage- and tissue-specific involvement in different human pathologies is only starting to be disclosed. This Special Issue provides original research and review articles focused on the involvement of specific MKPs and small-size atypical DUSPs in human disease, including relevant information on the use of different biological models to study the regulation and physiological functions of DUSP enzymes.

MKPs are also present in fungi, where they are major regulators of the different fungal MAPK adaptive pathways [12]. An update of the repertoire of MKPs in pathogenic and non-pathogenic fungi is presented, together with their functional effects on MAPK signaling in fungi and insights into their regulatory mechanisms of expression and function. The budding yeast *Saccharomyces cerevisiae* (with two active MKPs, Msg5 and Sdp1) and the fission yeast *Schyzosaccharomyces pombe* (with one active MKP, Pmp1) emerge as suitable models to study MKP regulation and function, with potential translation to other eukaryotic organisms [13].

The roles of MKPs and small-size atypical DUSPs in MAPK-dependent and -independent immune cell response have been reviewed [14,15]. In the context of airway epithelial signaling during viral infection in inflammatory airway processes, such as asthma and chronic obstructive pulmonary disease, several MKPs, including DUSP1, DUSP4 and DUSP10, arise as major negative regulators of inflammation due to their inhibitory effects on the major pro-inflammatory Jun N-terminal kinase (JNK), p38, and Extracellular signal-regulated kinase (ERK) MAPKs in the airway epithelium. Up-regulation or regulated activation of these MKPs constitute potential anti-inflammatory therapeutic approaches for inflammatory airway diseases [15]. Although the more frequently documented effects of MKPs in immunity involve direct dephosphorylation of MAPKs, the existence of non-MAPK MKP and small-size atypical DUSP substrates is also disclosed. The complexity and redundancy in DUSP signaling during immune cell response make dedicated studies and testing necessary, in appropriate biological models, for both the inhibition and activation of DUSPs as suitable therapeutic strategies for immune diseases [14]. In this regard, a comprehensive in silico study is presented analyzing the expression and functional interaction between DUSPs and protein kinases in hematopoietic cells, which unveils the interplay between DUSPs and novel non-MAPK protein kinases, including receptor tyrosine kinases (IGFR1, VEGF, FGF), AURKA, and LRRK2, among others [16]. In addition, the control of DUSPs protein stability by ubiquitination and phosphorylation has been reviewed as a major regulatory mechanism affecting most MKPs, whereas methylation-induced ubiquitination of DUSP14 is disclosed as a specific mechanism to activate the catalysis of this small-size atypical DUSP [17].

In a more specific context, the complex expression and regulation patterns of DUSP10, and its diverse functional roles in inflammation, immunity, and cancer, which could go beyond direct MAPK dephosphorylation, has been separately reviewed [18]. Li et al. presented their findings on the effects on gene expression and lipid metabolism of *Escherichia coli*-triggered sepsis, using a *Dusp1*$^{-/-}$ mouse model [19], whereas Neamatallah et al. described their findings on the macrophage gene expression profiles on *Dusp4*$^{-/-}$ mice [20]. Circulating tumor cells play a major role in tumor dissemination and metastasis, and Wu et al. reported the enrichment in nuclear localized DUSP6 in circulating tumor cells from triple negative breast cancers, as well as in brain metastases, suggesting a specific role for nuclear DUSP6 in cancer spreading [21]. Finally, Cao et al. provided new insights into the functions of the catalytically inactive MK-STYX pseudophosphatase as an indirect regulator of post-translational modifications from proteins regulating microtubule dynamics, including histone deacetylase isoform 6 and tubulin [22].

Neuroblastoma constitutes the most commonly diagnosed extracranial solid tumor in infants. The current knowledge on the involvement of MKPs and small-size atypical DUSPs in neuroblastoma cell growth and differentiation, in the context of Trks, STATs and ALK/RAS/MAPK signaling, has been covered. Highlights are made on the potential role of ERK1/2-specific DUSP5 and DUSP6 as neuroblastoma biomarkers, as well as on the potential of inhibition of other MKPs, such as DUSP1, DUSP8, DUSP10, or DUSP16 for therapy of neuroblastoma [23]. Further knowledge into the roles of MKPs in neuronal differentiation and nervous system development has also been reviewed by Perez-Sen et al., with emphasis in MKP-mediated neuroprotection upon genotoxic and ischemic neuron injury. Cell-specific signaling through neurotrophins, cannabinoids and nucleotides may have neuroprotective effects by up-regulation of specific MKPs. In addition, the dual anti- or pro-oncogenic role of DUSP1 and DUSP6 in glioblastoma, the most aggressive type of brain tumor, is discussed [24].

In summary, this Special Issue provides an updated overview on the complexity of DUSP biology at the physiological level, which is a prerequisite for the validation of DUSPs as useful biomarkers or drug-targetable proteins in human disease treatment. The direct functional relation between many of the DUSPs and MAPKs provides a high therapeutic potential for DUSP proteins, which is evident in the MKP DUSP subfamily. However, DUSPs redundancy, multiplicity in MAPK substrate specificity, and time-course and subcellular localization functional constraints, make it difficult to link unequivocally DUSPs expression and function with pathological manifestations. Well-defined biological models with precisely manipulated DUSP protein expression and function, as well as accurate molecular definition of functional DUSPs partners, with special emphasis on orphan small-size atypical DUSPs, will help in the future progress of the field.

We thank all the colleagues that contributed with their work and expertise to this Special Issue, and we hope that its content is of interest for clinicians and researchers aiming to explore and understand the role of DUSPs in human disease, as well as the potential benefits of their therapeutic manipulation.

Acknowledgments: Research on R.P. lab has been partially funded by BIO13/CI/001/BC from BIOEF (EITB maratoia), Basque Country, Spain; and SAF2016-79847-R from the Ministerio de Economía y Competitividad (Spain and Fondo Europeo de Desarrollo Regional). Research on DUSPs in the lab of R.L. has been supported by Deutsche Forschungsgemeinschaft (SFB 643) and by the Interdisziplinäres Zentrum für Klinische Forschung (IZKF) of the Medical Faculty at FAU.

Conflicts of Interest: The authors declare no conflicts of interest.

References

1. Alonso, A.; Pulido, R. The extended human PTPome: a growing tyrosine phosphatase family. *FEBS J.* **2016**, *283*, 1404–1429. [CrossRef] [PubMed]

2. Alonso, A.; Sasin, J.; Bottini, N.; Friedberg, I.; Friedberg, I.; Osterman, A.; Godzik, A.; Hunter, T.; Dixon, J.; Mustelin, T. Protein tyrosine phosphatases in the human genome. *Cell* **2004**, *117*, 699–711. [CrossRef] [PubMed]

3. Pulido, R.; Hooft van Huijsduijnen, R. Protein tyrosine phosphatases: Dual-specificity phosphatases in health and disease. *FEBS J.* **2008**, *275*, 848–866. [CrossRef] [PubMed]

4. Tonks, N.K. Protein tyrosine phosphatases: From genes, to function, to disease. *Nat. Rev. Mol. Cell Biol.* **2006**, *7*, 833–846. [CrossRef] [PubMed]

5. Pulido, R.; Stoker, A.W.; Hendriks, W.J. PTPs emerge as PIPs: Protein tyrosine phosphatases with lipid-phosphatase activities in human disease. *Hum. Mol. Genet.* **2013**, *22*, R66–R76. [CrossRef]

6. Hinton, S.D. The role of pseudophosphatases as signaling regulators. *Biochim. Biophys. Acta—Mol. Cell Res.* **2019**, *1866*, 167–174. [CrossRef] [PubMed]

7. Alonso, A.; Nunes-Xavier, C.E.; Bayon, Y.; Pulido, R. The Extended Family of Protein Tyrosine Phosphatases. *Methods Mol. Biol.* **2016**, *1447*, 1–23. [CrossRef] [PubMed]

8. Rios, P.; Nunes-Xavier, C.E.; Tabernero, L.; Kohn, M.; Pulido, R. Dual-specificity phosphatases as molecular targets for inhibition in human disease. *Antioxid. Redox Signal.* **2014**, *20*, 2251–2273. [CrossRef] [PubMed]

9. Kidger, A.M.; Keyse, S.M. The regulation of oncogenic Ras/ERK signalling by dual-specificity mitogen activated protein kinase phosphatases (MKPs). *Semin. Cell Dev. Biol.* **2016**, *50*, 125–132. [CrossRef]

10. Lang, R.; Hammer, M.; Mages, J. DUSP meet immunology: Dual specificity MAPK phosphatases in control of the inflammatory response. *J. Immunol.* **2006**, *177*, 7497–7504. [CrossRef]

11. Nunes-Xavier, C.; Roma-Mateo, C.; Rios, P.; Tarrega, C.; Cejudo-Marin, R.; Tabernero, L.; Pulido, R. Dual-specificity MAP kinase phosphatases as targets of cancer treatment. *Anti-Cancer Agents Med. Chem.* **2011**, *11*, 109–132. [CrossRef]

12. Martin, H.; Flandez, M.; Nombela, C.; Molina, M. Protein phosphatases in MAPK signalling: We keep learning from yeast. *Mol. Microbiol.* **2005**, *58*, 6–16. [CrossRef] [PubMed]

13. Gonzalez-Rubio, G.; Fernandez-Acero, T.; Martin, H.; Molina, M. Mitogen-Activated Protein Kinase Phosphatases (MKPs) in Fungal Signaling: Conservation, Function, and Regulation. *Int. J. Mol. Sci.* **2019**, *20*. [CrossRef] [PubMed]

14. Lang, R.; Raffi, F.A.M. Dual-Specificity Phosphatases in Immunity and Infection: An Update. *Int. J. Mol. Sci.* **2019**, *20*. [CrossRef] [PubMed]

15. Manley, G.C.A.; Parker, L.C.; Zhang, Y. Emerging Regulatory Roles of Dual-Specificity Phosphatases in Inflammatory Airway Disease. *Int. J. Mol. Sci.* **2019**, *20*. [CrossRef] [PubMed]

16. Subbannayya, Y.; Pinto, S.M.; Bosl, K.; Prasad, T.S.K.; Kandasamy, R.K. Dynamics of Dual Specificity Phosphatases and Their Interplay with Protein Kinases in Immune Signaling. *Int. J. Mol. Sci.* **2019**, *20*. [CrossRef] [PubMed]

17. Chen, H.F.; Chuang, H.C.; Tan, T.H. Regulation of Dual-Specificity Phosphatase (DUSP) Ubiquitination and Protein Stability. *Int. J. Mol. Sci.* **2019**, *20*. [CrossRef]

18. Jimenez-Martinez, M.; Stamatakis, K.; Fresno, M. The Dual-Specificity Phosphatase 10 (DUSP10): Its Role in Cancer, Inflammation, and Immunity. *Int. J. Mol. Sci.* **2019**, *20*. [CrossRef]

19. Li, J.; Wang, X.; Ackerman, W.E.T.; Batty, A.J.; Kirk, S.G.; White, W.M.; Wang, X.; Anastasakis, D.; Samavati, L.; Buhimschi, I.; et al. Dysregulation of Lipid Metabolism in Mkp-1 Deficient Mice during Gram-Negative Sepsis. *Int. J. Mol. Sci.* **2018**, *19*. [CrossRef]

20. Neamatallah, T.; Jabbar, S.; Tate, R.; Schroeder, J.; Shweash, M.; Alexander, J.; Plevin, R. Whole Genome Microarray Analysis of DUSP4-Deletion Reveals A Novel Role for MAP Kinase Phosphatase-2 (MKP-2) in Macrophage Gene Expression and Function. *Int. J. Mol. Sci.* **2019**, *20*. [CrossRef]

21. Wu, F.; McCuaig, R.D.; Sutton, C.R.; Tan, A.H.Y.; Jeelall, Y.; Bean, E.G.; Dai, J.; Prasanna, T.; Batham, J.; Malik, L.; et al. Nuclear-Biased DUSP6 Expression is Associated with Cancer Spreading Including Brain Metastasis in Triple-Negative Breast Cancer. *Int. J. Mol. Sci.* **2019**, *20*. [CrossRef] [PubMed]

22. Cao, Y.; Banks, D.A.; Mattei, A.M.; Riddick, A.T.; Reed, K.M.; Zhang, A.M.; Pickering, E.S.; Hinton, S.D. Pseudophosphatase MK-STYX Alters Histone Deacetylase 6 Cytoplasmic Localization, Decreases Its Phosphorylation, and Increases Detyrosination of Tubulin. *Int. J. Mol. Sci.* **2019**, *20*. [CrossRef] [PubMed]

23. Nunes-Xavier, C.E.; Zaldumbide, L.; Aurtenetxe, O.; Lopez-Almaraz, R.; Lopez, J.I.; Pulido, R. Dual-Specificity Phosphatases in Neuroblastoma Cell Growth and Differentiation. *Int. J. Mol. Sci.* **2019**, *20*. [CrossRef] [PubMed]

24. Perez-Sen, R.; Queipo, M.J.; Gil-Redondo, J.C.; Ortega, F.; Gomez-Villafuertes, R.; Miras-Portugal, M.T.; Delicado, E.G. Dual-Specificity Phosphatase Regulation in Neurons and Glial Cells. *Int. J. Mol. Sci.* **2019**, *20*. [CrossRef] [PubMed]

International Journal of
Molecular Sciences

MDPI

Review

Mitogen-Activated Protein Kinase Phosphatases (MKPs) in Fungal Signaling: Conservation, Function, and Regulation

Gema González-Rubio [†], Teresa Fernández-Acero [†], Humberto Martín * and María Molina *

Departamento de Microbiología y Parasitología. Facultad de Farmacia. Instituto Ramón y Cajal de Investigaciones Sanitarias (IRYCIS), Universidad Complutense de Madrid, Plaza de Ramón y Cajal s/n, 28040 Madrid, Spain; gemagonzalezrubio@ucm.es (G.G.-R.); teresafe@farm.ucm.es (T.F.-A.)
* Correspondence: humberto@ucm.es (H.M.); molmifa@ucm.es (M.M.); Tel.: +34-91-3941888 (H.M. & M.M.)
† These authors contributed equally to this work.

Received: 28 February 2019; Accepted: 3 April 2019; Published: 5 April 2019

Abstract: Mitogen-activated protein kinases (MAPKs) are key mediators of signaling in fungi, participating in the response to diverse stresses and in developmental processes. Since the precise regulation of MAPKs is fundamental for cell physiology, fungi bear dual specificity phosphatases (DUSPs) that act as MAP kinase phosphatases (MKPs). Whereas fungal MKPs share characteristic domains of this phosphatase subfamily, they also have specific interaction motifs and particular activation mechanisms, which, for example, allow some yeast MKPs, such as *Saccharomyces cerevisiae* Sdp1, to couple oxidative stress with substrate recognition. Model yeasts show that MKPs play a key role in the modulation of MAPK signaling flow. Mutants affected in *S. cerevisiae* Msg5 or in *Schizosaccharomyces pombe* Pmp1 display MAPK hyperactivation and specific phenotypes. MKPs from virulent fungi, such as *Candida albicans* Cpp1, *Fusarium graminearum* Msg5, and *Pyricularia oryzae* Pmp1, are relevant for pathogenicity. Apart from transcriptional regulation, MKPs can be post-transcriptionally regulated by RNA-binding proteins such as Rnc1, which stabilizes the *S. pombe* PMP1 mRNA. *P. oryzae* Pmp1 activity and *S. cerevisiae* Msg5 stability are regulated by phosphorylation and ubiquitination, respectively. Therefore, fungi offer a platform to gain insight into the regulatory mechanisms that control MKPs.

Keywords: fungal MKPs; MAPKs; signaling; Msg5; Sdp1; Pmp1; Cpp1

---

1. Fungi Respond to Distinct Stimuli Through MAPK Pathways

Cells communicate with the environment using an evolved machinery that allows them to interpret an external cue and translate it into an input message that permits the execution of cellular responses. Mitogen-activated protein kinase (MAPK) pathways are one of the main molecular systems involved in this process in eukaryotic organisms. Cells perceive extracellular stimuli, such as hormones, mitogens, and growth factors, through surface receptors attached to the plasma membrane, which transduce the external signal to intracellular proteins. This signal converges in the activation of a MAPK module that is conserved in eukaryotic cells and whose function is the amplification of such signal by sequential events of phosphorylation, making this system sensitive to subtle changes in the cell environment. MAPK pathways regulate a wide variety of cellular processes such as cell growth and division, metabolism, differentiation, and survival [1–3] (Figure 1).

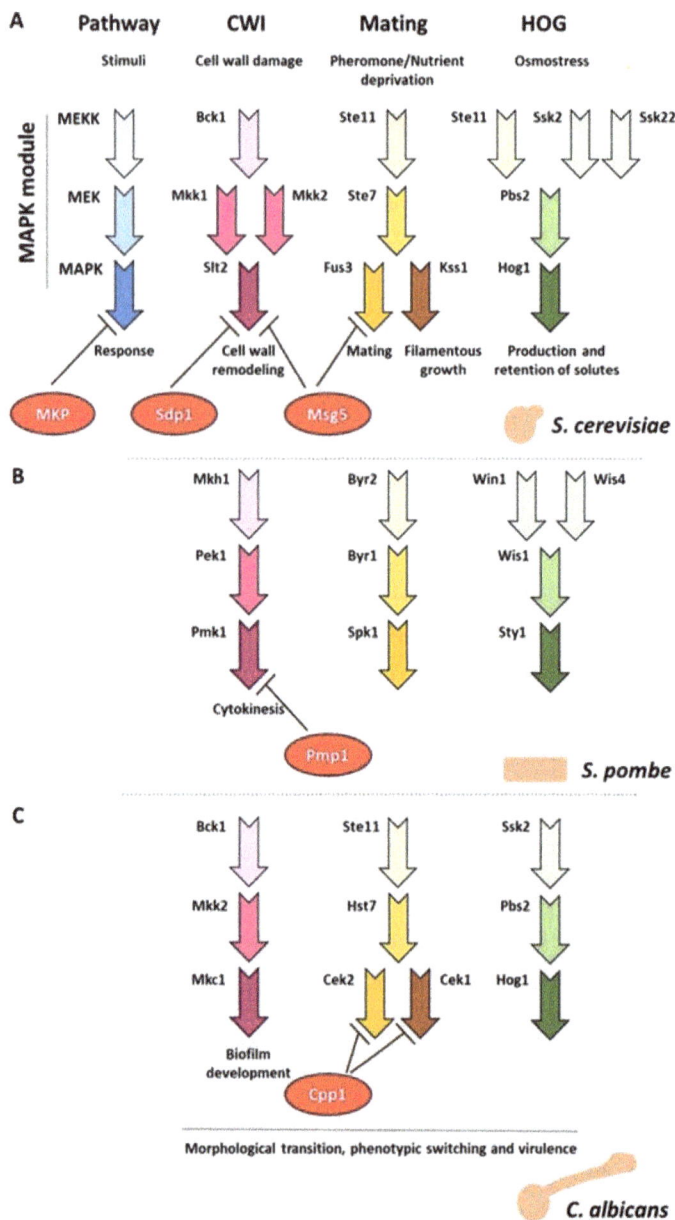

Figure 1. Mitogen-activated protein kinase (MAPK) signaling pathways in model fungi. At the uppermost left side, a schematic view of a MAPK pathway with the components of the MAPK module in blue and the MAP kinase phosphatase (MKP) in red. On the right side (head level), the major MAPK pathways described in fungi, Cell Wall Integrity (CWI), mating/filamentous growth and High Osmolarity Glycerol (HOG), and the stimuli that trigger their activation. The equivalent MAPK pathways are shown for the budding yeast *Saccharomyces cerevisiae* (**A**), the fission yeast *Schizosaccharomyces pombe* (**B**), and the dimorphic yeast *Candida albicans* (**C**). The same code of colors is shown in all cases, as indicated above.

MAPK modules are composed of three protein kinases acting in cascade. At the head level, the serine/threonine (Ser/Thr) kinase MAPKKK (MAP kinase kinase kinase), also known as MAP3K (mitogen-activated protein 3 kinase) or MEKK (MEK kinase), phosphorylates and activates its downstream effector, the MAPKK/MAP2K/MEK. The MEK, in turn, dually phosphorylates both the tyrosine and the threonine residues at the activation loop (Thr-X-Tyr) of the MAPK, which undergoes a conformational change that results in the full activation of the protein [3]. In higher eukaryotes, MAPKs are clustered into five classes: p38, ERK1/2, JNK, ERK5, and atypical MAPKs. The activated MAPK is the final component of the cascade and phosphorylates its substrates in a serine or threonine residue followed by a proline (Ser/Thr-Pro). Many of the MAPK substrates are transcription factors which, upon phosphorylation, adjust the transcriptional pattern of the cell to the particular condition determined by the stimulus. The activity of the MAPK is precisely regulated in the cell, and inappropriate modulation of these pathways is linked to several pathologies such as cancer, Parkinson's disease, inflammation, diabetes, memory dysfunction, and cardiac hypertrophy [4–6].

As eukaryotic organisms, fungi also process extracellular signals through MAPK cascades that conserve the architecture described above (Figure 1). These signaling pathways are specialized to face the different conditions that a fungus might encounter, such as high osmolarity concentrations, cell wall aggressions, mating pheromones, and, in certain cases, the presence of host factors or signals that trigger morphological transitions. Understanding the functioning of MAPK cascades in these organisms is particularly important since they are involved in the virulence of several human (e.g., *Candida albicans, Cryptococcus neoformans,* and *Aspergillus fumigatus*) and plant pathogens (e.g., *Ustilago maydis, Pyricularia oryzae*-sexual morph: *Magnaporthe oryzae*-, and *Ashbya gossypii*) [7,8]. The budding yeast model organism *Saccharomyces cerevisiae* has been the staple in the study of fungal MAPK signaling for its simplicity, easy handling, and genetic tractability. Many of the discoveries from research on the budding yeast have been translated not only to filamentous or dimorphic fungi, but also to higher eukaryotes. In *S. cerevisiae,* four different MAPKs were identified that regulate the high osmolarity response (Hog1), the pheromone response (Fus3), the pseudohyphal and invasive growth upon nutrient deprivation (Kss1), and the cell wall repair and integrity (Slt2). There is a fifth MAPK in *S. cerevisiae,* Smk1, which participates in spore wall formation, but no other elements of the MAPK module have been discovered yet [8–10]. In general, the main elements of the mating, high osmolarity (HOG), and cell wall integrity (CWI) MAPK pathways in fungi are conserved. These pathways are mediated by MAPKs Spk1, Sty1 and Pmk1 in the fission yeast *Schizosaccharomyces pombe,* and by Cek1/2, Hog1, and Mkc1 in the dimorphic model yeast *C. albicans* (Figure 1) [11,12]. The few compositional differences of the MAPK pathways between yeast and filamentous fungi were described in previous reviews [13,14].

2. General Structure and Essential Motifs of *S. cerevisiae* MKPs

The regulation of the signaling flow is executed on multiple levels of a MAPK cascade. Rapid downregulation of the stimulation generally occurs by receptor desensitization or direct dephosphorylation by phosphatases acting on the MAPKKK, the MAPKK, or predominantly the MAPK itself. Ser/Thr or Tyr phosphatases can dephosphorylate the Thr or Tyr, respectively, at the activation loop to inactivate the MAPK. Despite the general assumption that dephosphorylation of either of these two residues is sufficient for MAPK inactivation, recent evidence suggests that some monophosphorylated MAPKs retain some activity [15–17]. However, the main negative regulation is attributed to a particular type of phosphatases belonging to the family of dual specificity phosphatases (DSPs), the MAPK phosphatases (MKPs), which eliminate the phosphate group of both Thr and Tyr residues. MKPs regulate not only the magnitude and duration of MAPK signaling, but also the subcellular localization and substrate selectivity of MAPKs [18].

The general structure of MKPs includes a non-catalytic N-terminal domain and a C-terminal catalytic domain that contains a wide pocket with the critical Cys and Arg catalytic residues within the conserved signature HCXXGXXR. An aspartic residue upstream of this signature is also essential for catalysis. Within the N-terminal domain, a MAP kinase interaction motif (KIM), also called docking-domain or D-domain, is characteristic and defined by the presence of a cluster of basic residues followed by a hydrophobic sub-motif containing Leu, Ile, or Val separated by one residue: [K/R](1-3)-X(2–6)-[L/I/V]-X-[L/I/V] [19–21]. This domain is also found in other MAPK interactors, such as MAPKKs and MAPK substrates. The positive charge of the D-domain interacts with a negatively charged region at the MAPK called the common docking domain (CD).

Though mammalian cells contain at least 10 different MKPs, fungal cells only contain one or two. Among the putative or defined DSPs in *S. cerevisiae* (Yvh1, Cdc14, Pps1, Tep1, Msg5, Siw14/Oca3, Oca1, Oca2, Oca4, Oca6, and Sdp1), only Msg5 and Sdp1 have been shown to display MKP activity [22]. These two MKPs are encoded by paralogue genes likely originated from the ancient whole genome duplication that occurred in *S. cerevisiae* [23]. Msg5 is a 489 aa protein that negatively regulates Fus3 [24] and Slt2 [25] and presents the prototypical structure of MKPs, with a regulatory N-terminal domain and a catalytic phosphatase C-terminal domain (Figure 2). As an MKP, the cysteine 319 in its catalytic pocket is essential for its phosphatase activity. Msg5 possesses two different motifs that define its binding with the MAPKs. Msg5 bears a typical D-domain N-terminally located that mediates a canonical interaction with the CD domains of MAPKs Fus3 and Kss1. On the other hand, an unusual motif composed of Ile, Tyr, and Thr (IYT) located at positions 102–104 mediates the interaction with both Slt2 and the Slt2 pseudokinase paralogue Mlp1 through a CD-independent mechanism [26].

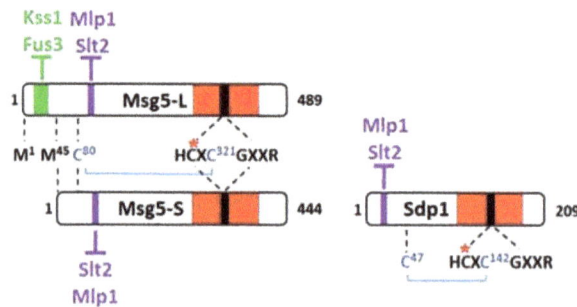

Figure 2. Diagram of the domain composition of *S. cerevisiae* MKPs Msg5, and Sdp1. Msg5-L corresponds to the long translational isoform of Msg5 and Msg5-S to the shorter one. The docking (D)-domain, IYT-motif, and phosphatase domain are drawn in green, purple, and red, respectively. The active site is represented in black and, inside this signature, the essential catalytic cysteine residue is labelled with a red star. The disulfide bond between the non-catalytic cysteine residue within the active site and its upstream cysteine partner is illustrated in blue. The target MAPKs are represented in green or purple, depending on the docking site involved in the binding.

Notably, Msg5 is not a single species, but is produced as two forms due to alternative translational initiation sites [25]. The short form lacks the first 44 amino acids and therefore does not contain the N-terminal D-domain, implying that full-length Msg5 would be able to act on both mating and CWI MAPKs, whereas the short form only would act on Slt2 (Figure 2). The physiological significance of the existence of these two forms of Msg5 remains to be established but it is tempting to speculate that it could constitute a mechanism for differential regulation of distinct MAPKs by the same MKP.

Sdp1 is only 209 amino acids long and its very short N-terminal domain presents an IYT motif that mediates interaction with Slt2 and Mlp1, but lacks the Msg5 D-domain counterpart [27]. The absence of this D-domain prevents the interaction of Sdp1 with Fus3 and Kss1, which explains why this MKP acts exclusively on the CWI pathway. The catalytic activity of both Msg5 and Sdp1 resides in the C-terminal part of the protein and, as in all MKPs, a cysteine residue at the active site is essential for its function (Figure 2).

Finally, both Msg5 and Sdp1 display enhanced catalytic activity under oxidative conditions. These phosphatases use an intramolecular disulfide bridge to recognize tyrosine-phosphorylated MAPK substrates. The bridge (Cys47–Cys142 in Sdp1 and Cys80–Cys321 in Msg5) involves a cysteine located two residues downstream of the conserved catalytic cysteine within the active site and an upstream cysteine partner out of the catalytic domain (Figure 2). This disulfide bond is critical for optimal activity of these MKPs and participates in a molecular mechanism that couples oxidative stress with substrate recognition [28].

3. Structural Conservation of Fungal MKPs

As mentioned above regarding MAPKs, orthologues of *S. cerevisiae* Msg5 are found across the fungal kingdom (Figure 3). We conducted a comparative analysis of 61 Msg5 orthologous protein sequences from a wide variety of fungi, selected from the genome databases National Center for Biotechnology Information (NCBI), Kyoto Encyclopedia of Genes and Genomes (KEGG), *Saccharomyces* Genome Database (SGD), *S. pombe* database (PomBase), *Candida* Genome Database (CGD), and *Aspergillus* Genome Database (AspGD) (Table S1). Twenty-eight representative proteins from four subphyla (Saccharomycotina, Pezizomycotina, and Taphrinomycotina from phylum Ascomycota, and Ustilaginomycotina from Basidiomycota) were chosen for a deeper analysis of structural diversity. Although the structure, function, and/or regulation of some of these fungal MKPs are already known, e.g. *S. cerevisiae* Msg5 and Sdp1, *U. maydis* Rok1, *C. albicans* Cpp1, *S. pombe* Pmp1, and *P. oryzae* Pmp1, most of these proteins have not yet been characterized.

As shown in Figure 3, the phylogram resulting from the multiple protein sequence alignment of these fungal MKPs indicates that proteins from species belonging to the same subphylum cluster together, with the exception of *N. crassa* NCU05049, which is distant from the other Pezizomycotina MKPs. This phylogram reflects the evolutionary branching of fungal MKPs. An important issue to highlight is that only one MKP has been found in most species, except in *S. cerevisiae*, *S. mikatae*, *S. paradoxus*, *U. maydis*, and *N. crassa*, which present two MKPs. In these cases, one of them is similar to ScMsg5 (*S. mikatae* smik406-g1.1, *S. paradoxus* spar252-g2.1, *U. maydis* Rok1, and *N. crassa* NCU06252), whereas the other one is closer to ScSdp1 (*S. mikatae* smik390-g11.1, *S. paradoxus* spar440-g11.1, *U. maydis* UMAG_02303, and *N. crassa* NCU05049) (Figure 3).

The size of fungal MKPs ranges from 177 (*S. mikatae* smik390-g11.1) to 1069 amino acids (*U. maydis* Rok1). In general, the larger fungal MKPs belong to Pezizomycotina, whereas the smaller ones are present in Saccharomycotina. All of them contain the distinctive active site signature motif HCXXGXXR within the typical dual specificity phosphatase catalytic region (DSPc), which is similar in size except for *U. maydis* Rok1, which displays a notably short DSPc (Figure 3). In agreement with previous observations [28], we found that the regulatory cysteine within the active site implicated in the formation of an intramolecular disulfide bridge required for full activity is conserved in all Saccharomycotina MKPs analyzed, but not in other subphyla excepting *U. maydis* Rok1 (Figure 3).

The scanning of sequences matching the consensus D-domain yielded several hits within each fungal MKP (Table S2). In the case of ScMsg5, only the one included between amino acids 29 and 38 has been proven to be responsible for its interaction with MAPKs Fus3 and Kss1 [26]. An equivalent D-domain was found in most Msg5-like MKPs of different subphyla, except in Taphrinomycotina (Figure 3), suggesting that a similar interaction mechanism with the corresponding MAPKs could be widely occurring in fungi. However, the presence of the IYT motif, known to mediate the non-canonical binding of ScMsg5 and ScSdp1 to the CWI MAPK Slt2 [27], seems to be restricted to yeast species of Saccharomycotina and Taphrinomycotina. Notably, some yeast MKPs only contain the IYT domain but not the D-domain, namely ScSdp1, SpPmp1, and SjPmp1. This could reflect their specialization in downregulating the CWI pathway.

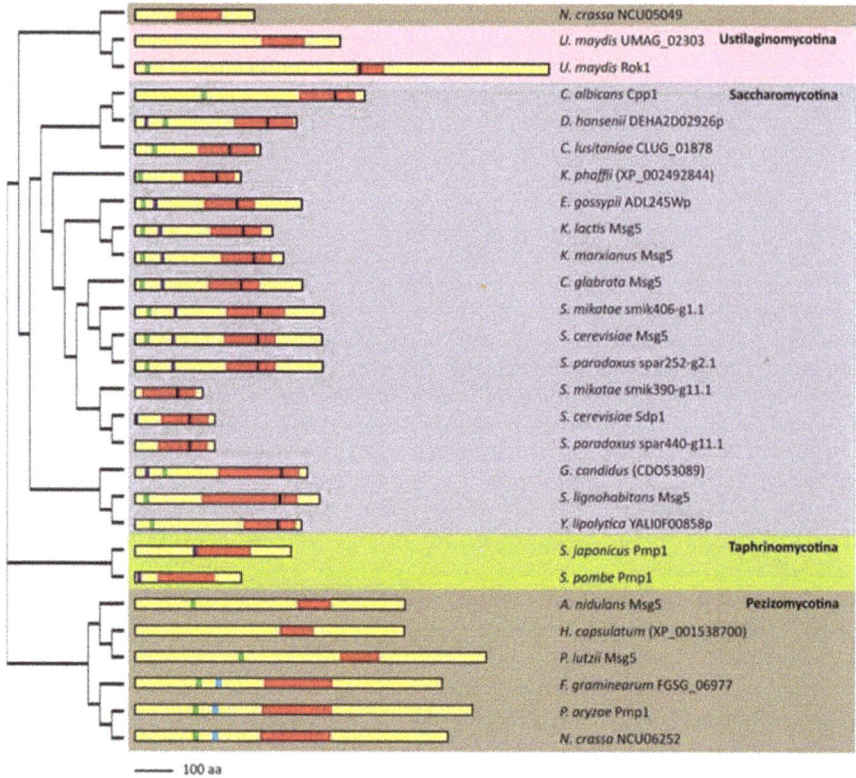

Figure 3. Phylogram and scaled scheme of the domain composition and taxonomic classification of the ScMsg5 orthologues from different fungal species. The phylogram was obtained by multiple protein sequence alignment, using Clustal Omega program (European Bioinformatics Institute (EMBL-EBI), Hinxton, UK) at default settings, of ScMsg5 orthologues selected from National Center for Biotechnology Information (NCBI), Kyoto Encyclopedia of Genes and Genomes (KEGG), or fungal genome databases [*Saccharomyces* Genome Database (SGD), *S. pombe* database (PomBase), *Candida* Genome Database (CGD), and *Aspergillus* Genome Database (AspGD)]. Proteins included are *Neurospora crassa* NCU05049 (XP_956423), *Ustilago maydis* UMAG_02303 (XP_011388628), *Ustilago maydis* UMAG_03701/Rok1 (XP_011390174), *Candida albicans* Cpp1 (XP_723551), *Debaryomyces hansenii* DEHA2D02926p (XP_458594), *Clavispora lusitaniae* CLUG_01878 (XP_002618419), *Komagataella phaffii* (formerly called *Pichia pastoris*) (XP_002492844), *Eremothecium gossypii* (also known as *Ashbya gossypii*) ADL245Wp (NP_983851), *Kluyveromyces lactis* KLLA0_F03597g/Msg5 (XP_455243), *Kluyveromyces marxianus* Msg5 (XP_022678215), *Candida glabrata* CAGL0G01320g/Msg5 (XP_446419), *Saccharomyces mikatae* (smik406-g1.1), *Saccharomyces cerevisiae* Msg5 (NP_014345), *Saccharomyces paradoxus* (spar252-g2.1), *Saccharomyces mikatae* (smik390-g11.1), *Saccharomyces cerevisiae* Sdp1 (NP_012153), *Saccharomyces paradoxus* (spar440-g11.1), *Galactomyces candidus* (CDO53089), *Sugiyamaella lignohabitans* Msg5 (XP_018733395), *Yarrowia lipolytica* YALI0F00858p (XP_504838), *Schizosaccharomyces japonicus* Pmp1 (XP_002174422), *Schizosaccharomyces pombe* Pmp1 (NP_595205), *Aspergillus nidulans* AN4544.2/Msg5 (XP_662148), *Histoplasma capsulatum* (XP_001538700), *Paracoccidioides lutzii* Msg5 (XP_002791021), *Fusarium graminearum* FGSG_06977 (XP_011326656), *Pyricularia oryzae* Pmp1 (XP_003712767), and *Neurospora crassa* NCU06252 (XP_962856). The ScanProsite tool (Swiss Institute of Bioinformatics, Laussanne, Switzerland) was used to search for dual specificity phosphatase catalytic (DSPc) domains (red), IYT/S motifs (purple), and D-domains (green for ScMsg5-like D-domain and blue for PoPmp1-like D-domain). All the sequences contain a DSPc domain

in the C-terminal region, including the active site signature motif HCXXGXXR. When a second cysteine that could be involved in a regulatory disulfide bridge is present at the fourth position within the active site, a black line is drawn. Several putative D-domains, characterized by the [K/R](1-3)-X(2–6)-[L/I/V]-X-[L/I/V] signature, were found for each protein (Table S2), but only those that aligned with the described ScMsg5 D-domain or the PoPmp1-like D-domain are shown. Different background colors indicate the subphylum to which the species belong: Pezizomycotina (brown), Ustilaginomycotina (pink), Saccharomycotina (blue), and Taphrinomycotina (green).

Within Saccharomycotina, the only MKPs lacking both interaction domains are *S. mikatae* smik390-g11.1 and *S. paradoxus* spar440-g11.1, even though they contain other hypothetical D-domains (Table S2), suggesting the probable use of alternative motifs for binding to the corresponding MAPKs.

P. oryzae Pmp1 has been shown to directly downregulate CWI MAPK Mps1. The region responsible for the interaction with Mps1 is highly conserved in other filamentous fungi [29] (Figure 3). This region does not include the ScMsg5-like D-domain, but another putative D-domain (Table S2) that could mediate the interaction with MAPK.

In conclusion, this analysis evidences the wide variety of interaction strategies and regulatory mechanisms that fungal MKPs possess and suggests several points to be further explored and extended to the putative fungal MKPs not yet characterized.

4. Function of Fungal MKPs

4.1. Msg5 Regulates the S. cerevisiae Mating Pathway by Targeting Fus3

MSG5 was initially identified as a multicopy suppressor gene of the cell cycle arrest of haploid *gpa1* mutants [24], which bear a mutation that leads to the constitutive activation of the mating pathway. *MSG5* overexpression causes cells to overcome pheromone-induced arrest and reduces the expression of a typical transcriptional readout of the mating pathway. In this seminal work, Msg5 was also proven to act as a phosphatase that was able to inactivate Fus3 in vitro (Figure 1a). Since *MSG5* was observed to be transcriptionally induced in response to pheromone exposure and the loss of this phosphatase increased the sensitivity of cells to pheromones, Msg5 was proposed to promote adaptation to this stimulus [24]. A negative effect on the mating pathway was soon defined for the protein tyrosine phosphatases Ptp2 and Ptp3, which dephosphorylate the same target, the MAPK Fus3. However, whereas Ptp2 and Ptp3 are critical in maintaining the mating pathway at a low basal activity in the absence of pheromone, Msg5 primarily participates in the inactivation of Fus3 following stimulation [30]. By employing both mathematical models and experimental data, the feedback inhibition of Fus3 by Msg5 has been proposed to be the most significant in the mid-to-late time points after pheromone-induced activation of this pathway. In contrast, the feedback inhibition of Fus3 by Ptp3 is most impactful in the early time points after the pheromone exposure [31].

Msg5 also seems to promote adaptation by regulating the subcellular distribution of Fus3. Pheromone recovery is correlated with a decrease in nuclear Fus3 [32] and, although Msg5 is distributed evenly in both the nucleus and the cytoplasm in a pheromone-independent manner, it counteracts the nuclear accumulation of Fus3 by dephosphorylating this MAPK in both cellular compartments [33]. The mechanisms underlying the downregulation of Fus3 by Msg5 also includes the involvement of monophosphorylated forms of the MAPK. A substantial pool of Fus3, which is only phosphorylated in Tyr182—the conserved Tyr residue of the MAPK activation motif—was shown to be generated in pheromone-stimulated cells. This pool of Tyr-phosphorylated kinase shows an inhibitory effect on signaling and acts in opposition to the fully phosphorylated Fus3. Importantly, cells lacking Msg5 not only show an increase in dual-phosphorylated Fus3, but also a decrease in monophosphorylated Fus3, indicating that Msg5 is involved in maintaining a pool of this inhibitory MAPK form [16].

Although overexpression of Msg5 reduces the amount of phosphorylated Kss1, this MAPK does not seem to be directly regulated by the MKP [34]. Removal of Msg5 does not result in increased

Kss1 phosphorylation but in decreased Kss1-driven FRE(Tec1)-*lacZ* expression, probably due to a Fus3-mediated effect [35].

4.2. CWI MAPK Slt2 is Regulated by the MKPs Msg5 and Sdp1 in S. cerevisiae

The ability of Msg5 to suppress the lethality promoted by the overexpression of *MKK1P396*, coding a hyperactive version of the MAPKK Mkk1, suggests the possibility that this DSP also acts on the CWI pathway [36]. It was later clarified that Msg5 downregulates this route since overexpression of *MSG5* eliminates the high Slt2 phosphorylation of cells lacking the Rho1 GAP Sac7, whereas disruption of *MSG5* in wild-type cells resulted in increased phospho-Slt2 levels, both at basal conditions and under cell wall stress [37]. When Msg5 was shown to bind and dephosphorylate Slt2 in vitro, this MAPK was formally identified as the target of Msg5 within the CWI pathway [25] (Figure 1a).

As occurs with other negative regulators, *msg5Δ* cells display sensitivity to cell wall stress [25], a phenotype that is likely derived from an increased CWI signaling, similar to that occurring in cells lacking the serine/threonine phosphatase Ptc1 [38]. This is also likely the mechanism underlying the vacuolar fragmentation observed in double *ptp2 msg5* mutants [39], a phenotype displayed by *ptc1* mutants and dependent on an exacerbated Slt2-mediated signaling [38]. However, although Msg5 is involved in the activation of Slt2 in the response to genotoxic stress, *msg5* mutant cells do not show sensitivity but tolerance to this stress. This could be explained by the fact that, under this particular insult, the signal enters the CWI pathway at the Slt2 level due to the degradation of Msg5 (see below), without the participation of upstream components and likely involved different outputs [40].

In contrast to Msg5, Sdp1 does not seem to regulate Slt2 in basal conditions. This phosphatase has been shown to act only under heat stress [41] and to be required for efficient Slt2 dephosphorylation during heat shock adaptation [42]. Whereas Sdp1 localizes both in nucleus and cytoplasm in basal conditions, it becomes localized in punctate spots throughout the cells upon heat shock. Many of these puncta colocalize with mitochondria, and Sdp1 reappears in the nucleus after prolonged heat stress, concurrent with Slt2 dephosphorylation [41].

4.3. C. albicans MKP Cpp1 Participates in Phenotypic Transition, Mating, and Virulence

Further knowledge about the functional relevance of MKPs within the subphylum Saccharomycotina was provided by studies on the dimorphic fungus and opportunistic pathogen *C. albicans*. The importance of fungal MKPs in morphogenesis is also illustrated in the case of *C. albicans* Cpp1 protein. This MKP is a negative regulator of filamentous growth and its removal derepresses hyphal formation under non-inducing conditions. This phenotype is eliminated if the MAPK Cek1 (the ScKss1 orthologue) is removed [43]. As expected, soon after this work, the Cek1 MAPK pathway was shown to be negatively regulated by Cpp1 [44] (Figure 1c). Surprisingly, Cpp1 represses hyphal gene expression by acting through a Cek1p-independent mechanism [45]. Cpp1 has been proposed to act as a key element in the crosstalk between MAPKs Hog1 and Cek1, since *hog1* and *cpp1* deletion mutants share hyperfilamentation and opaque cell formation phenotypes and display increased Cek1 activation. Moreover, *CPP1* expression is positively regulated by Hog1 [46].

Loss of Cpp1 was reported to activate the pheromone response pathway in a Cek1-dependent manner, indicating an additional role of this MKP in the mating pathway. These authors also found that basal Cek2 phosphorylation is increased in *cpp1* mutants, suggesting that Cpp1 could be downregulating Cek2 as well (Figure 1c) [47].

Pathogenicity of a homozygotic *cpp1* mutant strain is reduced in murine models of candidiasis, clarifying that Cpp1 is a virulence determinant [43]. Accordingly, it is transcriptionally induced within the human host during thrush [48]. The role of Cpp1 as a virulence factor has been connected with its activity on the target MAPK. The constitutive activation of Cek1 by deletion of its phosphatase Cpp1 increases susceptibility of cells to the protein Hst5, a salivary fungicidal histidine-rich protein constitutively produced by human salivary gland cells. Cek1 activation likely enhances Hst5-mediated killing in part through exposure of cell wall β-1,3-glucans [49].

4.4. S. pombe Pmp1: The Only Characterized MKP Within the Subphylum Taphrinomycotina

The MKP Pmp1 is the main phosphatase acting on the fission yeast MAPK Pmk1, which is orthologous to the MAPK Slt2 [50]. Pmp1 is a cytoplasmic protein, suggesting that Pmk1 inactivation by this phosphatase most likely occurs in the cytoplasm [51]. Removal of Pmp1 promotes an increase in signaling through the mating pathway [52], although the MAPK Spk1 has not been formally proven to be regulated by this MKP.

Pmp1 plays an important role in Cl⁻ homeostasis. It was originally identified as a multicopy suppressor of the chloride hypersensitivity of a calcineurin mutant, since calcineurin and Pmk1 MAPK pathways play antagonistic roles in the regulation of the Cl⁻ homeostasis [50]. Thus, cells lacking Pmp1 display a strong sensitivity to this anion [50]. Pmp1 also plays a critical role in morphogenesis. Mutant cells lacking this MKP present an aberrant morphology, frequently round or pear-shaped cells [50], and multiseptated and defective cell separation phenotypes [53]. Tight regulation of Pmk1 activity through Pmp1 is thus important in *S. pombe* cytokinesis (Figure 1b).

4.5. Filamentous Fungi MKPs are Involved in Mycelial Growth and Virulence

MKPs have been identified in other fungi, including some belonging to the subphylum Pezizomycotina, such as *Neurospora crassa, Fusarium oxysporum, Madurella mycetomatis, Gaeumannomyces graminis, Verticillium dahliae, Colletotrichum graminicola*, and *Ustilaginoidea virens* [29] (Table S1). However, only functional characterization of Msg5 homologues from filamentous fungi *F. graminearum* and *P. oryzae* has been completed to date.

In the wheat scab fungus *F. graminearum*, the Mgv1 MAPK pathway is homologous to the cell wall integrity (CWI) pathway in budding yeast and is important for plant infection [54]. The Msg5 orthologue, FgMsg5, was shown to dephosphorylate Mgv1 [55], and its elimination led to mycelial growth defects [56].

A functional study with the rice blast fungus *P. oryzae* Pmp1 not only identified this MKP as a negative regulator of the MAPK Pmk1 (Fus3 orthologue) but also of the CWI MAPK Mps1 (Slt2 orthologue), both essential for pathogenesis. In contrast to that observed for Mps1, yeast two-hybrid analysis revealed the absence of direct interaction of Pmp1 with Pmk1, opening up the possibility of the existence of some adaptor between both proteins. Δ*pmp1* mutants presented a reduced mycelial growth and conidiation and, more importantly, reduced virulence in both rice and barley [29]. However, further research is needed to provide more information about additional substrates of Pmp1, since both Δ*pmk1* and Δ*mps1* mutants develop a normal mycelial growth [57,58].

4.6. Choose the Right Host: A Role for MKPs Within Basidiomycota

MKP activity has also been proven to play a role in morphogenetic programs and disease development in Basidiomycota. This is the case of the causative agent of corn smut, *Ustilago maydis*. In this fungus, the partially redundant MAPKs Kpp2 and Kpp6 are regulated by the MKP Rok1. The lack of Rok1 activity promotes increased filamentation and enhanced virulence, whereas overexpression of *rok1* induces the opposite effect. The reason for this fungus not losing Rok1 through evolution could be that *rok1* mutants can no longer differentiate whether they are on the correct plant host [59]. Albeit the apparent dispensability of this fungal MKP in phenotypic assays, the precise control of MAPK phosphorylation is likely an advantage for *U. maydis* cells in nature.

5. Regulatory Mechanisms of Fungal MKPs

Like in mammalian MKPs, different regulatory mechanisms operating on fungal MPKs provide distinct layers of control on MAPK pathways (Figure 4). One of these mechanisms relies on the MAPK-dependent transcriptional induction of MKPs under stimulating conditions. This generates a negative feedback loop that maintains an adequate MAPK activation level and/or promotes adaptation [60]. In *S. cerevisiae*, transcription of *MSG5* is regulated by both Msg5-regulated pathways:

mating and CWI [24,61]. Whereas Msg5 is primarily involved in adaptation to pheromone exposure in the mating pathway, its main role in the CWI pathway is the control of the maximum intensity of signaling flow through the pathway [22]. *U. maydis* Rok1 is induced in response to the activation of its specific MAPK pathway [59]. In contrast, the *S. cerevisiae* Slt2-selective MKP Sdp1 is transcriptionally induced by heat stress, but in a CWI-independent manner. This induction is partially mediated by the transcription factors Msn2/4 of the cAMP-PKA pathway [41], which suggests a linkage between these two pathways through Sdp1. Therefore, the transcriptional regulation of MKPs can be exerted not only by their substrate MAPKs but also by other signaling pathways not targeted by the MKP, allowing signaling cross-inhibition. This is also the case for Cpp1, the Cek1-specific MKP from *C. albicans* that has been reported to mediate the crosstalk between Hog1 and Cek1 MAPK pathways involved in both yeast-hyphae transition and white-opaque switching [46]. In that work, Hog1 was shown to downregulate Cek1 signaling by inducing the expression of Cpp1.

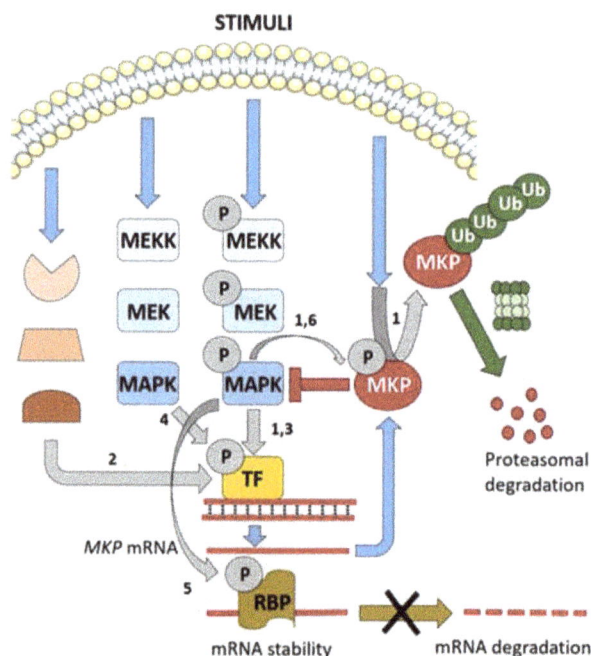

Figure 4. Multiple levels of MKP regulation in fungi. In response to stress, MKP gene expression can be transcriptionally induced by a transcription factor (TF) activated by either the MAPK target of the MKP, another related MAPK pathway, or another route. The stability of the MKP mRNA can be post-transcriptionally regulated by an RNA-binding protein (RBP), which increases the mRNA half-life and thereby the amount of MKP protein. The activity of the RBP can be in turn modulated by phosphorylation by the target MAPK. After translation, the MKP can be further regulated by different post-translational modifications, such as phosphorylation or ubiquitination. MKP phosphorylation is usually exerted by its target MAPK and ubiquitination generally leads to MKP degradation by the proteasome. Numbers designate the fungal MKPs known to be under the regulatory mechanisms indicated by grey arrows: (**1**) *S. cerevisiae* Msg5, (**2**) *S. cerevisiae* Sdp1, (**3**) *U. maydis* Rok1, (**4**) *C. albicans* Cpp1, (**5**) *S. pombe* Pmp1, and (**6**) *P. oryzae* Pmp1.

The increase in mRNA levels can be achieved not only by transcriptional upregulation but also by post-transcriptional mechanisms, including control of mRNA stability [62]. Such regulation has been reported for Pmp1, the MKP of the CWI MAPK Pmk1 in *S. pombe*. The RNA-binding protein (RBP) Rnc1 was shown to bind and stabilize an otherwise unstable Pmp1 mRNA [63]. A similar RBP-mediated post-transcriptional mRNA stabilization mechanism has also been reported for mammalian MKPs, such as MKP-1 [64]. *S. pombe* Rnc1 is phosphorylated at Thr^{50} by the MAPK Pmk1, mediating a novel type of negative-feedback regulatory loop in MAPK signaling pathways. This phosphorylation positively regulates the RNA-binding activity of Rnc1, resulting in stabilization of the Pmp1 mRNA and the subsequent increase in Pmk1 dephosphorylation [63].

Transcriptional and post-transcriptional regulation is often combined with post-translational mechanisms to ensure tight control of MKPs. Specific protein modifications, such as phosphorylation or ubiquitination, can alter the subcellular localization, stability, affinity, or catalytic activity of MKPs, resulting in either a reinforcement or a weakening of their action on MAPK signaling [60]. Phosphorylation of fungal MPKs was first described in *S. cerevisiae* Msg5, which is modified by its target MAPKs, Fus3 and Slt2. Phosphorylation of Msg5 by Fus3 was observed in vitro but its occurrence and functional significance in vivo remain to be established [24]. Slt2 was shown to phosphorylate Msg5 in yeast cells under cell wall stress conditions, alongside a decreased interaction between both proteins [25]. These results point to the existence of a positive feedback regulatory loop to reduce the activity of the phosphatase on the MAPK, allowing maximum activation of the CWI pathway. Subsequent *in vitro* kinase assays showed that both the N-terminal regulatory and the C-terminal catalytic domains of Msg5 are susceptible to phosphorylation by Slt2 [65]. However, the elimination of all the Ser/Thr-Pro putative MAPK target sites in Msg5 did not seem to alter its ability to dephosphorylate Slt2 in vivo, suggesting that this regulatory mechanism must be more subtle or complex than expected [65]. Regulatory phosphorylation has been revealed in the *P. oryzae* MKP Pmp1 by phosphoproteomic analysis [29]. Pmp1 was found phosphorylated at Ser^{240}, a position that is conserved only in filamentous fungi and lies just downstream of its region of interaction with the CWI MAPK Mps1. Pmp1 phosphorylation at this site was shown to be important for regulating MAPK phosphorylation of both Pmk1 (Fus3 orthologue) and Mps1 (Slt2 orthologue), but the involvement of such regulation in fungal virulence is not yet clear. The kinase responsible for Pmp1 phosphorylation is still unknown, but, curiously, Ser^{240} does not exactly match a canonical Ser/Thr-Pro MAPK target motif, although it is immediately followed by a Ser-Pro sequence [29].

Ubiquitination is another post-translational regulatory mechanism that allows cells to modulate signaling. Although ubiquitination can affect proteins in different ways, this modification usually results in the proteasomal degradation of the protein [60]. MAPKs can trigger proteolysis of their specific MKPs via the ubiquitin-proteasome pathway to enable long-term activation. This positive feedback loop is operating, for example, in the mammalian ERK1/2 pathway. MKP-1 stability has been reported to be regulated by ERK-mediated phosphorylation, which facilitates subsequent ubiquitination and degradation [66]. In *S. cerevisiae*, ubiquitin-mediated proteolysis of Msg5 has been proposed as the mechanism responsible for Slt2 activation in response to genotoxic stress [40]. In contrast to cell wall stress, which involves the stimulation of the whole CWI pathway, the phosphorylation of Slt2 by DNA damage does not require the activation of upstream protein kinases of the MAPK module. The degradation of Msg5 triggered by genotoxic stress depends on the presence of Slt2 but is independent of its kinase activity. Therefore, this mechanism does not seem to fit with the classical MAPK-mediated positive feedback loop described above. Slt2 activation by DNA damage does not lead to the general transcriptional pattern associated with the CWI pathway, suggesting the existence of unknown genotoxic stress-specific targets.

In summary, fungi display a similar complexity in the regulation of MPKs as other eukaryotic organisms. The regulation of MPKs can be exerted at multiple steps: Transcriptional, post-transcriptional, and post-translational, allowing a fine-tuning multifaceted modulation of MPK levels, thereby determining the duration and strength of MAPK signaling.

6. Concluding Remarks

So far, most studies on fungal MKPs focused on model yeast species, namely, *S. cerevisiae, S. pombe,* and *C. albicans.* However, orthologues of MKPs are found in all fungal proteomes examined, indicating broad conservation of this type of protein phosphatases in both pathogenic and environmental fungi. Although MKPs are very important in fungal biology as key negative regulators of MAPKs, most of them are yet uncharacterized. Hence their study will be crucial in future research. MKPs exert wide and complex functions on cellular signaling due to their ability not only to modulate but also to be regulated by several MAPK pathways. Therefore, they play an essential role in coordinating different routes and providing response plasticity to diverse environmental stimuli.

Fungal MKPs share structural and regulatory features with mammalian MKPs, such as typical docking and catalytic domains, and common transcriptional, post-transcriptional, and post-translational mechanisms that control MKP protein abundance and activity. Due to this conservation across evolution, knowledge generated in model yeasts is very helpful to mammalian MKPs researchers and vice versa. For example, the finding in yeast of a mechanism of MAPK activation through degradation of its MKP, triggered by a specific stress that does not stimulate the MAPK cascade, could inspire the search for similar cases in higher eukaryotic organisms. Nevertheless, fungal MKPs also display unique characteristics, such as disulfide bridge-mediated catalytic activation and distinctive MAPK interaction domains, which are even specific to particular fungal subphyla. These peculiarities likely reflect the different environments that fungal species have faced throughout evolution. Involvement of MKPs in fungal virulence leads us to speculate that differences between fungi and mammalian MKPs might be exploited for developing antifungal drugs with high selective toxicity.

Supplementary Materials: Supplementary Materials can be found at http://www.mdpi.com/1422-0067/20/7/1709/s1.

Author Contributions: G.G.-R., T.F.-A., H.M., and M.M. wrote and edited the manuscript.

Funding: This research was supported by grant BIO2016-75030-P, funded by Ministerio de Economía, Industria y Competitividad, and by grant S2017/BMD-3691-InGEMICS-CM, funded by Comunidad de Madrid and European Structural and Investment Funds. T.F.-A. was supported by a contract from BIO2016-75030-P, and G.G.-R. by a predoctoral contract from Universidad Complutense de Madrid.

Acknowledgments: We thank people from U3 for constructive comments.

Conflicts of Interest: The authors declare no conflict of interest.

Abbreviations

MAPK   Mitogen Activated Protein Kinase
MKP   MAP Kinase phosphatase
DUSP   Dual Specificity Phosphatase

References

1. Papa, S.; Choy, P.M.; Bubici, C. The ERK and JNK pathways in the regulation of metabolic reprogramming. *Oncogene* **2018**, *10*, 0582. [CrossRef]
2. Sun, Y.; Liu, W.Z.; Liu, T.; Feng, X.; Yang, N.; Zhou, H.F. Signaling pathway of MAPK/ERK in cell proliferation, differentiation, migration, senescence and apoptosis. *J. Recept. Signal. Transduct. Res.* **2015**, *35*, 600–604. [CrossRef] [PubMed]
3. Cargnello, M.; Roux, P.P. Activation and function of the MAPKs and their substrates, the MAPK-activated protein kinases. *Microbiol. Mol. Biol. Rev.* **2011**, *75*, 50–83. [CrossRef]
4. Low, H.B.; Zhang, Y. Regulatory Roles of MAPK Phosphatases in Cancer. *Immune. Netw.* **2016**, *16*, 85–98. [CrossRef] [PubMed]
5. Bohush, A.; Niewiadomska, G.; Filipek, A. Role of Mitogen Activated Protein Kinase Signaling in Parkinson's Disease. *Int. J. Mol. Sci.* **2018**, *19*, 2973. [CrossRef] [PubMed]
6. Kim, E.K.; Choi, E.J. Pathological roles of MAPK signaling pathways in human diseases. *Biochim. Biophys. Acta* **2010**, *1802*, 396–405. [CrossRef]

7.   Perez-Nadales, E.; Nogueira, M.F.; Baldin, C.; Castanheira, S.; El Ghalid, M.; Grund, E.; Lengeler, K.; Marchegiani, E.; Mehrotra, P.V.; Moretti, M.; et al. Fungal model systems and the elucidation of pathogenicity determinants. *Fungal. Genet. Biol.* **2014**, *70*, 42–67. [CrossRef]

8.   Hamel, L.P.; Nicole, M.C.; Duplessis, S.; Ellis, B.E. Mitogen-activated protein kinase signaling in plant-interacting fungi: Distinct messages from conserved messengers. *Plant Cell.* **2012**, *24*, 1327–1351. [CrossRef]

9.   Chen, R.E.; Thorner, J. Function and regulation in MAPK signaling pathways: Lessons learned from the yeast Saccharomyces cerevisiae. *Biochim. Biophys. Acta* **2007**, *1773*, 1311–1340. [CrossRef]

10.   Engelberg, D.; Perlman, R.; Levitzki, A. Transmembrane signaling in Saccharomyces cerevisiae as a model for signaling in metazoans: State of the art after 25 years. *Cell Signal.* **2014**, *26*, 2865–2878. [CrossRef]

11.   Perez, P.; Cansado, J. Cell integrity signaling and response to stress in fission yeast. *Curr. Protein Pept. Sci.* **2010**, *11*, 680–692. [CrossRef] [PubMed]

12.   Roman, E.; Arana, D.M.; Nombela, C.; Alonso-Monge, R.; Pla, J. MAP kinase pathways as regulators of fungal virulence. *Trends Microbiol.* **2007**, *15*, 181–190. [CrossRef] [PubMed]

13.   Tong, S.M.; Feng, M.G. Insights into regulatory roles of MAPK-cascaded pathways in multiple stress responses and life cycles of insect and nematode mycopathogens. *Appl. Microbiol. Biotechnol.* **2019**, *103*, 577–587. [CrossRef] [PubMed]

14.   Rispail, N.; Soanes, D.M.; Ant, C.; Czajkowski, R.; Grünler, A.; Huguet, R.; Perez-Nadales, E.; Poli, A.; Sartorel, E.; Valiante, V.; et al. Comparative genomics of MAP kinase and calcium-calcineurin signalling components in plant and human pathogenic fungi. *Fungal. Genet. Biol.* **2009**, *46*, 287–298. [CrossRef] [PubMed]

15.   Vazquez, B.; Soto, T.; del Dedo, J.E.; Franco, A.; Vicente, J.; Hidalgo, E.; Gacto, M.; Cansado, J.; Madrid, M. Distinct biological activity of threonine monophosphorylated MAPK isoforms during the stress response in fission yeast. *Cell Signal.* **2015**, *27*, 2534–2542. [CrossRef] [PubMed]

16.   Nagiec, M.J.; McCarter, P.C.; Kelley, J.B.; Dixit, G.; Elston, T.C.; Dohlman, H.G. Signal inhibition by a dynamically regulated pool of monophosphorylated MAPK. *Mol. Biol. Cell.* **2015**, *26*, 3359–3371. [CrossRef] [PubMed]

17.   Askari, N.; Beenstock, J.; Livnah, O.; Engelberg, D. p38alpha is active in vitro and in vivo when monophosphorylated at threonine 180. *Biochemistry* **2009**, *48*, 2497–2504. [CrossRef]

18.   Caunt, C.J.; Keyse, S.M. Dual-specificity MAP kinase phosphatases (MKPs): Shaping the outcome of MAP kinase signalling. *FEBS J.* **2013**, *280*, 489–504. [CrossRef]

19.   Pulido, R.; Zúñiga, A.; Ullrich, A. PTP-SL and STEP protein tyrosine phosphatases regulate the activation of the extracellular signal-regulated kinases ERK1 and ERK2 by association through a kinase interaction motif. *EMBO J.* **1998**, *17*, 7337–7350. [CrossRef] [PubMed]

20.   Tanoue, T.; Nishida, E. Molecular recognitions in the MAP kinase cascades. *Cell Signal.* **2003**, *15*, 455–462. [CrossRef]

21.   Peti, W.; Page, R. Molecular basis of MAP kinase regulation. *Protein Sci.* **2013**, *22*, 1698–1710. [CrossRef] [PubMed]

22.   Martin, H.; Flandez, M.; Nombela, C.; Molina, M. Protein phosphatases in MAPK signalling: We keep learning from yeast. *Mol. Microbiol.* **2005**, *58*, 6–16. [CrossRef] [PubMed]

23.   Kellis, M.; Birren, B.W.; Lander, E.S. Proof and evolutionary analysis of ancient genome duplication in the yeast *Saccharomyces cerevisiae*. *Nature* **2004**, *428*, 617–624. [CrossRef] [PubMed]

24.   Doi, K.; Gartner, A.; Ammerer, G.; Errede, B.; Shinkawa, H.; Sugimoto, K.; Matsumoto, K. MSG5, a novel protein phosphatase promotes adaptation to pheromone response in *S. cerevisiae*. *EMBO J.* **1994**, *13*, 61–70. [CrossRef] [PubMed]

25.   Flandez, M.; Cosano, I.C.; Nombela, C.; Martin, H.; Molina, M. Reciprocal regulation between Slt2 MAPK and isoforms of Msg5 dual-specificity protein phosphatase modulates the yeast cell integrity pathway. *J. Biol. Chem.* **2004**, *279*, 11027–11034. [CrossRef]

26.   Palacios, L.; Dickinson, R.J.; Sacristan-Reviriego, A.; Didmon, M.P.; Marín, M.J.; Martín, H.; Keyse, S.M.; Molina, M. Distinct docking mechanisms mediate interactions between the Msg5 phosphatase and mating or cell integrity mitogen-activated protein kinases (MAPKs) in *Saccharomyces cerevisiae*. *J. Biol. Chem.* **2011**, *286*, 42037–42050. [CrossRef] [PubMed]

27. Sacristan-Reviriego, A.; Madrid, M.; Cansado, J.; Martin, H.; Molina, M. A conserved non-canonical docking mechanism regulates the binding of dual specificity phosphatases to cell integrity mitogen-activated protein kinases (MAPKs) in budding and fission yeasts. *PLoS ONE* **2014**, *9*, e85390. [CrossRef] [PubMed]

28. Fox, G.C.; Shafiq, M.; Briggs, D.C.; Knowles, P.P.; Collister, M.; Didmon, M.J.; Makrantoni, V.; Dickinson, R.J.; Hanrahan, S.; Totty, N.; et al. Redox-mediated substrate recognition by Sdp1 defines a new group of tyrosine phosphatases. *Nature* **2007**, *447*, 487–492. [CrossRef] [PubMed]

29. Wang, R.J.; Peng, J.; Li, Q.X.; Peng, Y.L. Phosphorylation-mediated Regulatory Networks in Mycelia of *Pyricularia oryzae* Revealed by Phosphoproteomic Analyses. *Mol. Cell Proteomics* **2017**, *16*, 1669–1682. [CrossRef] [PubMed]

30. Zhan, X.L.; Deschenes, R.J.; Guan, K.L. Differential regulation of FUS3 MAP kinase by tyrosine-specific phosphatases PTP2/PTP3 and dual-specificity phosphatase MSG5 in *Saccharomyces cerevisiae*. *Genes Dev.* **1997**, *11*, 1690–1702. [CrossRef]

31. Dyjack, N.; Azeredo-Tseng, C.; Yildirim, N. Mathematical modeling reveals differential regulation of MAPK activity by phosphatase proteins in the yeast pheromone response pathway. *Mol. Biosyst.* **2017**, *13*, 1323–1335. [CrossRef] [PubMed]

32. Blackwell, E.; Halatek, I.M.; Kim, H.J.; Ellicott, A.T.; Obukhov, A.A.; Stone, D.E. Effect of the pheromone-responsive G(alpha) and phosphatase proteins of *Saccharomyces cerevisiae* on the subcellular localization of the Fus3 mitogen-activated protein kinase. *Mol. Cell Biol.* **2003**, *23*, 1135–1150. [CrossRef] [PubMed]

33. Blackwell, E.; Kim, H.J.; Stone, D.E. The pheromone-induced nuclear accumulation of the Fus3 MAPK in yeast depends on its phosphorylation state and on Dig1 and Dig2. *BMC Cell Biol.* **2007**, *8*, 44–48. [CrossRef]

34. Marin, M.J.; Flandez, M.; Bermejo, C.; Arroyo, J.; Martin, H.; Molina, M. Different modulation of the outputs of yeast MAPK-mediated pathways by distinct stimuli and isoforms of the dual-specificity phosphatase Msg5. *Mol. Genet. Genom.* **2009**, *281*, 345–359. [CrossRef] [PubMed]

35. Andersson, J.; Simpson, D.M.; Qi, M.; Wang, Y.; Elion, E.A. Differential input by Ste5 scaffold and Msg5 phosphatase route a MAPK cascade to multiple outcomes. *EMBO J.* **2004**, *23*, 2564–2576. [CrossRef] [PubMed]

36. Watanabe, Y.; Irie, K.; Matsumoto, K. Yeast RLM1 encodes a serum response factor-like protein that may function downstream of the Mpk1 (Slt2) mitogen-activated protein kinase pathway. *Mol. Cell Biol.* **1995**, *15*, 5740–5749. [CrossRef]

37. Martin, H.; Rodriguez-Pachon, J.M.; Ruiz, C.; Nombela, C.; Molina, M. Regulatory mechanisms for modulation of signaling through the cell integrity Slt2-mediated pathway in *Saccharomyces cerevisiae*. *J. Biol. Chem.* **2000**, *275*, 1511–1519. [CrossRef]

38. Tatjer, L.; Sacristan-Reviriego, A.; Casado, C.; González, A.; Rodríguez-Porrata, B.; Palacios, L.; Canadell, D.; Serra-Cardona, A.; Martín, H.; Molina, M.; et al. Wide-Ranging Effects of the Yeast Ptc1 Protein Phosphatase Acting Through the MAPK Kinase Mkk1. *Genetics* **2016**, *202*, 141–156. [CrossRef]

39. Hermansyah; Sugiyama, M.; Kaneko, Y.; Harashima, S. Yeast protein phosphatases Ptp2p and Msg5p are involved in G1-S transition, CLN2 transcription, and vacuole morphogenesis. *Arch. Microbiol.* **2009**, *191*, 721–733. [CrossRef]

40. Liu, L.; Levin, D.E. Intracellular mechanism by which genotoxic stress activates yeast SAPK Mpk1. *Mol. Biol. Cell* **2018**, *29*, 2898–2909. [CrossRef]

41. Hahn, J.S.; Thiele, D.J. Regulation of the *Saccharomyces cerevisiae* Slt2 kinase pathway by the stress-inducible Sdp1 dual specificity phosphatase. *J. Biol. Chem.* **2002**, *277*, 21278–21284. [CrossRef] [PubMed]

42. Kuravi, V.K.; Kurischko, C.; Puri, M.; Luca, F.C. Cbk1 kinase and Bck2 control MAP kinase activation and inactivation during heat shock. *Mol. Biol. Cell* **2011**, *22*, 4892–4907. [CrossRef] [PubMed]

43. Csank, C.; Makris, C.; Meloche, S.; Schroppel, K.; Röllinghoff, M.; Dignard, D.; Meloche, S.; Thomas, D.Y.; Whiteway, M. Derepressed hyphal growth and reduced virulence in a VH1 family-related protein phosphatase mutant of the human pathogen *Candida albicans*. *Mol. Biol. Cell* **1997**, *8*, 2539–2551. [CrossRef] [PubMed]

44. Csank, C.; Schroppel, K.; Leberer, E.; Harcus, D.; Mohamed, O.; Meloche, S.; Thomas, D.Y.; Whiteway, M. Roles of the *Candida albicans* mitogen-activated protein kinase homolog, Cek1p, in hyphal development and systemic candidiasis. *Infect. Immun.* **1998**, *66*, 2713–2721. [PubMed]

45. Schroppel, K.; Sprosser, K.; Whiteway, M.; Thomas, D.Y.; Rollinghoff, M.; Csank, C. Repression of hyphal proteinase expression by the mitogen-activated protein (MAP) kinase phosphatase Cpp1p of Candida albicans is independent of the MAP kinase Cek1p. *Infect. Immun.* **2000**, *68*, 7159–7161. [CrossRef]

46. Deng, F.S.; Lin, C.H. Cpp1 phosphatase mediated signaling crosstalk between Hog1 and Cek1 mitogen-activated protein kinases is involved in the phenotypic transition in *Candida albicans*. *Med. Mycol.* **2018**, *56*, 242–252. [CrossRef]

47. Rastghalam, G.; Omran, R.P.; Alizadeh, M.; Fulton, D.; Mallick, J.; Whiteway, M. MAP Kinase Regulation of the *Candida albicans* Pheromone Pathway. *mSphere* **2019**, *4*, e00598-18. [CrossRef]

48. Cheng, S.; Clancy, C.J.; Checkley, M.A.; Handfield, M.; Hillman, J.D.; Progulske-Fox, A.; Lewin, A.S.; Fidel, P.L.; Nguyen, M.H. Identification of *Candida albicans* genes induced during thrush offers insight into pathogenesis. *Mol. Microbiol.* **2003**, *48*, 1275–1288. [CrossRef]

49. Li, R.; Puri, S.; Tati, S.; Cullen, P.J.; Edgerton, M. *Candida albicans* Cek1 mitogen-activated protein kinase signaling enhances fungicidal activity of salivary histatin 5. *Antimicrob. Agents Chemother.* **2015**, *59*, 3460–3468. [CrossRef]

50. Sugiura, R.; Toda, T.; Shuntoh, H.; Yanagida, M.; Kuno, T. pmp1+, a suppressor of calcineurin deficiency, encodes a novel MAP kinase phosphatase in fission yeast. *EMBO J.* **1998**, *17*, 140–148. [CrossRef]

51. Madrid, M.; Soto, T.; Khong, H.K.; Franco, A.; Vicente, J.; Pérez, P.; Gacto, M.; Cansado, J. Stress-induced response, localization, and regulation of the Pmk1 cell integrity pathway in *Schizosaccharomyces pombe*. *J. Biol. Chem.* **2006**, *281*, 2033–2043. [CrossRef] [PubMed]

52. Didmon, M.; Davis, K.; Watson, P.; Ladds, G.; Broad, P.; Davey, J. Identifying regulators of pheromone signalling in the fission yeast *Schizosaccharomyces pombe*. *Curr. Genet.* **2002**, *41*, 241–253. [CrossRef] [PubMed]

53. Madrid, M.; Nunez, A.; Soto, T.; Vicente-Soler, J.; Gacto, M.; Cansado, J. Stress-activated protein kinase-mediated down-regulation of the cell integrity pathway mitogen-activated protein kinase Pmk1p by protein phosphatases. *Mol. Biol. Cell* **2007**, *18*, 4405–4419. [CrossRef] [PubMed]

54. Wang, C.; Zhang, S.; Hou, R.; Zhao, Z.; Zheng, Q.; Xu, Q.; Zheng, D.; Wang, G.; Liu, H.; Gao, X.; et al. Functional analysis of the kinome of the wheat scab fungus *Fusarium graminearum*. *PLoS Pathog.* **2011**, *7*, e1002460. [CrossRef] [PubMed]

55. Yu, F.; Gu, Q.; Yun, Y.; Yin, Y.; Xu, J.R.; Shim, W.B.; Ma, Z. The TOR signaling pathway regulates vegetative development and virulence in *Fusarium graminearum*. *New Phytol.* **2014**, *203*, 219–232. [CrossRef] [PubMed]

56. Yun, Y.; Liu, Z.; Yin, Y.; Jiang, J.; Chen, Y.; Xu, J.R.; Ma, Z. Functional analysis of the *Fusarium graminearum* phosphatome. *New Phytol.* **2015**, *207*, 119–134. [CrossRef] [PubMed]

57. Xu, J.R.; Hamer, J.E. MAP kinase and cAMP signaling regulate infection structure formation and pathogenic growth in the rice blast fungus *Magnaporthe grisea*. *Genes Dev.* **1996**, *10*, 2696–2706. [CrossRef] [PubMed]

58. Xu, J.R.; Staiger, C.J.; Hamer, J.E. Inactivation of the mitogen-activated protein kinase Mps1 from the rice blast fungus prevents penetration of host cells but allows activation of plant defense responses. *Proc. Natl. Acad. Sci. USA* **1998**, *95*, 12713–12718. [CrossRef]

59. Di Stasio, M.; Brefort, T.; Mendoza-Mendoza, A.; Munch, K.; Kahmann, R. The dual specificity phosphatase Rok1 negatively regulates mating and pathogenicity in Ustilago maydis. *Mol. Microbiol.* **2009**, *73*, 73–88. [CrossRef] [PubMed]

60. Molina, M.; Cid, V.J.; Martin, H. Fine regulation of *Saccharomyces cerevisiae* MAPK pathways by post-translational modifications. *Yeast* **2010**, *27*, 503–511. [CrossRef]

61. Garcia, R.; Bermejo, C.; Grau, C.; Pérez, R.; Rodríguez-Peña, J.M.; Francois, J.; Nombela, C.; Arroyo, J. The global transcriptional response to transient cell wall damage in *Saccharomyces cerevisiae* and its regulation by the cell integrity signaling pathway. *J. Biol. Chem.* **2004**, *279*, 15183–15195. [CrossRef] [PubMed]

62. Sugiura, R.; Satoh, R.; Ishiwata, S.; Umeda, N.; Kita, A. Role of RNA-Binding Proteins in MAPK Signal Transduction Pathway. *J. Signal. Transduct.* **2011**, *2011*, 109746. [CrossRef] [PubMed]

63. Sugiura, R.; Kita, A.; Shimizu, Y.; Shuntoh, H.; Sio, S.O.; Kuno, T. Feedback regulation of MAPK signalling by an RNA-binding protein. *Nature* **2003**, *424*, 961–965. [CrossRef] [PubMed]

64. Kuwano, Y.; Gorospe, M. Protecting the stress response, guarding the MKP-1 mRNA. *Cell Cycle* **2008**, *7*, 2640–2642. [CrossRef] [PubMed]

65. Alonso-Rodriguez, E.; Fernandez-Pinar, P.; Sacristan-Reviriego, A.; Molina, M.; Martin, H. An Analog-sensitive Version of the Protein Kinase Slt2 Allows Identification of Novel Targets of the Yeast Cell Wall Integrity Pathway. *J. Biol. Chem.* **2016**, *291*, 5461–5472. [CrossRef]

66. Lin, Y.W.; Chuang, S.M.; Yang, J.L. ERK1/2 achieves sustained activation by stimulating MAPK phosphatase-1 degradation via the ubiquitin-proteasome pathway. *J. Biol. Chem.* **2003**, *278*, 21534–21541. [CrossRef] [PubMed]

International Journal of
Molecular Sciences

MDPI

Review

Dual-Specificity Phosphatases in Immunity and Infection: An Update

Roland Lang * and **Faizal A.M. Raffi**

Institute of Clinical Microbiology, Immunology and Hygiene, Universitätsklinikum Erlangen,
Friedrich-Alexander-Universität Erlangen-Nürnberg, 91054 Erlangen, Germany
* Correspondence: roland.lang@uk-erlangen.de; Tel.: +49-9131-85-22979

Received: 15 May 2019; Accepted: 30 May 2019; Published: 2 June 2019

Abstract: Kinase activation and phosphorylation cascades are key to initiate immune cell activation in response to recognition of antigen and sensing of microbial danger. However, for balanced and controlled immune responses, the intensity and duration of phospho-signaling has to be regulated. The dual-specificity phosphatase (DUSP) gene family has many members that are differentially expressed in resting and activated immune cells. Here, we review the progress made in the field of DUSP gene function in regulation of the immune system during the last decade. Studies in knockout mice have confirmed the essential functions of several DUSP-MAPK phosphatases (DUSP-MKP) in controlling inflammatory and anti-microbial immune responses and support the concept that individual DUSP-MKP shape and determine the outcome of innate immune responses due to context-dependent expression and selective inhibition of different mitogen-activated protein kinases (MAPK). In addition to the canonical DUSP-MKP, several small-size atypical DUSP proteins regulate immune cells and are therefore also reviewed here. Unexpected and complex findings in DUSP knockout mice pose new questions regarding cell type-specific and redundant functions. Another emerging question concerns the interaction of DUSP-MKP with non-MAPK binding partners and substrate proteins. Finally, the pharmacological targeting of DUSPs is desirable to modulate immune and inflammatory responses.

Keywords: MAPK phosphatase; atypical DUSP; macrophage; T cell; cytokines; inflammation

---

1. Introduction

Reversible phosphorylation plays a fundamental role in signal transduction in physiological and pathological processes. Kinases phosphorylate multiple substrates, including other kinases and transcription factors, allowing the rapid transmission of signaling information to the nucleus and cellular responses at the level of gene expression and post-transcriptional regulation. The kinases and their substrates required for key biological processes, such as cell proliferation, cytokine or antigen receptor signaling, have been intensively studied by biochemical, genetic and pharmacological methods. In fact, several kinase inhibitors have been successfully developed for clinical application in oncology and autoimmune diseases. In contrast, the function and specificity of the approximately 200 phosphatases encoded in the human and murine genomes are much less well characterized, although the dephosphorylation of proteins and other substrates is an equally vital molecular switch to initiate or terminate signaling. One reason for this apparent neglect of phosphatase research has been the notion that the specific pharmacological targeting of phosphatases is not possible. However, the recent development of specific allosteric inhibitors for SHP-1, SHP-2 [1] and several other phosphatases has led to a renewed interest in drug discovery for phosphatases [2,3].

Dual-specificity phosphatases (DUSPs) come in two flavors, first as MAPK phosphatases and second in the form of atypical DUSPs that can have diverse substrates. Given the vastly documented importance of MAPK signaling in innate and adaptive immune cells, it is not surprising that several members of the DUSP-MKP family play essential roles in the regulation of antigen and pattern recognition receptor-driven immune responses. The research into the immunological functions of DUSP-MKP began more than a decade ago with the demonstration of the non-redundant functions of DUSP1/MKP-1, DUSP2/PAC1 and DUSP10/MKP-5 in the immune system (reviewed at the time by us and others [4–6]. Since then, the involvement of other DUSP-MKP family members in immune responses to infection and in inflammation has been investigated quite intensely by many groups in the mouse and human system. The phenotypes of DUSP-deficient mice and the mechanistic studies performed established that most members of the DUSP-MKP contribute to immune regulation and, overall, corroborate the notion that regulation of MAPK signaling is the prime mechanism of this group of phosphatases. However, results from the studies performed in the last decade also point to several open questions regarding the spectrum of substrates and molecular interaction partners and the cell type-specific expression and function of DUSP-MKP. Hence, the major aim of this review will be to give an overview of the main findings from the literature on DUSP-MKP in the immune system. For several members of the large group of atypical DUSPs (lacking a specific kinase-interaction motif), evidence is emerging for a function in immunity. Therefore, the state of the literature on DUSP3, DUSP11, DUSP12, DUSP14 and DUSP22 is also reviewed here. We will conclude by highlighting, in our opinion, the most important questions for ongoing research on the role of DUSPs in the immune system.

2. DUSP in Immunity and Infection

2.1. DUSP Family Phosphatases: A General Overview

DUSP proteins belong to the extended family of tyrosine phosphatases (for an overview, see [7]). They have been grouped in class I of the cysteine-based phosphotyrosine phosphatases (PTP) that carry the CxxxxxR signature motif in the active site of the catalytic domain. Several classical PTPs such as PTP1B, SHP1 and CD45 are in subclass I of class I. Subclass II of dual-specificity/VH1-like PTPs is characterized by a HCxxGxxR signature motif and consists of 63 members. These include the DUSP-MKP (10 members), the small-size atypical DUSPs (15 members), other atypical DUSPs (five members), slingshot phosphatases, phosphatases of regenerating liver (PRL), CDC14s (four members) and PTEN-like phosphatases (eight members). All DUSP-MKP, most small-size atypical DUSPs, and several other subclass II members show specific phosphatase activity against phosphorylated Ser, Thr or Tyr. Some atypical DUSPs, PRLs and PTEN-like proteins have non-amino acid substrates, including triose-phosphorylated RNA (DUSP11), phosphatidylinositol phosphates (PRL3, PTEN, DUSP23), or their substrates are yet unknown. In a narrower sense, we consider here as DUSP family members the ten DUSP-MKP and the 20 small-size and atypical DUSPs (Figure 1).

2.1.1. MAPK Phosphatases

The specificity of the 10 members of this group for distinct MAPK family members is due to the presence of an N-terminal MAPK-binding domain (MKB) containing a kinase-interaction motif (KIM) (Figure 1). While DUSP6/MKP-3 specifically binds to and dephosphorylates extracellular signal-regulated kinases (ERK)1/2, but not p38 and JNK, DUSP10/MKP-5 binds and inactivates p38, but not ERK1/2, whereas DUSP1/MKP-1 can interact and dephosphorylate ERK1/2, p38 and JNK. Of note, substrate preferences for distinct MAPK in vivo are not always predicted by in vitro binding and catalytic activation studies. The KIM contains positively charged arginine residues which are essential for the specific interaction with the MAPK. As excellently reviewed by Caunt and Keyse [8], the structural details of the binding interactions between DUSP-MKPs and their MAPK partners differ and can explain the specific activities for MAPK family members. Importantly, for some DUSP-MKPs (e.g., DUSP1/MKP-1, DUSP4/MKP-2 and DUSP6/MKP-3), the catalytic activity robustly increases as a consequence of a conformational change in the DSP domain, whereas this is not the case for DUSP5, DUSP10/MKP-5 and DUSP16/MKP-7 [8].

The DUSP-MKPs have relatively diverse C-terminal domains that are important for the regulation of protein stability and subcellular localization. PEST sequences in DUSP16/MKP-7 promote ubiquitinylation and proteasomal degradation, whereas phosphorylation of the C-terminus by ERK1/2 increases protein stability [9]. Nuclear export signals in the C-terminal domain govern the subcellular compartmentalization of DUSP6, DUSP8 and DUSP9 in the cytoplasm, while the combined presence of NES and nuclear localization signals in DUSP16 can explain its shuttling between the cytoplasm and nucleus [10].

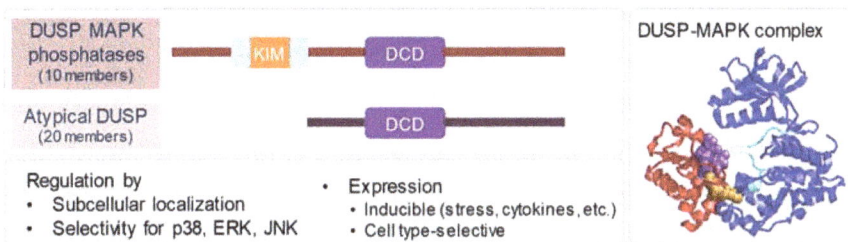

Figure 1. Dual-specificity phosphatases (DUSP): domain structure, mode of action, and levels of regulation. The 10 members of the classical DUSP-MAPK phosphatases contain a MAPK-binding kinase-interaction motif (KIM) conferring selective binding to ERK1/2, p38 or JNK1/2. Upon binding to MAPK, the DUSP catalytic domain (DCD) dephosphorylates the TXY motif in the activation loop (right panel). Atypical DUSPs lack a KIM and can have more diverse substrates, including phosphorylated RNA (DUSP11). Regulated expression between cell types and after stimulation, different compartmentalization of DUSP, and selectivity in binding to MAPK family members confers specificity of DUSP action in signaling.

2.1.2. Atypical DUSPs

Members of the 20 small-size and atypical DUSPs [7] lack a designated MKB domain and therefore dephosphorylate less well-defined substrates. Many of these DUSP dephosphorylate proteins with pSer, pThr and pTyr residues, and some of them interact with and regulate MAPK family members. However, activity towards other phosphoproteins is more commonly found in this family, and some members dephosphorylate non-protein substrates such as RNA or lipids. Thus, the field of atypical DUSP research is very diverse at the biochemical and functional level. In the context of this review, we will therefore focus on selected atypical DUSPs with a demonstrated role in the regulation of the immune response (DUSP3, DUSP11, DUSP12, DUSP14, and DUSP22).

2.2. Models of DUSP Regulation in the Immune System

Innate and adaptive immune cells rely on rapid kinase-driven signaling to translate the recognition of microbial structures (through TLR and other PRR) or the binding of specific antigen (by the B cell and T cell receptors) into massive transcriptional and post-transcriptional responses required for the swift production of cytokines, antimicrobial effector molecules and efficient clonal expansion in case of lymphocytes.

To achieve a balanced immune response, the intensity and duration of these cellular responses has to be controlled and, finally, to be terminated. Several DUSP-MKP and atypical DUSP are strongly upregulated in immune cells in response to the stimulation of PRR or antigen receptors, suggesting that they contribute to this regulatory process through the negative feedback regulation of MAPK and other kinases they interact with. Such negative feedback loops can also receive input from other cues, e.g., cytokines such as IL-10 that enhances expression of DUSP1 (Figure 2A). Prototypic examples of this type of regulation are the inducible nuclear DUSP-MKP DUSP1, DUSP2 and DUSP5, as described in detail below.

Triggering of different pattern recognition receptors on innate immune cells, such as TLR or C-type lectin receptors, induces overlapping yet distinct transcriptional responses. Ligand- or receptor-specific changes in gene expression are due to differences in signaling pathways and/or their spatio-temporal regulation, and, importantly, control the inflammatory reaction and direct the adaptive immune response. To date, the ligand- or receptor-specific induction of DUSP family genes and their kinetics in macrophages and dendritic cells has not been investigated in detail. However, while the TLR ligands LPS and CpG induce the strong upregulation of DUSP1, DUSP2 and DUSP16 [11], the mycobacterial cord factor TDM caused a marked induction of DUSP4 and DUSP5 [12]. Such ligand- or receptor-related differential expression of DUSP genes may contribute stimulus-specific transcriptional programs through the selective regulation of individual MAPK substrates and the subcellular localization and activity of the individual DUSP (Figure 2B).

The activation of different innate immune cells, e.g., macrophages, conventional or plasmacytoid DC, by the same stimulus can trigger remarkably different cellular outputs. For example, the activation of TLR9 by CpG ODN triggers inflammatory cytokine production in macrophages and cDC but high-level IFNα/β secretion in plasmacytoid DC. While these cell type-specific responses can have many reasons in the spatio-temporal regulation of signaling cascades, the constitutive or inducible expression of different sets of DUSP genes may strongly impact on the specification of the cellular reaction to exactly the same stimulus (Figure 2C). The selective constitutive expression of DUSP9 in plasmacytoid DC is one such example where a specific DUSP is associated with a cell type-specific response [13].

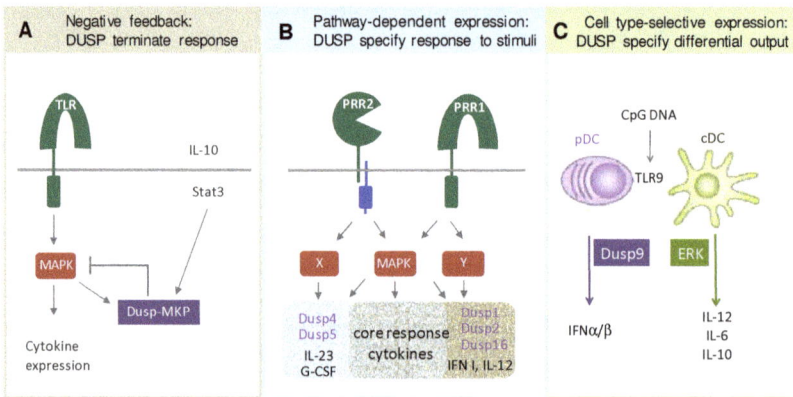

Figure 2. DUSP as regulators and specifiers of innate immune responses. (**A**) Negative feedback regulation of MAPK signaling by activation-induced DUSP upregulation, shown here as example TLR-induced DUSP1 that is enhanced by IL-10-STAT3 signaling. (**B**) Ligands for different pattern recognition receptors trigger overlapping yet distinct gene expression programs, including the differential expression of DUSP family members that then tune the amplitude and kinetics of MAPK activation. (**C**) Activation of the same pattern recognition receptor (here TLR9) in different cell types induces strikingly different cytokine outputs. Differential expression of DUSP genes (here DUSP9 in pDC, but not cDC) contributes to cell type-specificity of signaling and transcriptional responses.

2.3. Role of Specific DUSP in Immune Responses and Host Defence

2.3.1. DUSP-MKPs

DUSP1/MKP-1

This inducible nuclear MAPK phosphatase is expressed at high levels in many cell types in response to serum, growth factors, and environmental stresses (such as high salt and osmotic stress).

In cells of the innate immune system, DUSP1 is strongly induced by TLR ligands and further enhanced by concomitant treatment with IL-10 [11]. Dusp1 mRNA is highly unstable, with a $t_{1/2}$ in the range of 15–30 min [11,14]. In addition to regulation at the level of transcription and mRNA stability, DUSP1 levels are controlled by translational regulation, as shown by genome-wide profiling of polysomal mRNAs in resting and LPS-activated macrophages [15]. Similar to other negative feedback inhibitors of signaling, DUSP1 mRNA associated with polysomes was strongly increased after stimulation, indicating derepression of translational efficiency. Post-translationally, DUSP1 protein stability is strongly increased by ERK-mediated phosphorylation [16,17].

In addition to TLR ligands, several other microbial products have been shown to induce DUSP1 expression and to affect MAPK activation states. These include the helminth immunomodulatory cysteine protease inhibitor AvCystatin [18], the pore-forming protein pneumolysin of *Streptococcus (S.). pneumoniae* [19], and the hyphal form of *Candida albicans* [20]. Glucocorticoids are another strong stimulus of DUSP1 expression by macrophages [11,17], mast cells [21] and epithelial cells [22,23]. In fact, the anti-inflammatory effect of glucocorticoids is at least partially mediated by DUSP1, because inhibition of TNF, COX2 and IL-1 expression is resistant to dexamethasone in Dusp1$^{-/-}$ macrophages [24]. A strong phenotype of *Dusp1*$^{-/-}$ mice in the LPS challenge model was described by several groups, showing that prolonged and intensified p38 activation in response to TLR4 ligation led to the overshooting production of a subset of LPS target genes including chemokines (CCL3, CCL4), cytokines (TNF, IL-6, IL-10, IL-1), and other inflammatory mediators [25–28]. In several models of bacterial peritonitis (CLP, *Escherichia coli* injection) and sepsis (*Staphylococcus aureus*), a similar overproduction of cytokines, inflammatory damage to vital organs such as the lung, and increased lethality was found [29–32]. Lung infection of DUSP1$^{-/-}$ mice with *Chlamydia pneumoniae* led to increased bacterial loads associated with higher levels of IL-6 and chemokines in the lungs [33].

These results suggested that the TLR-induced expression of DUSP1 during infection is required to restrain damaging hyper-inflammation, partially mediating the effect of endogenous IL-10. Interestingly, the dramatically increased production of cytokines such as IL-6 and TNF was not efficient in achieving a reduction of pathogen burden. Mechanistically, THE increased activity of p38 in the absence of DUSP1 was shown to enhance inflammatory gene expression through signaling to MSK1/2 and the substrate transcription factors CREB and ATF1 [34] (Figure 3). In addition, the DUSP1-p38-MK2 regulatory module also controls cytokine levels via effects on mRNA stability through THE expression and post-translational modification of the RNA-binding protein TTP: first, TTP mRNA expression is upregulated in the absence of DUSP1; secondly, the phosphorylation of TTP by p38 inactivates its RNA-degrading capacity, thereby increasing the stability of mRNAs for TNF, IL-6, and multiple other inflammatory transcripts [35], including interferon beta [36]. Consequently, unleashed p38 activity in the absence of DUSP1 antagonizes TTP and increases inflammatory gene expression by prolonging stability ([37]; for review, see [38]).

While the phenotype of DUSP1$^{-/-}$ mice in infection and inflammation models was consistent with its role in innate immune cells in vitro, the broad expression in many cell types suggested that it may also be involved in the regulation of adaptive immune responses. Indeed, DUSP1-deficient mice developed pronounced Th17-biased cellular immunity after immunization through an indirect effect of APC-derived IL-6 and IL-1 [39]. In contrast, increased tissue inflammation in the T cell-dependent Experimental Autoimmune Encephalitis (EAE) model of multiple sclerosis was observed in DUSP1-deficient mice and depended on the dysregulated responses of astrocytes and fibroblasts to IL-17 receptor signaling [40]. Given its relatively broad tissue expression in multiple cell types, the clarification of the cell type-specific roles of DUSP1 in the regulation of inflammatory responses will require the use of conditional knockout mice [41,42].

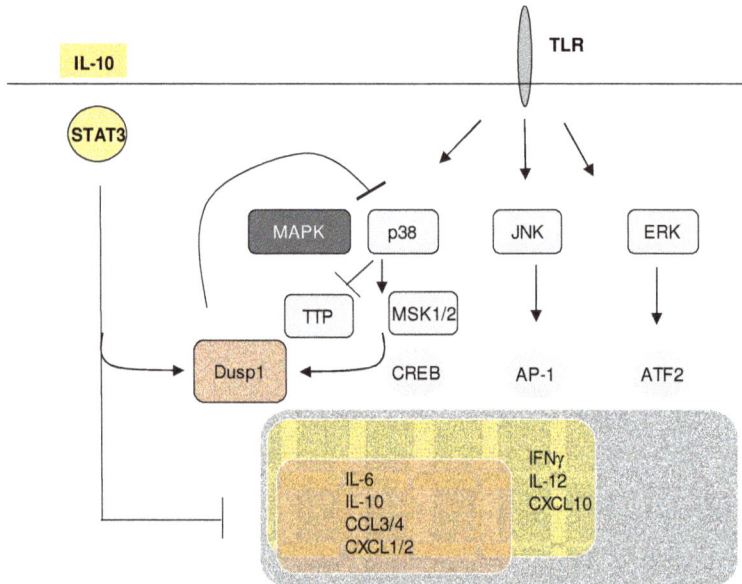

Figure 3. Mechanism of DUSP1 regulation of cytokine expression in Toll-like receptor (TLR)-stimulated macrophages. DUSP1 expression is induced after TLR triggering via MAPK activation and strongly enhanced by IL-10-STAT3 signaling. Preferential inhibition of p38 activity by DUSP1 down-regulates the expression of a subset of cytokines by interfering with MSK1/2-dependent transcription factors and through the control of mRNA decay via the post-translational regulation of TTP. See text for references and details.

DUSP2/PAC1

DUSP2 is, like DUSP1, an inducible nuclear protein [43]. It has been cloned from human T cells and is highly induced in lymphocytes after activation [44]. DUSP2 is a downstream target of the tumor suppressor p53 in signaling apoptosis and growth suppression [45]. Dusp2 mRNA was identified as one of the most highly induced transcripts in many activated leukocytes and acts as a positive regulator of inflammatory cell signaling and effector functions [46]. Dusp2$^{-/-}$ mice were protected from inflammation in the 'K/BxN' animal model of serum transfer-induced arthritis [46]. Protection was mediated by impairment in the effector responses by mast cells and macrophages that showed decreased phosphorylation of ERK and p38, but increased JNK phosphorylation, and impaired NFAT-AP1 and Elk1 transcriptional activation. Interestingly, JNK inhibition restored ERK activation showing crosstalk between MAPK pathways [46].

More recently, DUSP2 was shown to be a negative regulator of STAT3 signaling during the commitment of cells to the Th17 lineage in a T cell transfer model of colitis [47]. Here, the expression of DUSP1, DUSP2 and DUSP4 was upregulated in T cells upon stimulation with anti-CD3 and anti-CD28. Using a co-overexpression system in HEK293T cells, the repression of an IL2-luc reporter by DUSP1, DUSP2 and DUSP4 was found. DUSP2 and DUSP4 synergized in this assay. In this system, the authors also observed the association of DUSP2 with DUSP1 and DUSP4, but not with other co-expressed DUSP proteins. Further, DUSP2 directly catalyzed STAT3 dephosphorylation with physical association between DUSP2 and STAT3 [47]. In mouse models of inflammatory bowel disease (IBD), Dusp2$^{-/-}$ mice developed more severe inflammatory disease, with increased production of Th17 cells via STAT3 signaling. In human IBD, DUSP2 expression was down-regulated by methylation and failed to be induced during T cell activation in PBMCs from patients with ulcerative colitis (UC) [47].

Dusp2 expression was recently reported to be hampered in Batf$^{-/-}$ bone marrow-derived macrophages (BMM) stimulated with the TLR7 ligand R848 [48]. Interestingly, Dusp2 mRNA expression levels in macrophages correlated inversely with the amount of STAT3 phosphorylated on Tyr-705. Although these data are consistent with a dephosphorylation of STAT3 by DUSP2 as reported before [47], Kanemaru et al. did not demonstrate a direct interaction of DUSP2 with STAT3; increased phosphorylation of STAT3 in Batf$^{-/-}$ BMM expressing low levels of Dusp2 may therefore also be an indirect consequence of alterations in cytokine levels (e.g., higher concentrations of IL-10 would cause a similar phenotype). Together, more work is required to validate the unexpected and interesting observations made in Dusp2$^{-/-}$ mice and regarding its interaction with non-MAPK substrates. The further dissection of the function of DUSP2 would benefit from cell type-specific deletion in mice and further biochemical studies on its interaction with other proteins.

DUSP4/MKP-2

Dusp4 cDNA was cloned from PC12 cells and has significant homology with DUSP1/MKP-1 [49], but differs from MKP-1 in tissue distribution. DUSP4/MKP-2 is induced by growth factors, cellular stress, UV-light and by LPS [46,49]. DUSP4 is localized in the nucleus with two putative nuclear localization sequences (NLS1, NLS2) [50] and dephosphorylates selectively ERK and JNK in vitro [51].

DUSP4 deficiency in mice limited the Th1 immune response and increased the susceptibility to infection with the protozoan *Leishmania mexicana* [52]. DUSP4 deficiency in macrophages did not increase the phosphorylation of the previously described substrate ERK, but interestingly increased the LPS-induced phosphorylation of JNK and p38. Macrophages lacking DUSP4 produce increased amounts of the cytokines, IL-6, IL-12 and TNF, and of the prostaglandin PGE$_2$. They also express higher levels of the enzyme arginase-1 at baseline and after stimulation with LPS, and decreased levels of inducible nitric oxide synthase (iNOS) in response to LPS or IFNγ, resulting in the strongly reduced production of nitric oxide (NO) [52]. Consistent with the pivotal role for NO in control of *Leishmania* replication, DUSP4-deficient macrophages were unable to control *L. mexicana* parasite infection; however, the capacity for NO production and, in part, to suppress intracellular parasite proliferation could be restored by the arginase-inhibitor nor-NOHA. In vivo, DUSP4$^{-/-}$ mice had a strongly decreased ability to clear parasite growth with a limited Th1 and increased Th2 response [52]. Infection of DUSP4$^{-/-}$ mice with *Toxoplasma gondii*, another intracellular protozoal pathogen, mirrored the increased susceptibility to infection observed in the *L. mexicana* model in terms of increased parasite burden, reduced NO production and increased expression of arginase-1; however, in the case of *T. gondii* infection in Dusp4$^{-/-}$ mice, high arginase-1 levels contribute to parasite control, likely through depletion of L-arginine [53].

Two murine sepsis models revealed that DUSP4 is a positive regulator of inflammation, in contrast to the closely related DUSP1/MKP-1 [54]. Here, Dusp4$^{-/-}$ mice showed improved survival in the high-dose LPS and the cecal ligation and puncture (CLP) model of sepsis, with attenuated levels of systemic IL-1β, IL-6 and TNF. At variance with the findings of an earlier study [52], the phosphorylation of ERK was increased, whereas the phosphorylation of p38 and JNK were decreased in DUSP4-deficient macrophages. Interestingly, DUSP1 induction was increased in the absence of DUSP4 and siRNA knockdown of DUSP1 reversed the cytokine production [54]. As these results indicate an indirect phenotypic effect of the DUSP4 knockout, the combined deletion of both DUSP1 and DUSP4 is desirable to study DUSP redundancy and combinatorial effects during sepsis and inflammation.

DUSP4 knockout mice were also analyzed in the EAE mouse model of multiple sclerosis. Perhaps not surprisingly, given the phenotype of attenuated inflammation in the previous in vivo models, the severity of clinical disease and of immune cell infiltration in the CNS was attenuated in the absence of DUSP4. Antigen-specific cytokine production by immune cells from spleen and draining lymph nodes was reduced, which could be attributed to a CD4$^+$ T cell-intrinsic reduction of proliferation and production of IL-2 and IL-17, as well as to a contribution of impaired upregulation of MHC-II and of costimulatory molecules by DUSP4-deficient DC [55].

In CD4$^+$ T cells, the deletion of DUSP4 led to increased TCR-induced proliferation and expression of CD25 (the IL-2 receptor alpha chain) [56]. Contrary to expectations, the kinetics and intensity of the T cell receptor-triggered phosphorylation of MAPK were not altered in DUSP4$^{-/-}$ CD4$^+$ T cells, nor was the activation of IκB kinase β (IKKβ) affected. These findings indicated that DUSP4 plays a redundant role with regard to the dephosphorylation of its canonical substrates JNK and ERK. Interestingly, DUSP4$^{-/-}$ CD4$^+$ T cells responded to IL-2 stimulation with somewhat increased levels of STAT5 phosphorylation, which could explain the hyper-proliferative phenotype [56]. In a HEK 293T cell overexpression system, DUSP4 reduced the phosphorylation of STAT5, and DUSP4 protein could be immune-precipitated with STAT5 from thymocytes [56]. DUSP4 was subsequently shown by the same group to interact with STAT5, requiring both its substrate-interacting and the catalytic domain, which down-regulates STAT5 protein levels [57]. The CD4+ T cell response in the EAE model was characterized by a shift in Th cell differentiation from Th17 to regulatory T cells (Treg) in DUSP4-deficient mice, consistent with a reduced clinical severity in this Th17-dependent inflammation model [57]. Since DUSP4 is highly expressed in Treg, its potential contribution to dampened TCR-induced signaling was analyzed in another study. However, no increase in ERK phosphorylation was found in the absence of DUSP4 [58].

DUSP4 also appears to play a role in immunosenescence in human T cells, since it was found expressed at higher levels in the CD4$^+$ T cells of elderly donors and associated with weakened expression of the costimulatory molecules CD40L and ICOS-L, as well as the B-cell stimulating cytokines IL-21, IL-4 and IL-17. Importantly, the reduced vaccination-induced T cell-dependent B cell response in the elderly could be restored by silencing DUSP4 expression in CD4$^+$ T cells [59]. The defective TCR response in patients with idiopathic CD4 lymphopenia (ICL) has been reported to be due to the increased expression of DUSP4 [60]. ICL is a rare disorder characterized by very low numbers of T cells similar to T cell lymphopenia found in elderly people [61]. Gene-expression profiles in CD4$^+$ T cells from patients with ICL compared to healthy donors revealed a strong overexpression of DUSP4 in ICL patients. The repeated stimulation of T cells with anti-CD3 Abs in the ICL patients caused a senescent profile with gradual increase in DUSP4 expression and decreased ERK phosphorylation. However, the siRNA-mediated silencing of DUSP4 restored ERK activation and improved T cell activity.

Together, DUSP4 has important functions in both innate and adaptive immune cells, which appear to be only in part due to the regulation of its classical substrates JNK and ERK MAPK. Whereas in human T cells, the effects of DUSP4 overexpression and silencing could be explained by its control of ERK activation levels, in murine CD4$^+$ T cells, the mechanism of DUSP4 regulation of proliferation and Th differentiation appears to act at the level of STAT5 activation and stability. This mechanism may involve the direct binding and dephosphorylation of STAT5 by DUSP4 [56,57].

DUSP5/hVH-3

DUSP5 is a nuclear MKP and inducible by heat shock and growth factors in mammalian cells [62]. DUSP5 binds directly to and inactivates ERK MAPKs, but not other MAPK family members [63]. Its nuclear localization leads to the anchoring of ERK in the nucleus [63] and strict compartmentalization of ERK de-phosphorylation [64]. Paradoxically, in DUSP5$^{-/-}$ cells, a decrease in sustained cytoplasmic ERK phosphorylation was observed after serum stimulation, that could be explained by a relieving effect of DUSP5 on the ERK-mediated inhibition of upstream RAF kinases [65].

Computational modeling and molecular dynamics analysis of the binding of DUSP5 to phosphorylated ERK2 suggests that the N-terminal ERK-binding domain of DUSP5 first contacts the C-terminal lobe of dual phosphorylated ERK, followed by arranging the linear DUSP5 linker on the ERK2 groove, which brings the catalytic DUSP5 domain in close proximity to the phosphorylated tyrosine and threonine residues of ERK2 [66]. The DUSP5 linker appears to play a critical role in the specific binding to ERK, explaining substrate specificity for ERK [66].

The inducible expression of DUSP5 in the immune system was first identified in T cells after stimulation with cytokines signaling through the common gamma chain as negative feedback regulator of ERK activation [67]. During T cell development in the thymus, DUSP5 expression is increased during the transition from double-positive thymocytes to single-positive CD4$^+$ T cells [68]. The transgenic overexpression of DUSP5 blocked T cell development at the double-positive stage, whereas in mature CD4$^+$ T cells, it impaired the proliferative response to IL-2 stimulation [69]. Deletion of the Dusp5 gene did not cause an obvious phenotype in the murine immune system, but affected memory/effector CD8$^+$ T cell populations after viral infection [70]. DUSP5$^{-/-}$ T cells proliferated more strongly and underwent more cell death [70]. DUSP5 is a target of the DNA-damage induced transcription factor p53; whether a negative feedback loop between DUSP5 and p53 activity is operative in T cells remains to be established.

Eosinophilic granulocytes are induced by helminth infection and during the resolution phase of inflammatory responses. They express considerable levels of DUSP5 under basal conditions (www.immgen.org) and strongly upregulate DUSP5 mRNA and protein levels following stimulation with IL-33 [71]. Increased numbers of eosinophils were observed in bone marrow, blood and spleen of *Nippostrongylus brasiliensis*-infected DUSP5$^{-/-}$ mice compared to wild-type mice, with a decrease in apoptotic cells. Microarray analysis of IL-33-treated DUSP5$^{-/-}$ eosinophils revealed the increased expression of CD69 and Spred2, genes essential for eosinophil activation. DUSP5$^{-/-}$ eosinophils had increased ERK activation after IL-33 treatment, which enhanced the expression of anti-apoptotic Bcl-XL and cell survival [71].

In macrophages, DUSP5 expression was induced by the LPS stimulation of TLR4 and reduced cytokine production in RAW264.7 macrophages at the level of ERK1/2-dependent AP1 activation [72]. The stimulation of macrophages with the mycobacterial cord factor trehalose-6,6-dimycolate also upregulated DUSP5 expression, independent of the cord factor receptor Mincle (9). ERK1/2 signaling plays an important role in M-CSF-driven macrophage proliferation and differentiation. DUSP5 is upregulated by M-CSF in myeloid cells and acts as a negative feedback regulator of ERK1/2 activation. Its overexpression increased M-CSF driven proliferation and blocked macrophage differentiation [73]. Instead, DUSP5 directed progenitor cells to differentiate towards granulocytes in response to M-CSF [73], which is in line with data showing that the inhibition of ERK1/2 favors granulocyte over monocyte development [74].

Since DUSP5 is expressed and regulated in many innate and adaptive immune cell types, in vivo analysis of its function in specific immune responses would be aided by conditional knockout mice, that have not been reported yet.

DUSP6/MKP-3

DUSP6 is a cytoplasmic MAPK phosphatase with high selectivity for ERK1/2 due to high affinity binding of the DUSP6 N-terminal KIM to ERK1/2 and a binding-induced conformational change activating the catalytic domain [75]. DUSP6 is constitutively expressed in several immune cell types, including CD4$^+$ T cells [76], dendritic cells [13], macrophages [11] and microglia [77]. DUSP6$^{-/-}$ mice have increased ERK phosphorylation in the heart, spleen, kidney, brain and fibroblasts, but are otherwise healthy and fertile [78].

DUSP6$^{-/-}$ CD4$^+$ T cells responded to TCR stimulation with stronger ERK1/2 phosphorylation, proliferation and IFNγ production, but on the other hand made less IL-17, showed increased apoptosis and a reduced Treg function. In the IL-10 knockout mouse model of spontaneous colitis, additional DUSP6 deficiency exacerbated disease symptoms, indicating a regulatory function of DUSP6, which was confirmed by increased IFNγ production from colonic and mesenteric lymph node CD4$^+$ T cells [79]. DUSP6 was required for activation-induced glycolysis in CD4$^+$ T cells and for IL-21 production by follicular T helper cells (Tfh) [80]. DUSP6 expression in human CD4$^+$ T cells increases with age, due to declining levels of miR-181a, which reduces the sensitivity of TCR-triggering and results in suboptimal expression of activation markers and reduced proliferation [81]. The allosteric DUSP6 inhibitor BCI restores full responsiveness to TCR stimulation, suggesting that DUSP6 could be a promising target for improving cellular immune responses to vaccination in the elderly [81].

In macrophages, DUSP6 is also constitutively expressed and attenuates ERK phosphorylation. Hyperoxia inactivates DUSP6 catalytic activity in macrophages, leading to increased ERK function and the induction of a pro-survival expression program [82]. The pharmacological inhibition of DUSP6 in macrophages by BCI led to attenuated cytokine production through Nrf2-signaling and decreased NFκB activation [83]. In macrophages infected with *Leishmania major*, triggering of the activating receptor CD40 reciprocally regulates DUSP1 and DUSP6, and overexpression of DUSP6 promotes control of leishmanial replication [84]. Thus, depending on the cell type (CD4+ T cells versus macrophages), interfering with DUSP6 activity appears to result in immune restoration or the impairment of innate cytokine production.

DUSP9/MKP-4

DUSP9 is an ERK1/2-selective cytoplasmic MKP with an essential role in placental function during gestation [85,86]. Tetraploid rescue experiments showed that despite high-level DUSP9 expression in tubular epithelial cells in the ascending loop of Henle, kidney function in the absence of DUSP9 is normal [85]. DUSP9 is highly expressed in embryonic stem cells in a BMP4-dependent manner and appears to contribute to stemness [87]. Polymorphisms in the human DUSP9 gene were repeatedly associated with increased risk for diabetes in genome-wide association studies (GWAS) [88,89], consistent with a protective function of DUSP9 in the development of insulin resistance in a transgenic mouse model [90].

In the immune system, DUSP9 expression is weak in most cells types, but constitutively strong at the mRNA and protein level in murine plasmacytoid DC (pDC) [13]. This cell type is specialized in producing high-level type I IFN in response to the stimulation of TLR7/9 by nucleic acids, which has been attributed to constitutive high IRF7 levels, but is mechanistically incompletely understood. Compared to conventional DC (cDC), triggering TLR9 in pDC induces very little ERK1/2 activation, correlating with high DUSP9 levels. In addition, retrovirally enforced DUSP9 expression in cDC attenuated MAPK activation and increased IFNβ expression, showing that DUSP9 shapes cytokine/interferon production in DC by controlling ERK1/2 activation [13]. However, the conditional deletion of DUSP9 in CD11c-expressing DC did not restore ERK1/2 activation in pDC and only weakly reduced IFNβ and IL-12 expression in response to TLR9 stimulation [13]. These results indicate that either the lack of ERK activation in pDC is caused by intrinsic differences in signaling between pDC and cDC, or, alternatively, other phosphatases compensate for the absence of DUSP9, which may include DUSP5, DUSP6 or other DUSP family members expressed in pDC. Testing these hypotheses will require the simultaneous deletion of multiple DUSP genes in pDC, which is difficult to achieve by breeding conditional knockout mouse lines and may be facilitated by CRISPR/Cas9 technology.

DUSP10/MKP-5

DUSP10 is the only DUSP protein carrying an extended N-terminal domain of unknown function. DUSP10 can be found in the cytoplasm and in the nucleus, and shows selective phosphatase activity towards JNK and p38 MAPK [91,92].

DUSP10 was already defined in 2004 by Dong and coworkers as a regulator of the innate and adaptive immune system [93]. In T cells, DUSP10 down-regulates TCR-induced IFNγ-production [93], which was also observed in a murine malaria model during infection of DUSP10−/− mice with *Plasmodium yoelii* [94]. In this case, the enhanced killing of parasites and survival correlated with increased IFNγ production by T cells, which was attributable to enhanced stimulatory capacity of Dusp10−/− splenic DC [94]. DUSP10 plays a host-protective role in LPS-induced vascular injury by limiting ROS production [95] and in sepsis-induced lung injury [96]. On the other hand, DUSP10 limits production of type I IFN by macrophages infected with influenza virus, leading to a more rigorous IFN response to influenza infection in DUSP10 −/− mice, better control of viral replication and increased survival [97]. While MAPK phosphorylation was largely unaltered in DUSP10−/− cells following stimulation with the virus, the high-level induction of IFN I was associated with the increased phosphorylation of IRF-3, suggesting that IRF-3 may be a non-MAPK substrate of DUSP10 [97]. Similarly, in an earlier study, an interaction of DUSP10 (and DUSP1) with IRF-3 was shown by proximity ligation assay (PLA) in macrophages stimulated with a combination of TLR4 and TLR9 ligands (LPS and B-DNA), or infected with *Listeria monocytogenes* [98]. Since high-level type I IFN has a detrimental role in listeriosis, the inhibition of IRF-3 activation by TLR-induced DUSP10 or DUSP1 may have a beneficial role in control of bacterial replication [98].

Manley et al. have analyzed changes in DUSP gene expression in primary bronchial epithelial cells in response to rhinovirus infection and observed that DUSP10 was transiently increased followed by down-regulation after 8 h [99]. Functionally, DUSP10 attenuated chemokine expression in response to virus-induced IL-1β production, as demonstrated by siRNA knockdown of DUSP10 [99].

DUSP10 has recently been shown to constrain IL-33-mediated cytokine production by memory-type pathogenic Th2 cells [100]. The expression of DUSP10 was high in these Th2 cells but low in ILC2, which responded to IL-33 with much stronger induction of IL-5 and IL-13. This differential expression was not observed for other DUSP family members and was accompanied by an attenuated activation of p38 MAPK in Th2 cells compared to ILC2. Importantly, the CRISPR/Cas9-mediated deletion of functional DUSP10 in Th2 cells increased both p38 phosphorylation and IL-5/IL-13 production after stimulation with IL-33, whereas the overexpression of DUSP10, but not its phosphatase-dead mutant, inhibited IL-5 and IL-13 in ILC2 cells [100].

DUSP16/MKP-7

DUSP16 was identified as an LPS-inducible MKP in RAW264.7 macrophages with selectivity for JNK already in 2001 [10,101,102]. DUSP16 possesses a long C-terminal domain containing sequence motifs that control the protein's stability (PEST sequence) and localization (NES and NLS). In addition, the phosphorylation of DUSP16 at Ser446 increases the half-life of the protein [9]. DUSP16 can shuttle between cytoplasm and nucleus [10] and may therefore inhibit JNK in a compartmentalized fashion. DUSP16 also can bind to activated ERK in the cytoplasm and thereby interfere with nuclear ERK activation [103]. In addition to the full-length DUSP16 isoform A1, an alternative transcript encoding isoform B1 is generated by alternative splicing at comparable levels of steady state mRNA [104]. Due to the lack of high-quality antibodies for the detection of endogenous levels of DUSP16, the relative protein levels of both isoforms in different cell types have not been determined. Overexpressed DUSP16 A1 and B1 are readily detected in 293T cells, indicating that both are reasonably stable.

The interaction of the N-terminal domain of DUSP16 with the MAPK JNK1 has recently been analyzed in great detail by crystal structures and biochemical assays [105]. Interestingly, JNK1 binds to the 285FNFL288 segment of the DUSP16 catalytic domain (which is absent in the B1 isoform). This interaction was also critical for the dephosphorylation of JNK1 by DUSP16 in vitro and for the anti-apoptotic activity of DUSP16 after UV irradiation [105]. In contrast, the D-motif (or KIM) in the MKB-domain of DUSP16 was not required for binding to JNK1. While this work revealed an important function for the FXF motif, that is conserved in other DUSP, for the binding and regulation of JNK1, the D-motif (=KIM) appears to control the binding of DUSP16 and of DUSP10 to p38 [106].

Two labs have made use of an ES cell line with an insertion of a β-Galactosidase/Neomycin gene-trap in the fourth intron of the DUSP16 gene to generate DUSP16-deficient mice. Consistently, these mice died in the perinatal period for an as yet unknown reason [104,107]. Recently, it was demonstrated that DUSP16 deficiency in these gene-trap mice strongly enhances cell death in sensory neurons upon NGF withdrawal in vitro and a loss of sensory innervations in DUSP16-deficient embryos in vivo [108]. Moreover, DUSP16-deficient embryos develop hydrocephalus due to expansion of neural progenitors and blockade of cerebrospinal fluid circulation through the midbrain aqueduct [109]. The generation of radiation chimeras showed a largely normal reconstitution of the hematopoietic system by DUSP16-deficient homozygous gene-trap fetal liver cells and allowed the characterization of the role of DUSP16 in immune cells. DUSP16-deficient CD4+ T cells have a cell-intrinsic defect in Th17 polarization, whereas IFNγ-producing Th1 cells and IL-4-producing Th2 cells are not affected [107]. This effect is dependent on increased ERK1/2 MAPK activation in DUSP16-deficient T cells following TCR triggering, as the MEK1-inhibitor U0126 restored Th17 differentiation [107]. Analysis of innate immune activation by the TLR4 ligand LPS showed a largely comparable cytokine and chemokine output as in WT mice, with the exception of IL-12p40 which was overproduced in DUSP16-deficient mice in vivo, an effect that could be attributed to macrophages, but not DC, in vitro and was dependent on JNK activity [104]. Together, the data in the DUSP16 gene-trap mice indicate important cell type-specific roles of this MKP in the proliferation, differentiation and cytokine response of macrophages, DC and T cells (Figure 4). However, the perinatal lethality of the DUSP16 gene-trap mice and the need to generate radiation chimeras impede the straightforward experimental analysis. In addition, the expression and function of DUSP16 in multiple immune cell types can result in complex phenotypes. Therefore, the generation of conditional DUSP16 knockout mice is needed to overcome these obstacles and to investigate the cell type-specific functions of DUSP16 in detail.

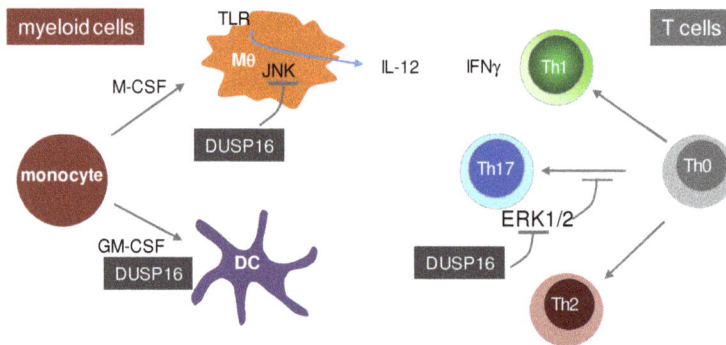

Figure 4. Role of DUSP16 in innate and adaptive immune cells. Analysis of CD4$^+$ T cell compartment and myeloid cells derived from radiation chimeras with a DUSP16-deficient hematopoietic system was performed by two labs [104,107].

2.3.2. Atypical DUSP with an Emerging Function in the Immune System

DUSP3/VHR

The *DUSP3* gene was cloned in 1992; DUSP3 is also known as Vaccinia-H1-related phosphatase (VHR) and is one of the smallest known phosphatases (21 kDa) [110]. The structure, function and regulation of DUSP3 have recently been reviewed [111]. VHR has been reported to be constitutively expressed in T cells, where it dephosphorylates ERK and JNK, but not p38 [112]. As atypical DUSP, DUSP3/VHR lacks a MAPK-binding motif. Still, the reported substrates of DUSP3 include the MAPK family members ERK1/2 [112], p38 [113] and JNK [114]. Non-MAPK substrates of DUSP3 include STAT5 (see below) and several nucleolar proteins involved in the DNA damage response and in DNA/RNA regulation [115].

DUSP3 mRNA expression levels were associated with resistance or susceptibility in infectious disease both in murine models and in humans. First, the down-regulated expression of *DUSP3* was linked to susceptibility to *Staphylococcus aureus* sepsis in A/J mice and in humans with *S. aureus* bloodstream infection [116]. Knockdown of DUSP3 in mouse macrophages led to increased cytokine production in response to *S. aureus* through the enhancement of NFκB activation [116]. Second, *DUSP3* mRNA was identified as part of a transcriptional signature for the diagnosis of tuberculosis from human peripheral blood in two independent studies [117,118]. Remarkably, *DUSP3* belonged to a set of three genes (with *GBP5* and *KLF2*) that were diagnostic for active tuberculosis in eight independent datasets from ten countries [119]. This diagnostic correlation of *DUSP3* expression in peripheral blood cells with active tuberculosis has not been linked to a function of this phosphatase in the host yet.

DUSP3 can inhibit IFNβ-induced STAT5 phosphorylation on Tyr-694/699 when overexpressed in HEK293T cells [120]. Similar to its function as a MAPK phosphatase for ERK/JNK, the dephosphorylation of Stat5 by DUSP3 required its phosphorylation at position Tyr-138 [120]. The binding specificity of DUSP3 for STAT5 has been investigated in a bioinformatics approach combining molecular modeling, docking and molecular dynamics simulations, which revealed a binding interface at the catalytic domain of DUSP3 that interacts with the SH2-domain of STAT5 and supported a model where DUSP3 binding at one SH2-domain facilitates the release of the flexible STAT5 C-terminus as a prerequisite for its dephosphorylation [121].

Loss of DUSP3 has been shown to be associated with cell-cycle arrest and senescence. The expression of *DUSP3* was undetected in G1 and was gradually increased during the progression of HeLa cells to S phase [122]. *DUSP3*$^{-/-}$ mice are fertile and healthy. However, DUSP3 deficiency led to a decrease in neovascularization and angiogenesis, likely due its expression in endothelial cells [123].

DUSP3 was shown by Rahmouni and colleagues to be a non-redundant regulator of the innate immune response to endotoxin or in polymicrobial infection [124,125]. The expression of *DUSP3* was higher in macrophages and monocytes than neutrophils, B cells and T cells. The authors elucidated that protection of DUSP3-deficient mice from septic shock involves polarization to M2-like macrophages and the decreased production of TNF. Bone marrow transfer of *DUSP3*$^{-/-}$ to irradiated WT mice showed LPS resistance in the recipients, demonstrating that the phenotype is indeed due to the deletion of *DUSP3* in hematopoietic cells, especially in monocytes [124]. Mechanistically, DUSP3-deficient macrophages unexpectedly showed reduced LPS-triggered ERK1/2 phosphorylation, whereas p38 and JNK activation kinetics were unaltered [124]. In a follow-up study, the protection against sepsis in DUSP3-deficient mice was revealed to be sex-specific for female mice and to involve estrogen-dependent reduction in ERK phosphorylation and enhanced M2 macrophage polarization [125]. The same group investigated the function of DUSP3 in tumor-associated macrophages in a Lewis Lung carcinoma (LLC) metastasis model, where *DUSP3*$^{-/-}$ macrophages displayed enhanced recruitment to LLC-bearing lungs and promoted metastatic tumor growth [126].

DUSP3 is highly expressed in human and mouse platelets. *DUSP3*$^{-/-}$ platelets displayed reduced activation and aggregate formation in response to collagen exposure in vitro, and DUSP3-deficient mice showed impaired pulmonary thromboembolism in vivo [127]. Mechanistically, the phosphorylation of the kinase SYK induced by platelet stimulation with collagen-related peptide (binding to Glycoprotein VI) or by the CLEC2-agonist Rhodocytin was selectively and strongly reduced in the absence of DUSP3. In fact, DUSP3 deficiency phenocopies the knockout of CLEC2 and GPVI [128], suggesting the functional relevance of SYK-regulation by DUSP3. It remains to be elucidated how DUSP3 participates in the activation of SYK. No hyperphosphorylation of tyrosine residues was detected by immunoblot. On the other hand, selective inhibitors of DUSP3 phosphatase activity replicated the impaired activation of SYK and platelet aggregation, indicating that DUSP3 may act by dephosphorylating phosphothreonine or serine residues on proteins in the CLEC2 or GPVI signaling complex [127].

In addition to the compounds identified and used in the platelet study by Rahmouni and colleagues, several groups have reported the development of DUSP3 inhibitors [129,130]. While these compounds have not been used in vivo yet, their availability demonstrates that the drug-targeting of DUSP3 is feasible. The identification of yet unknown substrates of this atypical DUSP should greatly facilitate the elucidation of the biological responses and signaling pathways it regulates.

DUSP11

This atypical DUSP is to date unique in its substrate specificity for phosphorylated RNA. Already in 1998 Yuan et al. identified DUSP11/PIR (for Phosphatase that interacts with RNP-complex-1) as a nuclear protein with nuclear RNA-ribonucleoprotein complexes [131]. Subsequently, DUSP11/PIR was shown to remove two phosphate groups from 5′-triphosphorylated RNA, with a selectivity of binding to phosphorylated RNA that was orders of magnitude higher over binding to phosphoprotein substrates [132]. 5′-PPP-RNA is characteristic for viral RNAs, including pre-miRNAs of several retroviruses, generated by RNA polymerase III. For proper binding of the 5p-miRNA to the RISC complex, the removal of the 5′-triphosphates is required. Burke et al. could show that the levels of BLV and AdV 5p miRNA were strongly reduced in human and sheep *DUSP11*-null cells, due to the lack of DUSP11 phosphatase activity, leading to inefficient loading onto the Argonaute proteins of the RISC complex [133,134]. These data raised the question whether DUSP11 is specific for viral RNA or also modifies endogenous cellular RNA. No consistent difference in expression of protein-coding RNA was observed when DUSP11 was deleted in two different parental cell lines. However, several RNAP-III-dependent non-coding RNAs of <500 nt length, including Alu-derived elements, were increased in samples treated with Terminator to digest 5′ monophosphorylated RNAs, indicating that DUSP11 affects phosphorylation status and steady state levels of diverse RNAP-III small nc RNA transcripts [134].

If DUSP11 regulates the level of 5′-triphosphorylated endogenous and viral RNAs, their accumulation in the absence of DUSP11 may trigger activation of pattern recognition receptors for viral RNA, especially of RIG-I that binds to 5′-triphosphorylated RNA. A recent study demonstrates that such a mechanism is indeed operating in cells during infection with the Kaposi sarcoma virus (KSV), a herpes virus [135]. Lytic infection with KSV substantially down-regulated the mRNA and protein levels of DUSP11, which was accompanied by an increase in triphosphorylated cellular non-coding vault RNA and RIG-I activity [135]. Thus, the reduced activity of DUSP11 during KSV infection results in host RNA recognition by RIG-I, which results in enhanced antiviral interferon gene expression.

A more direct anti-viral activity of DUSP11 was found toward the hepatitis C virus (HCV) [136]. Cellular XRN exonucleases can attack HCV RNA, provided that the viral transcripts are not protected by 5′-triphosphorylation. Using different approaches to delete DUSP11 from human cells, Kincaid et al. showed that dephosphorylation by DUSP11 is required to antagonize HCV replication by enabling access of XRN exonucleases to 5′ monophosphorylated HCV RNA [136]. The DNA damage response to irradiation is strongly dependent on the transcription factor p53. Caprara et al. recently identified Dusp11 as an irradiation-induced target gene of p53, which was shown to directly regulate DUSP11 expression by binding to its promoter [137].

In a genetic screening approach for phosphatases involved in the control of the replication of *Salmonella typhimurium* in a human cell line, DUSP11 (and Dusp27) were identified as candidates required for efficient inhibition of replication [138]. It is currently unclear, by which mechanism DUSP11 operates in this bacterial infection, and whether intracellular bacteria affect DUSP11 RNA expression similar to what was reported for the herpes virus KSV.

DUSP12 (=hYVH1)

This small atypical DUSP protein was initially identified in yeast [139] and is conserved across many species [140]. DUSP12 expression affects DNA content [141] and prevented cell death in mammalian cell lines [142], whereas it is involved in ribosome biosynthesis in yeast [143]. In a recent interactome analysis in human osteosarcoma cells, DUSP12 association with multiple ribonucleoproteins of the 60S and the 40S ribosomal subunits was found [144]. In addition, DUSP12 also bound to the stress granule proteins FMRP and YB1, and functionally contributed to the disassembly of stress granules [144].

Although DUSP12 lacks a MKB-domain, it bound to all three MAPK family members when overexpressed in HEK293T cells [145]. Upon overexpression in RAW264.7 macrophage cells, DUSP12 inhibited LPS-induced p38 and JNK phosphorylation, AP1-dependent promoter activation, and the expression of proinflammatory cytokines (TNF, IL-1, IL-6), whereas IL-10 production was increased [145]. When infected with *Mycobacterium bovis* BCG or *Listeria monocytogenes*, DUSP12-expressing macrophages showed less activation of p38 MAPK, a similar reduction in proinflammatory cytokine levels as observed after LPS, and increased bacterial loads [145]. The scaffold protein STAP-2 binds to both DUSP12 and p38 via its pleckstrin homology domain, which likely is required for the activity of DUSP12 towards p38 in macrophages [145]. To date, no loss-of-function experiments have been reported for DUSP12 in murine or human macrophages. Very recently, the generation of conditional DUSP12 knockout mice by CRISPR/Cas9 has been reported with liver-specific inactivation [146]. These transgenic mice will be an important resource to investigate the function of this broadly expression atypical DUSP in different immune cells during infection and inflammation.

DUSP14/MKP6

DUSP14 was identified in 2001 as a CD28-interacting protein in human T cells [147]. Its expression is strongly induced by the CD3/CD28 stimulation of T cells, suggesting it may function as a negative feedback regulator of TCR signaling. Indeed, a dominant negative DUSP14 mutant protein with a cysteine to serine mutation at position 111 in the active site (C111S) induced overshooting JNK and ERK activation after CD3/CD28 stimulation [147]. The phenotype of DUSP14-deficient mice confirmed the notion that this atypical small-size DUSP is a regulator of T cell activation: while the development of lymphoid cells in thymus and spleen was unaltered, a strong hyper-proliferation of DUSP14-deficient T cells in response to CD3 stimulation was observed [148]. Mechanistically, DUSP14 associated with TAB1 after anti-CD3 stimulation, leading to the dephosphorylation and inactivation of TAB1. Consequently, in DUSP14-deficient T cells, the activation of TAK1 and IKK was strongly enhanced, as well as that of the MAPK ERK and JNK [148]. More recently, the same group showed that the phosphatase activity of DUSP14 is dependent on TRAF2-mediated Lys63-ubiquitinylation [149]. Regulation of the NFκB pathway by DUSP14 was also described by another group in HEK293T and HeLa cells, although in this setting, DUSP14 interacted with TAK1 and regulated its phosphorylation, but not with TAB1, in response to IL-1 and TNF [150]. These differences in the molecular binding partners of DUSP14 may reflect species differences. In mice, DUSP14 is required to restrain T cell responses in vivo, since immunization with the model antigen KLH induced enhanced T cell responses and EAE induction with MOG peptide in Complete Freund's Adjuvant (CFA) induced more severe clinical disease in the mice and overshooting Th1 and Th17 cytokine production [148].

In the context of human infectious diseases, DUSP14 was identified in a large genome-wide expression quantitative trait loci (eQTL) analysis of human DC before and after infection with *M. tuberculosis* and linked to genetic susceptibility to tuberculosis by integration with genome-wide association studies (GWAS) [151]. A more recent study on the role of the genetic polymorphism in DUSP14 found that a low-expression genotype of DUSP14 was accompanied by high transcript levels of IFNGR2 and STAT1 and may thus protect against early TB development [152].

DUSP22/JKAP

This atypical DUSP is also known as MKP-X, VHX, JSP1 and LMW-DSP2. The first description of DUSP22 was by Aoyama et al. who cloned it as LMW-DSP2 from mouse testis cDNA [153]. DUSP22 was then identified as JNK-associated phosphatase (JKAP) that is required for full JNK activation in response to TNF [154]. The interaction of DUSP22 with JNK was indirect and the phosphatase activity of DUSP22 was required for JNK activation [154]. In contrast, more recent work reported that DUSP22 acts as a scaffold protein in the ASK1–MKK7–JNK pathway and that its phosphatase activity is not needed for JNK activation [155]. DUSP22 can also bind to STAT3 and inhibit IL-6-induced responses in cell lines (HEK 293T cells) [156]. The expression of DUSP22 is found in many tissues and cell types, with myeloid cells showing higher mRNA expression among immune cells (see ImmGen database). At the protein level, the highest DUSP22 levels were reported in heart and muscle [157]. Over the years, studies in immune cells have provided accumulating evidence for a function of DUSP22 in T cells, in which DUSP22 inactivates the SRC-kinase LCK and thereby inhibits TCR signaling [158]. As a consequence of enhanced TCR signaling, increased T cell proliferation and cytokine production, DUSP22-deficient mice are more susceptible in models of autoimmune disease such as EAE and develop spontaneous inflammation and autoimmunity in old age [158]. In human patients with inflammatory bowel disease (IBD), DUSP22 expression was inversely correlated with disease activity and with the levels of IL-17 and TNF in the inflamed mucosa [159]. The lentiviral overexpression of DUSP22 suppressed CD4$^+$ T cell activation status and proliferation, whereas the lentiviral siRNA knockdown of DUSP22 increased Th1/Th17 differentiation [159]. Since the down-regulation of DUSP22 expression in CD4+ T cells was also correlated with the presence and activity of Systemic Lupus Erythematodes (SLE) in human patients, DUSP22 is emerging as a biomarker for several autoimmune diseases [160]. In addition, the reduced expression of DUSP22 has been associated with T cell lymphoma subtypes, where silenced DUSP22 expression and hypomethylation of the DUSP22 promoter were found [161,162]. Interestingly, DUSP22-hypomethylated anaplastic large cell lymphomas with DUSP22 rearrangements are characterized by better prognosis which appears to be linked to the higher expression of costimulatory molecules and reduced PD1/PD1-L activity, and hence stronger immunogenicity [161].

3. Conclusions and Open Questions

In the last decade, the investigations of DUSP gene expression and function in different cells of the immune system and during immune responses in mouse models and in human disease have greatly expanded our knowledge about the role of this phosphatase family in immunity and inflammation. It is now also well established that in addition to the classical DUSP-MKP, several atypical DUSPs make important contributions to the signaling processes that control innate and adaptive immune cell functions. The field has benefitted tremendously from the generation of knockout mice for DUSP family members, enabling researchers to scrutinize the relevance of individual DUSP genes in immune responses and inflammation in vivo. Much has been learned about the amazing complexity of how the expression and function of different DUSP family members are regulated at the level of transcription, mRNA stability and post-translational modifications. In turn, the investigations into the mechanisms of the DUSP-mediated control of signaling processes have revealed a much more detailed and multi-faceted picture than the simple model of MAPK-binding and dephosphorylation. Elegant studies have (1) shown the importance of the compartmentalization of DUSP activity [65,163], (2) defined the molecular details of DUSP interaction with specific substrates [66,105], and (3) suggested the emerging role of non-MAPK substrates and binding partners of several DUSP [47]. Moreover, the potential of targeting DUSPs by pharmacological means has been demonstrated in several examples identifying specific inhibitors.

Yet, despite this progress, and in some cases because of newly made unexpected findings, there are several open questions and hurdles the DUSP field is facing. These need to be overcome to fully understand the role of the DUSP family in immune cell development, activation and regulation, and to potentially employ this knowledge for the manipulation of DUSP functions in infection, pathological inflammation or vaccination-triggered immunity:

- *Obtain better genetic tools to investigate cell type-specific and redundant functions of DUSP family members* While the power of knockout mouse models has been instrumental to define an immunological role for many DUSP, the expression of several DUSP in diverse immune and non-hematopoietic cell types can produce complex phenotypes and impedes the clear definition of the contribution made by DUSP deficiency in a specific cell type. To date, conditional knockout models have been published for *DUSP1, DUSP9,* and *DUSP12*. The generation and wide availability of additional conditional knockout lines is eagerly awaited for more family members. In cases where the genetic abrogation of a specific DUSP gene does not produce a phenotypic change, the possibility of functional compensation by one of the other family members is difficult to exclude. To tackle this potential redundancy, cells or mice lacking more than one DUSP gene would be very helpful. The crossing of multiple DUSP-deficient mice to obtain combined knockouts is possible but very time-consuming and expensive. Hence, the application of CRISPR/Cas9 technology to target several DUSP genes of interest in a combinatorial manner appears very tempting and may help to overcome this technical hurdle.
- *Comprehensive identification of the interaction partners and substrates of DUSP proteins* Beyond the paradigm of classical MKP-DUSP functions as regulators of MAPK activity, evidence for the direct binding and dephosphorylation of other molecular interaction partners has been emerging. The

massive advances in proteomics and phospho-proteomics methods in the last decade now provide the opportunity for an unbiased, in-depth investigation of DUSP interaction partners and to quantitatively assess the impact of DUSP overexpression or deletion on protein phosphorylation in cells. Such datasets can then also be employed for bioinformatics analysis of signaling pathways regulated by DUSP proteins. In addition, the increasing availability of structural information on DUSP proteins will facilitate the validation of candidate interactors by the computational modelling of their binding modes, an important prerequisite for the design and development of novel allosteric inhibitors of DUSPs. Depending on the specific DUSP, and on the cell type, such inhibitors may enhance immune responses and anti-microbial mechanisms, or conversely lead to the attenuation of inflammation. Therefore, the opposite approach to increase the expression or activity of certain DUSPs could be an attractive complementary or alternative strategy for the modulation of signaling and immune activation states.

Author Contributions: The conceptualization, literature search and writing of this review article were performed by R.L. and F.A.M.R.

Funding: The research on DUSP in immunity in the Lang lab was funded by Deutsche Forschungsgemeinschaft (SFB643, TPA10 to R.L.) and the Interdisziplinäres Zentrum für Klinische Forschung of the Medical Faculty at the Friedrich-Alexander-Universität Erlangen-Nürnberg (IZKF grant F1-11 to R.L.).

Conflicts of Interest: The authors declare no conflict of interest.

Abbreviations

CLP	Cecal ligation and puncture
DC	Dendritic cell
DUSP	Dual-specificity phosphatase
EAE	Experimental autoimmune encephalomyelitis
ICL	Idiopathic CD4 lymphopenia
ILC	Innate lymphoid cell
MAPK	Mitogen activated protein kinase
MKB	MAPK-binding domain
MKP	MAPK phosphatase
PRRPTP	Pattern recognition receptorProtein tyrosine phosphatase
TCR	T cell receptor
TLR	Toll-like receptor

References

1. Chen, Y.N.; LaMarche, M.J.; Chan, H.M.; Fekkes, P.; Garcia-Fortanet, J.; Acker, M.G.; Antonakos, B.; Chen, C.H.; Chen, Z.; Cooke, V.G.; et al. Allosteric inhibition of SHP2 phosphatase inhibits cancers driven by receptor tyrosine kinases. *Nature* **2016**, *535*, 148–152. [CrossRef] [PubMed]

2. Mullard, A. Phosphatases start shedding their stigma of undruggability. *Nat. Rev. Drug Discov.* **2018**, *17*, 847–849. [CrossRef] [PubMed]

3. Ran, H.; Tsutsumi, R.; Araki, T.; Neel, B.G. Sticking It to Cancer with Molecular Glue for SHP2. *Cancer Cell* **2016**, *30*, 194–196. [CrossRef]

4. Lang, R.; Hammer, M.; Mages, J. DUSP Meet Immunology: Dual Specificity MAPK Phosphatases in Control of the Inflammatory Response. *J. Immunol.* **2006**, *177*, 7497–7504. [CrossRef] [PubMed]

5. Liu, Y.; Shepherd, E.G.; Nelin, L.D. MAPK phosphatases–regulating the immune response. *Nat. Rev. Immunol.* **2007**, *7*, 202–212. [CrossRef]

6. Jeffrey, K.L.; Camps, M.; Rommel, C.; Mackay, C.R. Targeting dual-specificity phosphatases: Manipulating MAP kinase signalling and immune responses. *Nat. Rev. Drug Discov.* **2007**, *6*, 391–403. [CrossRef] [PubMed]

7. Alonso, A.; Pulido, R. The extended human PTPome: A growing tyrosine phosphatase family. *FEBS J.* **2016**, *283*, 2197–2201. [CrossRef]

8. Caunt, C.J.; Keyse, S.M. Dual-specificity MAP kinase phosphatases (MKPs): Shaping the outcome of MAP kinase signalling. *FEBS J.* **2013**, *280*, 489–504. [CrossRef]

9. Katagiri, C.; Masuda, K.; Urano, T.; Yamashita, K.; Araki, Y.; Kikuchi, K.; Shima, H. Phosphorylation of Ser-446 determines stability of MKP-7. *J. Biol. Chem.* **2005**, *280*, 14716–14722. [CrossRef]

10. Masuda, K.; Shima, H.; Watanabe, M.; Kikuchi, K. MKP-7, a novel mitogen-activated protein kinase phosphatase, functions as a shuttle protein. *J. Biol. Chem.* **2001**, *276*, 39002–39011. [CrossRef]

11. Hammer, M.; Mages, J.; Dietrich, H.; Schmitz, F.; Striebel, F.; Murray, P.J.; Wagner, H.; Lang, R. Control of dual-specificity phosphatase-1 expression in activated macrophages by IL-10. *Eur. J. Immunol.* **2005**, *35*, 2991–3001. [CrossRef] [PubMed]

12. Hansen, M.; Peltier, J.; Killy, B.; Amin, B.; Bodendorfer, B.; Hartlova, A.; Uebel, S.; Bosmann, M.; Hofmann, J.; Buttner, C.; et al. Macrophage phosphoproteome analysis reveals MINCLE-dependent and -independent mycobacterial cord factor signaling. *Mol. Cell. Proteom.* **2019**. [CrossRef] [PubMed]

13. Niedzielska, M.; Raffi, F.A.; Tel, J.; Muench, S.; Jozefowski, K.; Alati, N.; Lahl, K.; Mages, J.; Billmeier, U.; Schiemann, M.; et al. Selective Expression of the MAPK Phosphatase Dusp9/MKP-4 in Mouse Plasmacytoid Dendritic Cells and Regulation of IFN-beta Production. *J. Immunol.* **2015**, *195*, 1753–1762. [CrossRef]

14. Kuwano, Y.; Kim, H.H.; Abdelmohsen, K.; Pullmann, R., Jr.; Martindale, J.L.; Yang, X.; Gorospe, M. MKP-1 mRNA stabilization and translational control by RNA-binding proteins HuR and NF90. *Mol. Cell. Biol.* **2008**, *28*, 4562–4575. [CrossRef] [PubMed]

15. Schott, J.; Reitter, S.; Philipp, J.; Haneke, K.; Schafer, H.; Stoecklin, G. Translational regulation of specific mRNAs controls feedback inhibition and survival during macrophage activation. *PLoS Genet.* **2014**, *10*, e1004368. [CrossRef] [PubMed]

16. Brondello, J.M.; Pouyssegur, J.; McKenzie, F.R. Reduced MAP kinase phosphatase-1 degradation after p42/p44MAPK-dependent phosphorylation. *Science* **1999**, *286*, 2514–2517. [CrossRef] [PubMed]

17. Chen, P.; Li, J.; Barnes, J.; Kokkonen, G.C.; Lee, J.C.; Liu, Y. Restraint of proinflammatory cytokine biosynthesis by mitogen-activated protein kinase phosphatase-1 in lipopolysaccharide-stimulated macrophages. *J. Immunol.* **2002**, *169*, 6408–6416. [CrossRef] [PubMed]

18. Klotz, C.; Ziegler, T.; Figueiredo, A.S.; Rausch, S.; Hepworth, M.R.; Obsivac, N.; Sers, C.; Lang, R.; Hammerstein, P.; Lucius, R.; et al. A helminth immunomodulator exploits host signaling events to regulate cytokine production in macrophages. *PLoS Pathog.* **2011**, *7*, e1001248. [CrossRef] [PubMed]

19. Shin, H.S.; Yoo, I.H.; Kim, Y.J.; Kim, H.B.; Jin, S.; Ha, U.H. MKP1 regulates the induction of inflammatory response by pneumococcal pneumolysin in human epithelial cells. *FEMS Immunol. Med. Microbiol.* **2010**, *60*, 171–178. [CrossRef]

20. Moyes, D.L.; Runglall, M.; Murciano, C.; Shen, C.; Nayar, D.; Thavaraj, S.; Kohli, A.; Islam, A.; Mora-Montes, H.; Challacombe, S.J.; et al. A biphasic innate immune MAPK response discriminates between the yeast and hyphal forms of Candida albicans in epithelial cells. *Cell Host Microbe* **2010**, *8*, 225–235. [CrossRef]

21. Kassel, O.; Sancono, A.; Kratzschmar, J.; Kreft, B.; Stassen, M.; Cato, A.C. Glucocorticoids inhibit MAP kinase via increased expression and decreased degradation of MKP-1. *EMBO J.* **2001**, *20*, 7108–7116. [CrossRef] [PubMed]

22. Lasa, M.; Abraham, S.M.; Boucheron, C.; Saklatvala, J.; Clark, A.R. Dexamethasone causes sustained expression of mitogen-activated protein kinase (MAPK) phosphatase 1 and phosphatase-mediated inhibition of MAPK p38. *Mol. Cell. Biol.* **2002**, *22*, 7802–7811. [CrossRef] [PubMed]

23. Sakai, A.; Han, J.; Cato, A.C.; Akira, S.; Li, J.D. Glucocorticoids synergize with IL-1beta to induce TLR2 expression via MAP Kinase Phosphatase-1-dependent dual Inhibition of MAPK JNK and p38 in epithelial cells. *BMC Mol. Biol.* **2004**, *5*, 2. [CrossRef] [PubMed]

24. Abraham, S.M.; Lawrence, T.; Kleiman, A.; Warden, P.; Medghalchi, M.; Tuckermann, J.; Saklatvala, J.; Clark, A.R. Antiinflammatory effects of dexamethasone are partly dependent on induction of dual specificity phosphatase 1. *J. Exp. Med.* **2006**, *203*, 1883–1889. [CrossRef] [PubMed]

25. Hammer, M.; Mages, J.; Dietrich, H.; Servatius, A.; Howells, N.; Cato, A.C.; Lang, R. Dual specificity phosphatase 1 (DUSP1) regulates a subset of LPS-induced genes and protects mice from lethal endotoxin shock. *J. Exp. Med.* **2006**, *203*, 15–20. [CrossRef]

26. Chi, H.; Barry, S.P.; Roth, R.J.; Wu, J.J.; Jones, E.A.; Bennett, A.M.; Flavell, R.A. Dynamic regulation of pro- and anti-inflammatory cytokines by MAPK phosphatase 1 (MKP-1) in innate immune responses. *Proc. Natl. Acad. Sci. USA* **2006**, *103*, 2274–2279. [CrossRef]

27. Salojin, K.V.; Owusu, I.B.; Millerchip, K.A.; Potter, M.; Platt, K.A.; Oravecz, T. Essential role of MAPK phosphatase-1 in the negative control of innate immune responses. *J. Immunol.* **2006**, *176*, 1899–1907. [CrossRef]

28. Zhao, Q.; Wang, X.; Nelin, L.D.; Yao, Y.; Matta, R.; Manson, M.E.; Baliga, R.S.; Meng, X.; Smith, C.V.; Bauer, J.A.; et al. MAP kinase phosphatase 1 controls innate immune responses and suppresses endotoxic shock. *J. Exp. Med.* **2006**, *203*, 131–140. [CrossRef]

29. Frazier, W.J.; Wang, X.; Wancket, L.M.; Li, X.A.; Meng, X.; Nelin, L.D.; Cato, A.C.; Liu, Y. Increased inflammation, impaired bacterial clearance, and metabolic disruption after gram-negative sepsis in Mkp-1-deficient mice. *J. Immunol.* **2009**, *183*, 7411–7419. [CrossRef]

30. Hammer, M.; Echtenachter, B.; Weighardt, H.; Jozefowski, K.; Rose-John, S.; Mannel, D.N.; Holzmann, B.; Lang, R. Increased inflammation and lethality of Dusp1-/- mice in polymicrobial peritonitis models. *Immunology* **2010**, *131*, 395–404. [CrossRef]

31. Li, J.; Wang, X.; Ackerman, W.E.T.; Batty, A.J.; Kirk, S.G.; White, W.M.; Wang, X.; Anastasakis, D.; Samavati, L.; Buhimschi, I.; et al. Dysregulation of Lipid Metabolism in Mkp-1 Deficient Mice during Gram-Negative Sepsis. *Int. J. Mol. Sci.* **2018**, *19*, 3904. [CrossRef]

32. Wang, X.; Meng, X.; Kuhlman, J.R.; Nelin, L.D.; Nicol, K.K.; English, B.K.; Liu, Y. Knockout of Mkp-1 enhances the host inflammatory responses to gram-positive bacteria. *J. Immunol.* **2007**, *178*, 5312–5320. [CrossRef] [PubMed]

33. Rodriguez, N.; Dietrich, H.; Mossbrugger, I.; Weintz, G.; Scheller, J.; Hammer, M.; Quintanilla-Martinez, L.; Rose-John, S.; Miethke, T.; Lang, R. Increased inflammation and impaired resistance to Chlamydophila pneumoniae infection in Dusp1(-/-) mice: Critical role of IL-6. *J. Leukoc. Biol.* **2010**, *88*, 579–587. [CrossRef]

34. Ananieva, O.; Darragh, J.; Johansen, C.; Carr, J.M.; McIlrath, J.; Park, J.M.; Wingate, A.; Monk, C.E.; Toth, R.; Santos, S.G.; et al. The kinases MSK1 and MSK2 act as negative regulators of Toll-like receptor signaling. *Nat. Immunol.* **2008**, *9*, 1028. [CrossRef] [PubMed]

35. Schaljo, B.; Kratochvill, F.; Gratz, N.; Sadzak, I.; Sauer, I.; Hammer, M.; Vogl, C.; Strobl, B.; Muller, M.; Blackshear, P.J.; et al. Tristetraprolin is required for full anti-inflammatory response of murine macrophages to IL-10. *J. Immunol.* **2009**, *183*, 1197–1206. [CrossRef]

36. McGuire, V.A.; Rosner, D.; Ananieva, O.; Ross, E.A.; Elcombe, S.E.; Naqvi, S.; van den Bosch, M.M.; Monk, C.E.; Ruiz-Zorrilla Diez, T.; Clark, A.R.; et al. Beta Interferon Production Is Regulated by p38 Mitogen-Activated Protein Kinase in Macrophages via both MSK1/2- and Tristetraprolin-Dependent Pathways. *Mol. Cell. Biol.* **2017**, *37*, e00454-16. [CrossRef]

37. Smallie, T.; Ross, E.A.; Ammit, A.J.; Cunliffe, H.E.; Tang, T.; Rosner, D.R.; Ridley, M.L.; Buckley, C.D.; Saklatvala, J.; Dean, J.L.; et al. Dual-Specificity Phosphatase 1 and Tristetraprolin Cooperate To Regulate Macrophage Responses to Lipopolysaccharide. *J. Immunol.* **2015**, *195*, 277–288. [CrossRef]

38. Clark, A.R.; Dean, J.L. The control of inflammation via the phosphorylation and dephosphorylation of tristetraprolin: A tale of two phosphatases. *Biochem. Soc. Trans.* **2016**, *44*, 1321–1337. [CrossRef]

39. Huang, G.; Wang, Y.; Shi, L.Z.; Kanneganti, T.D.; Chi, H. Signaling by the phosphatase MKP-1 in dendritic cells imprints distinct effector and regulatory T cell fates. *Immunity* **2011**, *35*, 45–58. [CrossRef]

40. Huang, G.; Wang, Y.; Vogel, P.; Chi, H. Control of IL-17 receptor signaling and tissue inflammation by the p38alpha-MKP-1 signaling axis in a mouse model of multiple sclerosis. *Sci. Signal.* **2015**, *8*, ra24. [CrossRef]

41. Lawan, A.; Min, K.; Zhang, L.; Canfran-Duque, A.; Jurczak, M.J.; Camporez, J.P.G.; Nie, Y.; Gavin, T.P.; Shulman, G.I.; Fernandez-Hernando, C.; et al. Skeletal Muscle-Specific Deletion of MKP-1 Reveals a p38 MAPK/JNK/Akt Signaling Node That Regulates Obesity-Induced Insulin Resistance. *Diabetes* **2018**, *67*, 624–635. [CrossRef] [PubMed]

42. Lawan, A.; Zhang, L.; Gatzke, F.; Min, K.; Jurczak, M.J.; Al-Mutairi, M.; Richter, P.; Camporez, J.P.; Couvillon, A.; Pesta, D.; et al. Hepatic mitogen-activated protein kinase phosphatase 1 selectively regulates glucose metabolism and energy homeostasis. *Mol. Cell. Biol.* **2015**, *35*, 26–40. [CrossRef] [PubMed]

43. Rohan, P.J.; Davis, P.; Moskaluk, C.A.; Kearns, M.; Krutzsch, H.; Siebenlist, U.; Kelly, K. PAC-1: A mitogen-induced nuclear protein tyrosine phosphatase. *Science* **1993**, *259*, 1763–1766. [CrossRef]

44. Grumont, R.J.; Rasko, J.E.; Strasser, A.; Gerondakis, S. Activation of the mitogen-activated protein kinase pathway induces transcription of the PAC-1 phosphatase gene. *Mol. Cell. Biol.* **1996**, *16*, 2913–2921. [CrossRef] [PubMed]

45. Yin, Y.; Liu, Y.X.; Jin, Y.J.; Hall, E.J.; Barrett, J.C. PAC1 phosphatase is a transcription target of p53 in signalling apoptosis and growth suppression. *Nature* **2003**, *422*, 527–531. [CrossRef] [PubMed]

46. Jeffrey, K.L.; Brummer, T.; Rolph, M.S.; Liu, S.M.; Callejas, N.A.; Grumont, R.J.; Gillieron, C.; Mackay, F.; Grey, S.; Camps, M.; et al. Positive regulation of immune cell function and inflammatory responses by phosphatase PAC-1. *Nat. Immunol.* **2006**, *7*, 274–283. [CrossRef] [PubMed]

47. Lu, D.; Liu, L.; Ji, X.; Gao, Y.; Chen, X.; Liu, Y.; Liu, Y.; Zhao, X.; Li, Y.; Li, Y.; et al. The phosphatase DUSP2 controls the activity of the transcription activator STAT3 and regulates TH17 differentiation. *Nat. Immunol.* **2015**, *16*, 1263–1273. [CrossRef]

48. Kanemaru, H.; Yamane, F.; Tanaka, H.; Maeda, K.; Satoh, T.; Akira, S. BATF2 activates DUSP2 gene expression and up-regulates NF-kappaB activity via phospho-STAT3 dephosphorylation. *Int. Immunol.* **2018**, *30*, 255–265. [CrossRef]

49. Misra-Press, A.; Rim, C.S.; Yao, H.; Roberson, M.S.; Stork, P.J. A novel mitogen-activated protein kinase phosphatase. Structure, expression, and regulation. *J. Biol. Chem.* **1995**, *270*, 14587–14596. [CrossRef]

50. Sloss, C.M.; Cadalbert, L.; Finn, S.G.; Fuller, S.J.; Plevin, R. Disruption of two putative nuclear localization sequences is required for cytosolic localization of mitogen-activated protein kinase phosphatase-2. *Cell Signal.* **2005**, *17*, 709–716. [CrossRef]

51. Chu, Y.; Solski, P.A.; Khosravi-Far, R.; Der, C.J.; Kelly, K. The mitogen-activated protein kinase phosphatases PAC1, MKP-1, and MKP-2 have unique substrate specificities and reduced activity in vivo toward the ERK2 sevenmaker mutation. *J. Biol. Chem.* **1996**, *271*, 6497–6501. [CrossRef] [PubMed]

52. Al-Mutairi, M.S.; Cadalbert, L.C.; McGachy, H.A.; Shweash, M.; Schroeder, J.; Kurnik, M.; Sloss, C.M.; Bryant, C.E.; Alexander, J.; Plevin, R. MAP kinase phosphatase-2 plays a critical role in response to infection by Leishmania mexicana. *PLoS Pathog.* **2010**, *6*, e1001192. [CrossRef] [PubMed]

53. Woods, S.; Schroeder, J.; McGachy, H.A.; Plevin, R.; Roberts, C.W.; Alexander, J. MAP kinase phosphatase-2 plays a key role in the control of infection with Toxoplasma gondii by modulating iNOS and arginase-1 activities in mice. *PLoS Pathog.* **2013**, *9*, e1003535. [CrossRef]

54. Cornell, T.T.; Rodenhouse, P.; Cai, Q.; Sun, L.; Shanley, T.P. Mitogen-activated protein kinase phosphatase 2 regulates the inflammatory response in sepsis. *Infect. Immun.* **2010**, *78*, 2868–2876. [CrossRef] [PubMed]

55. Barbour, M.; Plevin, R.; Jiang, H.R. MAP kinase phosphatase 2 deficient mice develop attenuated experimental autoimmune encephalomyelitis through regulating dendritic cells and T cells. *Sci. Rep.* **2016**, *6*, 38999. [CrossRef]

56. Huang, C.Y.; Lin, Y.C.; Hsiao, W.Y.; Liao, F.H.; Huang, P.Y.; Tan, T.H. DUSP4 deficiency enhances CD25 expression and CD4+ T-cell proliferation without impeding T-cell development. *Eur. J. Immunol.* **2012**, *42*, 476–488. [CrossRef]

57. Hsiao, W.Y.; Lin, Y.C.; Liao, F.H.; Chan, Y.C.; Huang, C.Y. Dual-Specificity Phosphatase 4 Regulates STAT5 Protein Stability and Helper T Cell Polarization. *PLoS ONE* **2015**, *10*, e0145880. [CrossRef]

58. Yan, D.; Farache, J.; Mingueneau, M.; Mathis, D.; Benoist, C. Imbalanced signal transduction in regulatory T cells expressing the transcription factor FoxP3. *Proc. Natl. Acad. Sci. USA* **2015**, *112*, 14942–14947. [CrossRef]

59. Yu, M.; Li, G.; Lee, W.W.; Yuan, M.; Cui, D.; Weyand, C.M.; Goronzy, J.J. Signal inhibition by the dual-specific phosphatase 4 impairs T cell-dependent B-cell responses with age. *Proc. Natl. Acad. Sci. USA* **2012**, *109*, E879–E888. [CrossRef]

60. Bignon, A.; Regent, A.; Klipfel, L.; Desnoyer, A.; de la Grange, P.; Martinez, V.; Lortholary, O.; Dalloul, A.; Mouthon, L.; Balabanian, K. DUSP4-mediated accelerated T-cell senescence in idiopathic CD4 lymphopenia. *Blood* **2015**, *125*, 2507–2518. [CrossRef]

61. Ahmad, D.S.; Esmadi, M.; Steinmann, W.C. Idiopathic CD4 Lymphocytopenia: Spectrum of opportunistic infections, malignancies, and autoimmune diseases. *Avicenna J. Med.* **2013**, *3*, 37–47.

62. Kwak, S.P.; Dixon, J.E. Multiple dual specificity protein tyrosine phosphatases are expressed and regulated differentially in liver cell lines. *J. Biol. Chem.* **1995**, *270*, 1156–1160. [CrossRef] [PubMed]

63. Mandl, M.; Slack, D.N.; Keyse, S.M. Specific inactivation and nuclear anchoring of extracellular signal-regulated kinase 2 by the inducible dual-specificity protein phosphatase DUSP5. *Mol. Cell. Biol.* **2005**, *25*, 1830–1845. [CrossRef] [PubMed]

64. Rushworth, L.K.; Kidger, A.M.; Delavaine, L.; Stewart, G.; van Schelven, S.; Davidson, J.; Bryant, C.J.; Caddye, E.; East, P.; Caunt, C.J.; et al. Dual-specificity phosphatase 5 regulates nuclear ERK activity and suppresses skin cancer by inhibiting mutant Harvey-Ras (HRasQ61L)-driven SerpinB2 expression. *Proc. Natl. Acad. Sci. USA* **2014**, *111*, 18267–18272. [CrossRef] [PubMed]

65. Kidger, A.M.; Rushworth, L.K.; Stellzig, J.; Davidson, J.; Bryant, C.J.; Bayley, C.; Caddye, E.; Rogers, T.; Keyse, S.M.; Caunt, C.J. Dual-specificity phosphatase 5 controls the localized inhibition, propagation, and transforming potential of ERK signaling. *Proc. Natl. Acad. Sci. USA* **2017**, *114*, E317–E326. [CrossRef] [PubMed]

66. Kutty, R.G.; Talipov, M.R.; Bongard, R.D.; Lipinski, R.A.J.; Sweeney, N.L.; Sem, D.S.; Rathore, R.; Ramchandran, R. Dual Specificity Phosphatase 5-Substrate Interaction: A Mechanistic Perspective. *Compr. Physiol.* **2017**, *7*, 1449–1461. [PubMed]

67. Kovanen, P.E.; Rosenwald, A.; Fu, J.; Hurt, E.M.; Lam, L.T.; Giltnane, J.M.; Wright, G.; Staudt, L.M.; Leonard, W.J. Analysis of gamma c-family cytokine target genes. Identification of dual-specificity phosphatase 5 (DUSP5) as a regulator of mitogen-activated protein kinase activity in interleukin-2 signaling. *J. Biol. Chem.* **2003**, *278*, 5205–5213. [CrossRef]

68. Tanzola, M.B.; Kersh, G.J. The dual specificity phosphatase transcriptome of the murine thymus. *Mol. Immunol.* **2006**, *43*, 754–762. [CrossRef]

69. Kovanen, P.E.; Bernard, J.; Al-Shami, A.; Liu, C.; Bollenbacher-Reilley, J.; Young, L.; Pise-Masison, C.; Spolski, R.; Leonard, W.J. T-cell development and function are modulated by dual specificity phosphatase DUSP5. *J. Biol. Chem.* **2008**, *283*, 17362–17369. [CrossRef]

70. Kutty, R.G.; Xin, G.; Schauder, D.M.; Cossette, S.M.; Bordas, M.; Cui, W.; Ramchandran, R. Dual Specificity Phosphatase 5 Is Essential for T Cell Survival. *PLoS ONE* **2016**, *11*, e0167246. [CrossRef]

71. Holmes, D.A.; Yeh, J.H.; Yan, D.; Xu, M.; Chan, A.C. Dusp5 negatively regulates IL-33-mediated eosinophil survival and function. *EMBO J.* **2015**, *34*, 218–235. [CrossRef] [PubMed]

72. Seo, H.; Cho, Y.C.; Ju, A.; Lee, S.; Park, B.C.; Park, S.G.; Kim, J.H.; Kim, K.; Cho, S. Dual-specificity phosphatase 5 acts as an anti-inflammatory regulator by inhibiting the ERK and NF-kappaB signaling pathways. *Sci. Rep.* **2017**, *7*, 17348. [CrossRef]

73. Grasset, M.F.; Gobert-Gosse, S.; Mouchiroud, G.; Bourette, R.P. Macrophage differentiation of myeloid progenitor cells in response to M-CSF is regulated by the dual-specificity phosphatase DUSP5. *J. Leukoc. Biol.* **2010**, *87*, 127–135. [CrossRef]

74. Hu, N.; Qiu, Y.; Dong, F. Role of Erk1/2 signaling in the regulation of neutrophil versus monocyte development in response to G-CSF and M-CSF. *J. Biol. Chem.* **2015**, *290*, 24561–24573. [CrossRef] [PubMed]

75. Arkell, R.S.; Dickinson, R.J.; Squires, M.; Hayat, S.; Keyse, S.M.; Cook, S.J. DUSP6/MKP-3 inactivates ERK1/2 but fails to bind and inactivate ERK5. *Cell Signal.* **2008**, *20*, 836–843. [CrossRef]

76. Gonzalez-Navajas, J.M.; Fine, S.; Law, J.; Datta, S.K.; Nguyen, K.P.; Yu, M.; Corr, M.; Katakura, K.; Eckman, L.; Lee, J.; et al. TLR4 signaling in effector CD4+ T cells regulates TCR activation and experimental colitis in mice. *J. Clin. Investig.* **2010**, *120*, 570–581. [CrossRef] [PubMed]

77. Ham, J.E.; Oh, E.K.; Kim, D.H.; Choi, S.H. Differential expression profiles and roles of inducible DUSPs and ERK1/2-specific constitutive DUSP6 and DUSP7 in microglia. *Biochem. Biophys. Res. Commun.* **2015**, *467*, 254–260. [CrossRef] [PubMed]

78. Maillet, M.; Purcell, N.H.; Sargent, M.A.; York, A.J.; Bueno, O.F.; Molkentin, J.D. DUSP6 (MKP3) null mice show enhanced ERK1/2 phosphorylation at baseline and increased myocyte proliferation in the heart affecting disease susceptibility. *J. Biol. Chem.* **2008**, *283*, 31246–31255. [CrossRef] [PubMed]

79. Bertin, S.; Lozano-Ruiz, B.; Bachiller, V.; Garcia-Martinez, I.; Herdman, S.; Zapater, P.; Frances, R.; Such, J.; Lee, J.; Raz, E.; et al. Dual-specificity phosphatase 6 regulates CD4+ T-cell functions and restrains spontaneous colitis in IL-10-deficient mice. *Mucosal Immunol.* **2015**, *8*, 505–515. [CrossRef] [PubMed]

80. Hsu, W.C.; Chen, M.Y.; Hsu, S.C.; Huang, L.R.; Kao, C.Y.; Cheng, W.H.; Pan, C.H.; Wu, M.S.; Yu, G.Y.; Hung, M.S.; et al. DUSP6 mediates T cell receptor-engaged glycolysis and restrains TFH cell differentiation. *Proc. Natl. Acad. Sci. USA* **2018**, *115*, E8027–E8036. [CrossRef]

81. Li, G.; Yu, M.; Lee, W.W.; Tsang, M.; Krishnan, E.; Weyand, C.M.; Goronzy, J.J. Decline in miR-181a expression with age impairs T cell receptor sensitivity by increasing DUSP6 activity. *Nat. Med.* **2012**, *18*, 1518–1524. [CrossRef] [PubMed]

82. Nyunoya, T.; Monick, M.M.; Powers, L.S.; Yarovinsky, T.O.; Hunninghake, G.W. Macrophages survive hyperoxia via prolonged ERK activation due to phosphatase down-regulation. *J. Biol. Chem.* **2005**, *280*, 26295–26302. [CrossRef]

83. Zhang, F.; Tang, B.; Zhang, Z.; Xu, D.; Ma, G. DUSP6 Inhibitor (E/Z)-BCI Hydrochloride Attenuates Lipopolysaccharide-Induced Inflammatory Responses in Murine Macrophage Cells via Activating the Nrf2 Signaling Axis and Inhibiting the NF-kappaB Pathway. *Inflammation* **2018**. [CrossRef] [PubMed]

84. Srivastava, N.; Sudan, R.; Saha, B. CD40-modulated dual-specificity phosphatases MAPK phosphatase (MKP)-1 and MKP-3 reciprocally regulate Leishmania major infection. *J. Immunol.* **2011**, *186*, 5863–5872. [CrossRef]

85. Christie, G.R.; Williams, D.J.; Macisaac, F.; Dickinson, R.J.; Rosewell, I.; Keyse, S.M. The dual-specificity protein phosphatase DUSP9/MKP-4 is essential for placental function but is not required for normal embryonic development. *Mol. Cell. Biol.* **2005**, *25*, 8323–8333. [CrossRef] [PubMed]

86. Dickinson, R.J.; Eblaghie, M.C.; Keyse, S.M.; Morriss-Kay, G.M. Expression of the ERK-specific MAP kinase phosphatase PYST1/MKP3 in mouse embryos during morphogenesis and early organogenesis. *Mech. Dev.* **2002**, *113*, 193–196. [CrossRef]

87. Li, Z.; Fei, T.; Zhang, J.; Zhu, G.; Wang, L.; Lu, D.; Chi, X.; Teng, Y.; Hou, N.; Yang, X.; et al. BMP4 Signaling Acts via dual-specificity phosphatase 9 to control ERK activity in mouse embryonic stem cells. *Cell Stem Cell* **2012**, *10*, 171–182. [CrossRef]

88. Fukuda, H.; Imamura, M.; Tanaka, Y.; Iwata, M.; Hirose, H.; Kaku, K.; Maegawa, H.; Watada, H.; Tobe, K.; Kashiwagi, A.; et al. A single nucleotide polymorphism within DUSP9 is associated with susceptibility to type 2 diabetes in a Japanese population. *PLoS ONE* **2012**, *7*, e46263. [CrossRef]

89. Bao, X.Y.; Peng, B.; Yang, M.S. Replication study of novel risk variants in six genes with type 2 diabetes and related quantitative traits in the Han Chinese lean individuals. *Mol. Biol. Rep.* **2012**, *39*, 2447–2454. [CrossRef]

90. Emanuelli, B.; Eberle, D.; Suzuki, R.; Kahn, C.R. Overexpression of the dual-specificity phosphatase MKP-4/DUSP-9 protects against stress-induced insulin resistance. *Proc. Natl. Acad. Sci. USA* **2008**, *105*, 3545–3550. [CrossRef]

91. Tanoue, T.; Moriguchi, T.; Nishida, E. Molecular cloning and characterization of a novel dual specificity phosphatase, MKP-5. *J. Biol. Chem.* **1999**, *274*, 19949–19956. [CrossRef] [PubMed]

92. Theodosiou, A.; Smith, A.; Gillieron, C.; Arkinstall, S.; Ashworth, A. MKP5, a new member of the MAP kinase phosphatase family, which selectively dephosphorylates stress-activated kinases. *Oncogene* **1999**, *18*, 6981–6988. [CrossRef] [PubMed]

93. Zhang, Y.; Blattman, J.N.; Kennedy, N.J.; Duong, J.; Nguyen, T.; Wang, Y.; Davis, R.J.; Greenberg, P.D.; Flavell, R.A.; Dong, C. Regulation of innate and adaptive immune responses by MAP kinase phosphatase 5. *Nature* **2004**, *430*, 793–797. [CrossRef] [PubMed]

94. Cheng, Q.; Zhang, Q.; Xu, X.; Yin, L.; Sun, L.; Lin, X.; Dong, C.; Pan, W. MAPK phosphotase 5 deficiency contributes to protection against blood-stage Plasmodium yoelii 17XL infection in mice. *J. Immunol.* **2014**, *192*, 3686–3696. [CrossRef] [PubMed]

95. Qian, F.; Deng, J.; Cheng, N.; Welch, E.J.; Zhang, Y.; Malik, A.B.; Flavell, R.A.; Dong, C.; Ye, R.D. A non-redundant role for MKP5 in limiting ROS production and preventing LPS-induced vascular injury. *EMBO J.* **2009**, *28*, 2896–2907. [CrossRef] [PubMed]

96. Qian, F.; Deng, J.; Gantner, B.N.; Flavell, R.A.; Dong, C.; Christman, J.W.; Ye, R.D. Map kinase phosphatase 5 protects against sepsis-induced acute lung injury. *Am. J. Physiol. Lung Cell. Mol. Physiol.* **2012**, *302*, L866–L874. [CrossRef]

97. James, S.J.; Jiao, H.; Teh, H.Y.; Takahashi, H.; Png, C.W.; Phoon, M.C.; Suzuki, Y.; Sawasaki, T.; Xiao, H.; Chow, V.T.; et al. MAPK Phosphatase 5 Expression Induced by Influenza and Other RNA Virus Infection Negatively Regulates IRF3 Activation and Type I Interferon Response. *Cell Rep.* **2015**. [CrossRef]

98. Negishi, H.; Matsuki, K.; Endo, N.; Sarashina, H.; Miki, S.; Matsuda, A.; Fukazawa, K.; Taguchi-Atarashi, N.; Ikushima, H.; Yanai, H.; et al. Beneficial innate signaling interference for antibacterial responses by a Toll-like receptor-mediated enhancement of the MKP-IRF3 axis. *Proc. Natl. Acad. Sci. USA* **2013**, *110*, 19884–19889. [CrossRef]

99. Manley, G.C.A.; Stokes, C.A.; Marsh, E.K.; Sabroe, I.; Parker, L.C. DUSP10 Negatively Regulates the Inflammatory Response to Rhinovirus through Interleukin-1beta Signaling. *J. Virol.* **2019**, *93*, e01659-18.

100. Yamamoto, T.; Endo, Y.; Onodera, A.; Hirahara, K.; Asou, H.K.; Nakajima, T.; Kanno, T.; Ouchi, Y.; Uematsu, S.; Nishimasu, H.; et al. DUSP10 constrains innate IL-33-mediated cytokine production in ST2(hi) memory-type pathogenic Th2 cells. *Nat. Commun.* **2018**, *9*, 4231. [CrossRef]

101. Matsuguchi, T.; Musikacharoen, T.; Johnson, T.R.; Kraft, A.S.; Yoshikai, Y. A novel mitogen-activated protein kinase phosphatase is an important negative regulator of lipopolysaccharide-mediated c-Jun N-terminal kinase activation in mouse macrophage cell lines. *Mol. Cell. Biol.* **2001**, *21*, 6699–7009. [CrossRef] [PubMed]

102. Tanoue, T.; Yamamoto, T.; Maeda, R.; Nishida, E. A Novel MAPK phosphatase MKP-7 acts preferentially on JNK/SAPK and p38 alpha and beta MAPKs. *J. Biol. Chem.* **2001**, *276*, 26629–26639. [CrossRef] [PubMed]

103. Masuda, K.; Katagiri, C.; Nomura, M.; Sato, M.; Kakumoto, K.; Akagi, T.; Kikuchi, K.; Tanuma, N.; Shima, H. MKP-7, a JNK phosphatase, blocks ERK-dependent gene activation by anchoring phosphorylated ERK in the cytoplasm. *Biochem. Biophys. Res. Commun.* **2010**, *393*, 201–206. [CrossRef] [PubMed]

104. Niedzielska, M.; Bodendorfer, B.; Munch, S.; Eichner, A.; Derigs, M.; da Costa, O.; Schweizer, A.; Neff, F.; Nitschke, L.; Sparwasser, T.; et al. Gene trap mice reveal an essential function of dual specificity phosphatase Dusp16/MKP-7 in perinatal survival and regulation of Toll-like receptor (TLR)-induced cytokine production. *J. Biol. Chem.* **2014**, *289*, 2112–2126. [CrossRef] [PubMed]

105. Liu, X.; Zhang, C.S.; Lu, C.; Lin, S.C.; Wu, J.W.; Wang, Z.X. A conserved motif in JNK/p38-specific MAPK phosphatases as a determinant for JNK1 recognition and inactivation. *Nat. Commun.* **2016**, *7*, 10879. [CrossRef]

106. Zhang, Y.Y.; Wu, J.W.; Wang, Z.X. A distinct interaction mode revealed by the crystal structure of the kinase p38alpha with the MAPK binding domain of the phosphatase MKP5. *Sci. Signal.* **2011**, *4*, ra88. [CrossRef]

107. Zhang, Y.; Nallaparaju, K.C.; Liu, X.; Jiao, H.; Reynolds, J.M.; Wang, Z.X.; Dong, C. MAPK phosphatase 7 regulates T cell differentiation via inhibiting ERK-mediated IL-2 expression. *J. Immunol.* **2015**, *194*, 3088–3095. [CrossRef]

108. Maor-Nof, M.; Romi, E.; Shalom, H.S.; Ulisse, V.; Raanan, C.; Nof, A.; Leshkowitz, D.; Lang, R.; Yaron, A. Axonal Degeneration Is Regulated by a Transcriptional Program that Coordinates Expression of Pro- and Anti-degenerative Factors. *Neuron* **2016**, *92*, 991–1006. [CrossRef]

109. Zega, K.; Jovanovic, V.M.; Vitic, Z.; Niedzielska, M.; Knaapi, L.; Jukic, M.M.; Partanen, J.; Friedel, R.F.; Lang, R.; Brodski, C. Dusp16 Deficiency Causes Congenital Obstructive Hydrocephalus and Brain Overgrowth by Expansion of the Neural Progenitor Pool. *Front. Mol. Neurosci.* **2017**, *10*, 372. [CrossRef]

110. Ishibashi, T.; Bottaro, D.P.; Chan, A.; Miki, T.; Aaronson, S.A. Expression cloning of a human dual-specificity phosphatase. *Proc. Natl. Acad. Sci. USA* **1992**, *89*, 12170–12174. [CrossRef]

111. Pavic, K.; Duan, G.; Kohn, M. VHR/DUSP3 phosphatase: Structure, function and regulation. *FEBS J.* **2015**, *282*, 1871–1890. [CrossRef] [PubMed]

112. Alonso, A.; Saxena, M.; Williams, S.; Mustelin, T. Inhibitory role for dual specificity phosphatase VHR in T cell antigen receptor and CD28-induced Erk and Jnk activation. *J. Biol. Chem.* **2001**, *276*, 4766–4771. [CrossRef] [PubMed]

113. Schumacher, M.A.; Todd, J.L.; Rice, A.E.; Tanner, K.G.; Denu, J.M. Structural basis for the recognition of a bisphosphorylated MAP kinase peptide by human VHR protein Phosphatase. *Biochemistry* **2002**, *41*, 3009–3017. [CrossRef] [PubMed]

114. Todd, J.L.; Rigas, J.D.; Rafty, L.A.; Denu, J.M. Dual-specificity protein tyrosine phosphatase VHR down-regulates c-Jun N-terminal kinase (JNK). *Oncogene* **2002**, *21*, 2573–2583. [CrossRef] [PubMed]

115. Panico, K.; Forti, F.L. Proteomic, cellular, and network analyses reveal new DUSP3 interactions with nucleolar proteins in HeLa cells. *J. Proteome Res.* **2013**, *12*, 5851–5866. [CrossRef] [PubMed]

116. Yan, Q.; Sharma-Kuinkel, B.K.; Deshmukh, H.; Tsalik, E.L.; Cyr, D.D.; Lucas, J.; Woods, C.W.; Scott, W.K.; Sempowski, G.D.; Thaden, J.T.; et al. Dusp3 and Psme3 are associated with murine susceptibility to Staphylococcus aureus infection and human sepsis. *PLoS Pathog.* **2014**, *10*, e1004149. [CrossRef] [PubMed]

117. Kaforou, M.; Wright, V.J.; Oni, T.; French, N.; Anderson, S.T.; Bangani, N.; Banwell, C.M.; Brent, A.J.; Crampin, A.C.; Dockrell, H.M.; et al. Detection of tuberculosis in HIV-infected and -uninfected African adults using whole blood RNA expression signatures: A case-control study. *PLoS Med.* **2013**, *10*, e1001538. [CrossRef]

118. Sweeney, T.E.; Braviak, L.; Tato, C.M.; Khatri, P. Genome-wide expression for diagnosis of pulmonary tuberculosis: A multicohort analysis. *Lancet Respir. Med.* **2016**, *4*, 213–224. [CrossRef]

119. Singhania, A.; Wilkinson, R.J.; Rodrigue, M.; Haldar, P.; O'Garra, A. The value of transcriptomics in advancing knowledge of the immune response and diagnosis in tuberculosis. *Nat. Immunol.* **2018**, *19*, 1159–1168. [CrossRef]

120. Hoyt, R.; Zhu, W.; Cerignoli, F.; Alonso, A.; Mustelin, T.; David, M. Cutting Edge: Selective Tyrosine Dephosphorylation of Interferon-Activated Nuclear STAT5 by the VHR Phosphatase. *J. Immunol.* **2007**, *179*, 3402–3406. [CrossRef]

121. Jardin, C.; Sticht, H. Identification of the structural features that mediate binding specificity in the recognition of STAT proteins by dual-specificity phosphatases. *J. Biomol. Struct. Dyn.* **2012**, *29*, 777–792. [CrossRef] [PubMed]

122. Rahmouni, S.; Cerignoli, F.; Alonso, A.; Tsutji, T.; Henkens, R.; Zhu, C.; Louis-dit-Sully, C.; Moutschen, M.; Jiang, W.; Mustelin, T. Loss of the VHR dual-specific phosphatase causes cell-cycle arrest and senescence. *Nat. Cell Biol.* **2006**, *8*, 524–531. [CrossRef]

123. Amand, M.; Erpicum, C.; Bajou, K.; Cerignoli, F.; Blacher, S.; Martin, M.; Dequiedt, F.; Drion, P.; Singh, P.; Zurashvili, T.; et al. DUSP3/VHR is a pro-angiogenic atypical dual-specificity phosphatase. *Mol. Cancer* **2014**, *13*, 108. [CrossRef]

124. Singh, P.; Dejager, L.; Amand, M.; Theatre, E.; Vandereyken, M.; Zurashvili, T.; Singh, M.; Mack, M.; Timmermans, S.; Musumeci, L.; et al. DUSP3 Genetic Deletion Confers M2-like Macrophage-Dependent Tolerance to Septic Shock. *J. Immunol.* **2015**, *194*, 4951–4962. [CrossRef]

125. Vandereyken, M.M.; Singh, P.; Wathieu, C.P.; Jacques, S.; Zurashvilli, T.; Dejager, L.; Amand, M.; Musumeci, L.; Singh, M.; Moutschen, M.P.; et al. Dual-Specificity Phosphatase 3 Deletion Protects Female, but Not Male, Mice from Endotoxemia-Induced and Polymicrobial-Induced Septic Shock. *J. Immunol.* **2017**, *199*, 2515–2527. [CrossRef]

126. Vandereyken, M.; Jacques, S.; Van Overmeire, E.; Amand, M.; Rocks, N.; Delierneux, C.; Singh, P.; Singh, M.; Ghuysen, C.; Wathieu, C.; et al. Dusp3 deletion in mice promotes experimental lung tumour metastasis in a macrophage dependent manner. *PLoS ONE* **2017**, *12*, e0185786. [CrossRef] [PubMed]

127. Musumeci, L.; Kuijpers, M.J.; Gilio, K.; Hego, A.; Theatre, E.; Maurissen, L.; Vandereyken, M.; Diogo, C.V.; Lecut, C.; Guilmain, W.; et al. Dual-specificity phosphatase 3 deficiency or inhibition limits platelet activation and arterial thrombosis. *Circulation* **2015**, *131*, 656–668. [CrossRef]

128. Bender, M.; May, F.; Lorenz, V.; Thielmann, I.; Hagedorn, I.; Finney, B.A.; Vogtle, T.; Remer, K.; Braun, A.; Bosl, M.; et al. Combined in vivo depletion of glycoprotein VI and C-type lectin-like receptor 2 severely compromises hemostasis and abrogates arterial thrombosis in mice. *Arterioscler. Thromb. Vasc. Biol.* **2013**, *33*, 926–934. [CrossRef]

129. Shi, Z.; Tabassum, S.; Jiang, W.; Zhang, J.; Mathur, S.; Wu, J.; Shi, Y. Identification of a potent inhibitor of human dual-specific phosphatase, VHR, from computer-aided and NMR-based screening to cellular effects. *Chembiochem. A Eur. J. Chem. Biol.* **2007**, *8*, 2092–2099. [CrossRef] [PubMed]

130. Wu, S.; Vossius, S.; Rahmouni, S.; Miletic, A.V.; Vang, T.; Vazquez-Rodriguez, J.; Cerignoli, F.; Arimura, Y.; Williams, S.; Hayes, T.; et al. Multidentate small-molecule inhibitors of vaccinia H1-related (VHR) phosphatase decrease proliferation of cervix cancer cells. *J. Med. Chem.* **2009**, *52*, 6716–6723. [CrossRef]

131. Yuan, Y.; Li, D.M.; Sun, H. PIR1, a novel phosphatase that exhibits high affinity to RNA. ribonucleoprotein complexes. *J. Biol. Chem.* **1998**, *273*, 20347–20353. [CrossRef] [PubMed]

132. Deshpande, T.; Takagi, T.; Hao, L.; Buratowski, S.; Charbonneau, H. Human PIR1 of the protein-tyrosine phosphatase superfamily has RNA 5′-triphosphatase and diphosphatase activities. *J. Biol. Chem.* **1999**, *274*, 16590–16594. [CrossRef] [PubMed]

133. Burke, J.M.; Sullivan, C.S. DUSP11—An RNA phosphatase that regulates host and viral non-coding RNAs in mammalian cells. *RNA Biol.* **2017**, *14*, 1457–1465. [CrossRef] [PubMed]

134. Burke, J.M.; Kincaid, R.P.; Nottingham, R.M.; Lambowitz, A.M.; Sullivan, C.S. DUSP11 activity on triphosphorylated transcripts promotes Argonaute association with noncanonical viral microRNAs and regulates steady-state levels of cellular noncoding RNAs. *Genes Dev.* **2016**, *30*, 2076–2092. [CrossRef]

135. Zhao, Y.; Ye, X.; Dunker, W.; Song, Y.; Karijolich, J. RIG-I like receptor sensing of host RNAs facilitates the cell-intrinsic immune response to KSHV infection. *Nat. Commun.* **2018**, *9*, 4841. [CrossRef] [PubMed]

136. Kincaid, R.P.; Lam, V.L.; Chirayil, R.P.; Randall, G.; Sullivan, C.S. RNA triphosphatase DUSP11 enables exonuclease XRN-mediated restriction of hepatitis C virus. *Proc. Natl. Acad. Sci. USA* **2018**, *115*, 8197–8202. [CrossRef] [PubMed]

137. Caprara, G.; Zamponi, R.; Melixetian, M.; Helin, K. Isolation and characterization of DUSP11, a novel p53 target gene. *J. Cell. Mol. Med.* **2009**, *13*, 2158–2170. [CrossRef]

138. Albers, H.M.; Kuijl, C.; Bakker, J.; Hendrickx, L.; Wekker, S.; Farhou, N.; Liu, N.; Blasco-Moreno, B.; Scanu, T.; den Hertog, J.; et al. Integrating chemical and genetic silencing strategies to identify host kinase-phosphatase inhibitor networks that control bacterial infection. *ACS Chem. Biol.* **2014**, *9*, 414–422. [CrossRef]

139. Guan, K.; Hakes, D.J.; Wang, Y.; Park, H.D.; Cooper, T.G.; Dixon, J.E. A yeast protein phosphatase related to the vaccinia virus VH1 phosphatase is induced by nitrogen starvation. *Proc. Natl. Acad. Sci. USA* **1992**, *89*, 12175–12179. [CrossRef]

140. Muda, M.; Manning, E.R.; Orth, K.; Dixon, J.E. Identification of the human YVH1 protein-tyrosine phosphatase orthologue reveals a novel zinc binding domain essential for in vivo function. *J. Biol. Chem.* **1999**, *274*, 23991–23995. [CrossRef]

141. Kozarova, A.; Hudson, J.W.; Vacratsis, P.O. The dual-specificity phosphatase hYVH1 (DUSP12) is a novel modulator of cellular DNA content. *Cell Cycle* **2011**, *10*, 1669–1678. [CrossRef] [PubMed]

142. Sharda, P.R.; Bonham, C.A.; Mucaki, E.J.; Butt, Z.; Vacratsis, P.O. The dual-specificity phosphatase hYVH1 interacts with Hsp70 and prevents heat-shock-induced cell death. *Biochem. J.* **2009**, *418*, 391–401. [CrossRef] [PubMed]

143. Lo, K.Y.; Li, Z.; Wang, F.; Marcotte, E.M.; Johnson, A.W. Ribosome stalk assembly requires the dual-specificity phosphatase Yvh1 for the exchange of Mrt4 with P0. *J. Cell Biol.* **2009**, *186*, 849–862. [CrossRef] [PubMed]

144. Geng, Q.; Xhabija, B.; Knuckle, C.; Bonham, C.A.; Vacratsis, P.O. The Atypical Dual Specificity Phosphatase hYVH1 Associates with Multiple Ribonucleoprotein Particles. *J. Biol. Chem.* **2017**, *292*, 539–550. [CrossRef] [PubMed]

145. Cho, S.S.L.; Han, J.; James, S.J.; Png, C.W.; Weerasooriya, M.; Alonso, S.; Zhang, Y. Dual-Specificity Phosphatase 12 Targets p38 MAP Kinase to Regulate Macrophage Response to Intracellular Bacterial Infection. *Front. Immunol.* **2017**, *8*, 1259. [CrossRef] [PubMed]

146. Huang, Z.; Wu, L.M.; Zhang, J.L.; Sabri, A.; Wang, S.J.; Qin, G.J.; Guo, C.Q.; Wen, H.T.; Du, B.B.; Zhang, D.H.; et al. DUSP12 Regulates Hepatic Lipid Metabolism through Inhibition of Lipogenesis and ASK1 Pathways. *Hepatology* **2019**. [CrossRef] [PubMed]

147. Marti, F.; Krause, A.; Post, N.H.; Lyddane, C.; Dupont, B.; Sadelain, M.; King, P.D. Negative-feedback regulation of CD28 costimulation by a novel mitogen-activated protein kinase phosphatase, MKP6. *J. Immunol.* **2001**, *166*, 197–206. [CrossRef]

148. Yang, C.Y.; Li, J.P.; Chiu, L.L.; Lan, J.L.; Chen, D.Y.; Chuang, H.C.; Huang, C.Y.; Tan, T.H. Dual-specificity phosphatase 14 (DUSP14/MKP6) negatively regulates TCR signaling by inhibiting TAB1 activation. *J. Immunol.* **2014**, *192*, 1547–1557. [CrossRef]

149. Yang, C.Y.; Chiu, L.L.; Tan, T.H. TRAF2-mediated Lys63-linked ubiquitination of DUSP14/MKP6 is essential for its phosphatase activity. *Cell Signal.* **2016**, *28*, 145–151. [CrossRef]

150. Zheng, H.; Li, Q.; Chen, R.; Zhang, J.; Ran, Y.; He, X.; Li, S.; Shu, H.B. The dual-specificity phosphatase DUSP14 negatively regulates tumor necrosis factor- and interleukin-1-induced nuclear factor-kappaB activation by dephosphorylating the protein kinase TAK1. *J. Biol. Chem.* **2013**, *288*, 819–825. [CrossRef]

151. Barreiro, L.B.; Tailleux, L.; Pai, A.A.; Gicquel, B.; Marioni, J.C.; Gilad, Y. Deciphering the genetic architecture of variation in the immune response to Mycobacterium tuberculosis infection. *Proc. Natl. Acad. Sci. USA* **2012**, *109*, 1204–1209. [CrossRef]

152. Hijikata, M.; Matsushita, I.; Le Hang, N.T.; Thuong, P.H.; Tam, D.B.; Maeda, S.; Sakurada, S.; Cuong, V.C.; Lien, L.T.; Keicho, N. Influence of the polymorphism of the DUSP14 gene on the expression of immune-related genes and development of pulmonary tuberculosis. *Genes Immun.* **2016**, *17*, 207–212. [CrossRef] [PubMed]

153. Aoyama, K.; Nagata, M.; Oshima, K.; Matsuda, T.; Aoki, N. Molecular cloning and characterization of a novel dual specificity phosphatase, LMW-DSP2, that lacks the cdc25 homology domain. *J. Biol. Chem.* **2001**, *276*, 27575–27583. [CrossRef]

154. Chen, A.J.; Zhou, G.; Juan, T.; Colicos, S.M.; Cannon, J.P.; Cabriera-Hansen, M.; Meyer, C.F.; Jurecic, R.; Copeland, N.G.; Gilbert, D.J.; et al. The dual specificity JKAP specifically activates the c-Jun N-terminal kinase pathway. *J. Biol. Chem.* **2002**, *277*, 36592–36601. [CrossRef] [PubMed]

155. Ju, A.; Cho, Y.C.; Kim, B.R.; Park, S.G.; Kim, J.H.; Kim, K.; Lee, J.; Park, B.C.; Cho, S. Scaffold Role of DUSP22 in ASK1-MKK7-JNK Signaling Pathway. *PLoS ONE* **2016**, *11*, e0164259. [CrossRef] [PubMed]

156. Sekine, Y.; Tsuji, S.; Ikeda, O.; Sato, N.; Aoki, N.; Aoyama, K.; Sugiyama, K.; Matsuda, T. Regulation of STAT3-mediated signaling by LMW-DSP2. *Oncogene* **2006**, *25*, 5801–5806. [CrossRef]

157. Hamada, N.; Mizuno, M.; Tomita, H.; Iwamoto, I.; Hara, A.; Nagata, K.I. Expression analyses of Dusp22 (Dual-specificity phosphatase 22) in mouse tissues. *Med. Mol. Morphol.* **2018**, *51*, 111–117. [CrossRef]

158. Li, J.P.; Yang, C.Y.; Chuang, H.C.; Lan, J.L.; Chen, D.Y.; Chen, Y.M.; Wang, X.; Chen, A.J.; Belmont, J.W.; Tan, T.H. The phosphatase JKAP/DUSP22 inhibits T-cell receptor signalling and autoimmunity by inactivating Lck. *Nat. Commun.* **2014**, *5*, 3618. [CrossRef]

159. Zhou, R.; Chang, Y.; Liu, J.; Chen, M.; Wang, H.; Huang, M.; Liu, S.; Wang, X.; Zhao, Q. JNK Pathway-Associated Phosphatase/DUSP22 Suppresses CD4(+) T-Cell Activation and Th1/Th17-Cell Differentiation and Negatively Correlates with Clinical Activity in Inflammatory Bowel Disease. *Front. Immunol.* **2017**, *8*, 781. [CrossRef]

160. Chuang, H.C.; Chen, Y.M.; Hung, W.T.; Li, J.P.; Chen, D.Y.; Lan, J.L.; Tan, T.H. Downregulation of the phosphatase JKAP/DUSP22 in T cells as a potential new biomarker of systemic lupus erythematosus nephritis. *Oncotarget* **2016**, *7*, 57593–57605. [CrossRef]

161. Luchtel, R.A.; Dasari, S.; Oishi, N.; Pedersen, M.B.; Hu, G.; Rech, K.L.; Ketterling, R.P.; Sidhu, J.; Wang, X.; Katoh, R.; et al. Molecular profiling reveals immunogenic cues in anaplastic large cell lymphomas with DUSP22 rearrangements. *Blood* **2018**, *132*, 1386–1398. [CrossRef] [PubMed]

162. Melard, P.; Idrissi, Y.; Andrique, L.; Poglio, S.; Prochazkova-Carlotti, M.; Berhouet, S.; Boucher, C.; Laharanne, E.; Chevret, E.; Pham-Ledard, A.; et al. Molecular alterations and tumor suppressive function of the DUSP22 (Dual Specificity Phosphatase 22) gene in peripheral T-cell lymphoma subtypes. *Oncotarget* **2016**, *7*, 68734–68748. [CrossRef] [PubMed]

163. Caunt, C.J.; Kidger, A.M.; Keyse, S.M. Visualizing and Quantitating the Spatiotemporal Regulation of Ras/ERK Signaling by Dual-Specificity Mitogen-Activated Protein Phosphatases (MKPs). *Methods Mol. Biol.* **2016**, *1447*, 197–215. [PubMed]

International Journal of
Molecular Sciences

MDPI

Review

Emerging Regulatory Roles of Dual-Specificity Phosphatases in Inflammatory Airway Disease

Grace C. A. Manley [1,2], Lisa C. Parker [3] and Yongliang Zhang [1,2,*]

[1] Department of Microbiology and Immunology, Yong Loo Lin School of Medicine, National University of Singapore, Singapore 117545, Singapore; micmgca@nus.edu.sg
[2] Immunology Programme, Life Science Institute, National University of Singapore, Singapore 117597, Singapore
[3] Department of Infection, Immunity and Cardiovascular Disease, University of Sheffield, Sheffield S10 2RX, UK; l.c.parker@sheffield.ac.uk
* Correspondence: miczy@nus.edu.sg; Tel.: +65-6516-6407

Received: 11 January 2019; Accepted: 1 February 2019; Published: 5 February 2019

Abstract: Inflammatory airway disease, such as asthma and chronic obstructive pulmonary disease (COPD), is a major health burden worldwide. These diseases cause large numbers of deaths each year due to airway obstruction, which is exacerbated by respiratory viral infection. The inflammatory response in the airway is mediated in part through the MAPK pathways: p38, JNK and ERK. These pathways also have roles in interferon production, viral replication, mucus production, and T cell responses, all of which are important processes in inflammatory airway disease. Dual-specificity phosphatases (DUSPs) are known to regulate the MAPKs, and roles for this family of proteins in the pathogenesis of airway disease are emerging. This review summarizes the function of DUSPs in regulation of cytokine expression, mucin production, and viral replication in the airway. The central role of DUSPs in T cell responses, including T cell activation, differentiation, and proliferation, will also be highlighted. In addition, the importance of this protein family in the lung, and the necessity of further investigation into their roles in airway disease, will be discussed.

Keywords: inflammation; asthma; COPD; MAPK; respiratory viruses; influenza; rhinovirus; RSV

---

1. Introduction

Inflammatory airway diseases are major causes of morbidity and mortality. The most common chronic respiratory diseases are asthma and chronic obstructive pulmonary disease (COPD), affecting around 300 million and 65 million people worldwide, respectively [1,2]. Both diseases are characterized by chronic inflammation of the respiratory tract, which is worsened in acute exacerbations, leading to airway obstruction, wheezing, and breathlessness [3]. The main cause of exacerbations is infection with respiratory viruses, including rhinovirus, respiratory syncytial virus (RSV), and influenza. Studies to determine the aetiology of exacerbations detected respiratory viruses in 65–82% of asthma exacerbations and 37–56% of COPD exacerbations [4–11].

The airway epithelium is the main target of respiratory viruses. Pattern recognition receptors (PRRs) on the surface and within epithelial cells recognize components of viruses and activate a range of signaling pathways, including the mitogen-activated protein kinase (MAPK) pathways [12,13]. The MAPK pathways consist of a three-tier kinase cascade, culminating in the dual-phosphorylation and activation of the MAPKs: extracellular signal-regulated kinase (ERK), Jun N-terminal kinase (JNK), and p38. These proteins translocate to the nucleus and activate a range of transcription factors, such as NF-κB and AP-1, leading to the production and release of many different molecules, including interferons, cytokines, and adhesion molecules [12,14], initiating inflammatory responses.

These responses are aberrant in patients with underlying airway disease. The reasons for this remain incompletely understood, but involve impaired control of viral infection [15,16], damaged epithelium [17,18], and altered lymphocyte responses [19,20]. This review will discuss the roles of the MAPK pathways in these processes and their regulation by a group of proteins known as dual-specificity phosphatases (DUSPs) or MAPK phosphatases (MKPs).

2. The Epithelial Response to Respiratory Viral Infection

Activation of PRRs in respiratory epithelial cells leads to induction of the MAPK pathways, as summarized in Figure 1 [21]. Respiratory viral infection of epithelial cells can also activate the MAPKs through other means; for example, p38 can be activated by infection with rhinovirus, through the protein kinase Syk [22–24], or influenza, through the endoplasmic-reticulum stress response [25]. Once activated, the MAPKs have roles in many different processes, with severe implications in airway disease. These roles are summarized in the following sections.

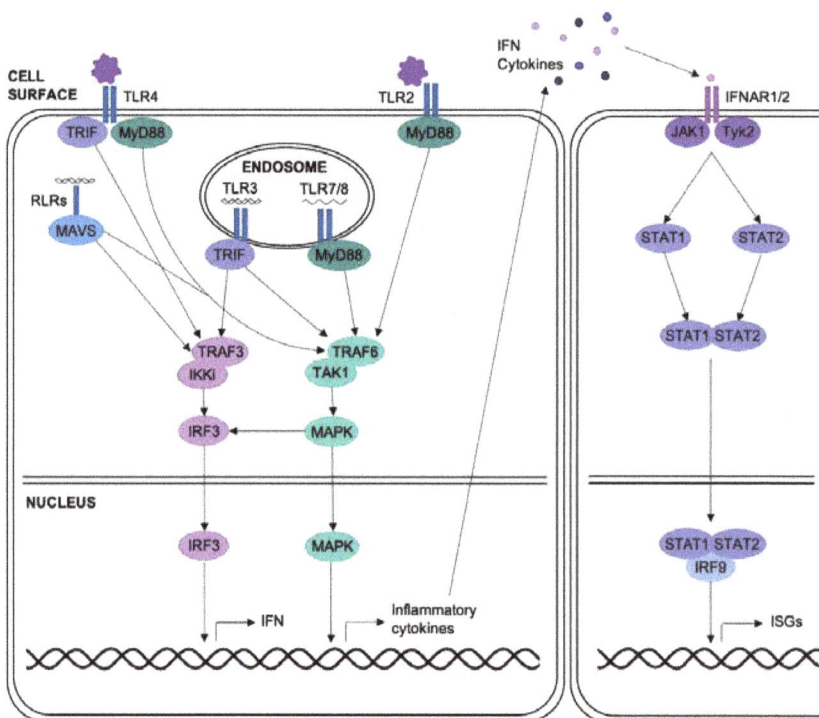

Figure 1. Activation of signaling pathways in respiratory epithelial cells upon viral infection. PRRs detect viral infection of the cell: TLRs 2 and 4 can bind components of the viral surface, TLR3 binds dsRNA, TLR7/8 bind ssRNA, and the RLRs bind dsRNA or 5′-triphosphorylated ssRNA. Adaptor proteins MyD88, TRIF, and MAVS mediate the activation of signaling pathways, including the MAPK pathways. The MAPKs translocate into the nucleus where they activate transcription factors, leading to the transcription of genes for inflammatory cytokines. TRIF and MAVS signaling activates IRF3, leading to interferon production. The MAPK pathways can also activate IRF3. Inflammatory cytokines and interferons are released by the cell and act upon surrounding cells. IFN binds to the IFN receptor complex IFNAR1/2, activating the JAK/STAT pathway. JAK1 and Tyk2 phosphorylate STAT1 and STAT2 which dimerize, translocate to the nucleus and bind IRF9, forming ISGF3, which induces transcription of interferon stimulated genes (ISGs).

2.1. The MAPKs and Cytokine Release

The specific roles of each MAPK pathway have been examined using small molecule inhibitors. Pyridinyl imidazole compounds inhibit p38 by competing with ATP for its binding site, blocking its catalytic activity [26]. Griego et al. used two pyridinyl imidazole inhibitors, SB203580 and SB239053, to examine the role of p38 in cytokine and chemokine production by the BEAS-2B human bronchial epithelial cell line in response to infection with rhinovirus [27]. They found that infection caused a time- and dose-dependent increase in p38 phosphorylation. Treatment with either inhibitor prior to infection led to a significant reduction in the secretion of all cytokines and chemokines examined, including CXCL8, growth-related oncogene-α (GRO-α), granulocyte colony-stimulating factor (G-CSF), and granulocyte-macrophage colony-stimulating factor (GM-CSF), all of which have important roles in neutrophilia [27]. Recent work has furthered this knowledge, showing reduced production of CXCL8 by primary bronchial epithelial cells when p38 signaling was inhibited prior to infection with rhinovirus [28].

Inhibitors of p38 have also been used to highlight its importance in inflammatory cytokine production in response to other respiratory viruses. Treatment of A549 cells with SB203580 decreased release of CCL5 in response to RSV infection, and CXCL8 in response to parainfluenza virus infection [29,30]. Supporting this, inhibition of p38 in primary bronchial epithelial cells reduced mRNA production of IL-1β and TNF-α in response to RSV infection [31]. This pro-inflammatory role of p38 has also been demonstrated in vivo, as treatment of influenza-infected BALB/c mice with SB203580 lowered the concentration of TNF-α, IL-1β and IL-6 protein in lung homogenates [32].

The ERK pathway also has roles in cytokine induction in epithelial cells in response to viral infection. Liu et al. and Newcomb et al. treated airway epithelial cell lines with U0126 prior to rhinoviral infection. U0126 inhibits the ERK pathway by blocking activation of upstream kinases MEK1/2 [33]. Treatment with this drug reduced the secretion of CXCL8 in response to rhinovirus to almost baseline levels [34,35]; however, this was not replicated in primary bronchial epithelial cells treated with the MEK inhibitor PD90859 [28]. This could be due to differences in potency between the two chemical inhibitors, or between primary and immortalized cells. ERK signaling also induces inflammatory cytokine release in response to infection with RSV, with decreased levels of CXCL8 and CCL5 in supernatants of infected A549 cells treated with PD98059 [29,36].

Less is known about the role of the JNK pathway in inflammatory cytokine production in viral infection of the airway. One study showed weaker production and release of CXCL8 in response to infection with two strains of rhinovirus in primary bronchial epithelial cells treated with the JNK inhibitor SP600125 [28]. Together, these studies illustrate the central role of the MAPKs in the inflammatory response to respiratory viral infection. The precise contribution of each pathway seems to depend on the specific virus and cell type studied, but together they induce a large proportion of inflammatory cytokine production.

Respiratory epithelial cells release type I and type III interferons in response to viral infection (Figure 1) [37–39]. Interferons limit replication of respiratory viruses; pre-treatment of airway epithelial cells with interferon-β (IFN-β) significantly reduced replication of rhinovirus or influenza virus [40,41]. Several viruses, including influenza and RSV, target components of the interferon pathway in order to limit the antiviral response [42,43], and highly pathogenic strains of influenza induce lower levels of interferon [41]. The MAPK pathways have previously been implicated in interferon induction in response to influenza infection. Infection of MDCK cells or chicken macrophages with avian influenza viruses in the presence of JNK inhibitors led to increased viral replication due to decreased activation of IRF3 [44,45]. Recently, a gene expression array compared the response of primary HUVECs infected with highly pathogenic avian influenza viruses with and without SB202190, a p38 inhibitor. In addition to diminished production of inflammatory mediators, p38 inhibition reduced expression of IFN-β [46]. Signaling by ERK has also been linked to interferon signaling in RSV infection; ERK inhibition in A549 cells lessened activation of STAT1 in response to RSV [47]. This identifies the MAPKs as key pathways in both the anti-viral and pro-inflammatory responses to viral infection (Figure 1).

2.2. The MAPKs and Viral Replication

In addition to regulating respiratory viral infection through the interferon response, the MAPKs may also have roles in the viral life cycle. Marchant et al. showed that inhibition of p38 using SB203580 in the bronchial epithelial cell line 1HAEo- reduced replication of a number of respiratory viruses, including: influenza, RSV, coxsackie virus B3, human parainfluenza virus 3, and adenovirus [48]. Influenza genome replication occurs within the nucleus, forming viral ribonucleoprotein (vRNP) complexes, which are then exported into the cytoplasm [49]. Inhibition of either p38 or ERK was found to decrease influenza virus replication in MDCK cells due to a reduction in vRNP export from the nucleus [49–51]. Nencioni et al. hypothesized this was due to phosphorylation of vRNP by p38, affecting its affinity for the viral surface protein M1 [51]. This was supported by co-localization of p38 and vRNP in the nucleus of MDCK cells, and a reduction of vRNP phosphorylation when p38 was inhibited [51].

The roles of the MAPK pathways in RSV replication have also been investigated. Inhibition of p38 or ERK diminished levels of viral RNA and progeny release in A549 cells [52,53]. In both cases, this was thought to be due to impaired transport of viral proteins through the secretory pathway. Inhibition of p38 in vero cells decreased phosphorylation of the SH protein, a viral membrane protein with unknown function [54]. This altered the cellular distribution of SH, increasing localization in the golgi, implying that phosphorylation of SH may be necessary for transport through the secretory pathway and thus, viral assembly [54]. A similar role was proposed for ERK, as treatment of A549 cells with U0126 reduced surface expression of the viral F protein [53].

RSV and influenza viruses can also successfully evade the immune response and antiviral therapies by direct cell to cell spread [55,56]. RSV forms syncytia in the airway epithelium by fusing the membranes of neighbor cells, leading to cytosol mixing and viral transfer. RSV can also induce the formation of long filaments to reach, and spread to, more distant cells. This process is dependent on actin rearrangement through RhoA and the Arp2/Arp3 complex [56,57]. In wound healing assays, inhibition of ERK in epithelial cell lines reduced Arp2/3 recruitment and actin polymerization, indicating a possible role for ERK in syncytia formation during RSV infection [58]. ERK has previously been implicated in syncytia formation in cancer, with U0126 treatment of a choriocarcinoma cell line mitigating syncytia formation [59]. One result of syncytia formation in RSV infection is disruption of the epithelium and decreased membrane barrier integrity, which can lead to pneumonia and secondary bacterial infection. This can be modelled in A549 cells, where RSV infection lowers trans-epithelial resistance and causes paracellular gap formation. Treatment of A549 cells with SB203580 lessened these effects of RSV infection on the cell monolayer [60]. This was associated with reduced phosphorylation of heat shock protein 27 (Hsp27), a protein involved in actin rearrangement [60], suggesting that p38 may also be involved in syncytia formation through Hsp27.

As the MAPKs are involved in a wide variety of processes, they may also be indirectly involved in viral replication. For example, enteroviruses, such as rhinovirus, utilize the host cell endocytosis machinery, mainly the protein Rab11, to traffic cholesterol to replication organelles [61]. p38 has been shown to phosphorylate and activate guanyl-nucelotide dissociation inhibitor, a protein which facilitates cycling of Rab proteins between the membrane and the cytosol in endocytosis [62]. Cholesterol plays an important role in viral polyprotein processing and genome synthesis, and inhibition of cholesterol trafficking blocks viral replication [61]. Thus, p38 activation of Rab protein cycling may facilitate viral replication.

Overall, the literature suggests that respiratory viruses hijack the MAPKs and their downstream targets for their own ends; utilizing them for protein trafficking, viral assembly, and cell to cell spread. This highlights the need for strict regulation of these pathways, in order to limit viral replication, and proposes the MAPKs as targets for antiviral therapies [63].

2.3. The MAPKs and Mucus Production

A defining feature of asthma and COPD is goblet cell hyperplasia and excessive mucus production. This can lead to blockage of the airway and contributes to asthma-associated deaths [64]. The predominant mucin in asthma and COPD is MUC5AC [65,66]. The T helper 2 cytokine IL-13 is thought to be the main inducer of goblet cell hyperplasia and MUC5AC production in murine models of asthma, through activation of STAT6 [67,68]. The MAPKs also participate in this process; inhibition of p38 or ERK in differentiated primary murine or human airway epithelial cell monolayers reduced the IL-13 induced upregulation of goblet cell numbers and MUC5AC expression [69–71]. Furthermore, activation of p38 in response to IL-13 is weaker in epithelial cells from STAT6 knock out mice, indicating STAT6-induced mucin production occurs via p38 [69].

Respiratory viral infection has also been shown to upregulate mucus production. Double-stranded RNA is a common component or replication intermediate of viruses. Stimulation of NCI-H292 cells with double-stranded RNA upregulated expression of mucin MUC2, and this could be reversed by treatment with a p38 inhibitor [72]. MUC5AC expression is raised in ovalbumin murine models of asthma, and is increased further by RSV infection [73]. This observation may be dependent of IL-33, as IL-33 levels are higher in the lungs of RSV-infected mice, leading to enhanced production of IL-13. In addition, treatment of RAW cells with MAPK inhibitors decreases the release of IL-33 in response to RSV [74]. Inhibition of p38 has also been shown to repress IL-33 production in primary nasal epithelial cells in response to TNF-α stimulation [75].

Another mechanism by which mucin production is upregulated is via activation of the EGF receptor (EGFR) and Ras-Raf-MEK-ERK pathway [76,77]. Rhinovirus infection of differentiated primary human tracheal epithelial cells upregulates MUC5AC RNA levels and protein release, as well as RNA for MUC2, MUC3, MUC5B and MUC6 [78,79]. This induction was mediated by the EGFR pathway, as treatment with MEK or EGFR inhibitors returned MUC5AC levels to baseline. The authors hypothesized this was due to an autocrine loop, where rhinoviral infection induced production and release of EGRF ligands, as shown in NCI-H292 cells, which activated EGRF on the cell surface, and thus activated the ERK pathway [79]. This highlights the roles of the MAPKs in viral induced mucus production and has substantial implications for airway disease, where mucus hyperplasia is a common symptom.

2.4. Regulation of the MAPKs by DUSPs in Respiratory Viral Infection

The above studies underline the importance of the MAPK pathways in respiratory viral infection and airway disease. Although the majority of these studies rely on small molecule inhibitors which have significant off-target effects [80], they do indicate roles for the MAPKs in many of the processes implicated in exacerbations of asthma or COPD, including inflammation, mucus production and elevated viral replication. Thus, regulation of the MAPKs is of extreme importance. These pathways are primarily inactivated by simultaneous dephosphorylation of the threonine and tyrosine residues within the MAPK activation motif by dual-specificity phosphatases (DUSPs) (Figure 2) [81].

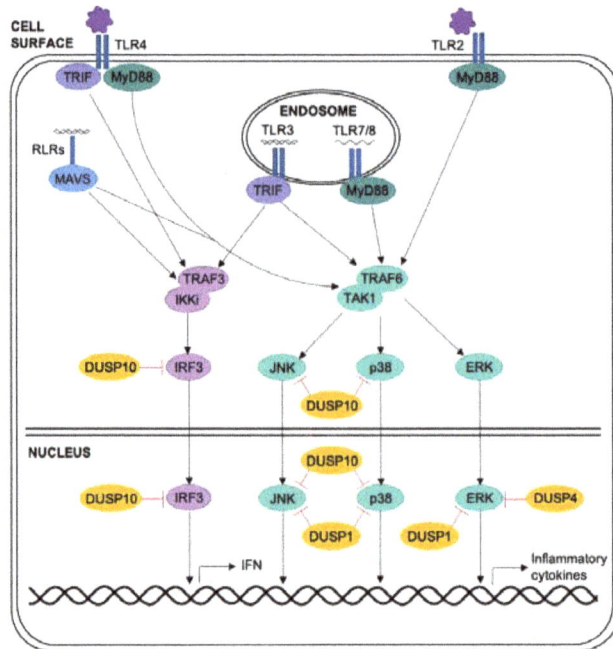

Figure 2. Regulation of the MAPK pathways by DUSPs in epithelial cells upon viral infection. PRR recognition of viruses or viral components activates the MAPK and IRF3 pathways. The MAPKs and IRF3 translocate to the nucleus and induce expression of inflammatory cytokines and interferon. These pathways are negatively regulated (red arrows) through dephosphorylation by DUSPs. DUSP1 is present in the nucleus and dephosphorylates all three MAPKs. DUSP4 is a nuclear protein, and is thought to dephosphorylate ERK. DUSP10 is present in both the nucleus and the cytoplasm and dephosphorylates JNK, p38 and IRF3. Black arrows represent activating interactions, red arrows represent inhibition.

2.4.1. DUSP1/MKP1

Much of the literature regarding DUSPs in innate immunity have focused on bacterial infection, and few studies have examined their roles in viral infection. DUSP1 (MKP1) is the archetype of the family and the most well studied. DUSP1 is a nuclear protein, capable of dephosphorylating p38, JNK and ERK, with substrate specificity depending on the stimuli and cell type [82,83]. It has been characterized as an early response gene, with undetectable expression at baseline, and rapid upregulation upon exposure to a variety of stimuli [82,84]. The airway epithelial cell line NCI-H292 upregulates DUSP1 mRNA within one hour in response to the synthetic double-stranded RNA molecule polyinosinic:polycytidylic acid (poly(I:C)) [85]. Poly(I:C) is a ligand for the PRRs toll-like receptor 3 (TLR3) and the RIG-I-like receptors (RLRs), which are predominantly activated by viral infection. Knock down of DUSP1 expression using small-interfering RNA (siRNA) in NCI-H292 cells amplified the release of two pro-inflammatory cytokines in response to poly(I:C) stimulation, CXCL8 and IL-6 [86]. A similar role for DUSP1 was seen in infection of the NCI-H1299 cell line with the avian coronavirus infectious bronchitis virus, with DUSP1 siRNA treatment increasing mRNA levels of CXCL8 in response to infection [87]. This augmented cytokine expression is likely to be due to elevated MAPK activation, with increased levels of phosphorylated p38 and JNK found in RSV infected A549 cells treated with DUSP1 siRNA [88].

DUSP1 has also been implicated in regulating the interferon response, with DUSP1 knock down in hepatocyte cell lines increasing STAT1 activation in response to hepatitis C virus or IFN-γ stimulation [89,90]. However, a yeast two-hybrid assay was unable to find an interaction between DUSP1 and STAT1, and overexpression of DUSP1 in COS-1 cells did not affect the level of STAT1 activation in response to IFN-γ [91]. Thus, the effects of DUSP1 knock down on the interferon response to hepatitis C infection may be indirect effects of increased MAPK activation (Section 2.1) rather than direct inactivation by DUSP1.

The inflammatory cytokine TNF-α is induced by respiratory viral infection, with higher expression in asthmatic patients [92,93], and elicits secondary inflammatory cytokine release from airway smooth muscle (ASM) cells [94]. Stimulation of primary ASM cells with TNF-α also caused the upregulation of DUSP1 mRNA and protein. When DUSP1 expression was knocked down in ASM cells, the release of CXCL8 increased in response to TNF-α stimulation [95]. CXCL8 is a neutrophil chemoattractant commonly detected in asthmatic airways [96]. Neutrophilia can harm the airway, causing epithelial cell damage and necrosis, and levels of CXCL8 correlate with asthma severity [96,97]. TNF-α stimulation of epithelial cells also induced the expression of mucins, and DUSP1 knock down in NCI-H292 cells further amplified the expression of airway mucin MUC5AC in response to TNF-α [98]. Taken together, this work suggests that DUSP1 has an important role in the response of the epithelium to insult, including regulation of inflammatory cytokine and mucin production.

2.4.2. DUSP10/MKP5

DUSP10 (MKP5) is expressed ubiquitously in the nucleus and cytoplasm [99], and is upregulated in response to viral infection: bone-marrow derived macrophages (BMDMs) infected with influenza virus or stimulated with poly(I:C) have enhanced DUSP10 mRNA and protein expression [100]. Knock down of DUSP10 in primary bronchial epithelial cells increased the release of the neutrophil chemoattractants CXCL8 and CXCL1 in response to stimulation with a key proinflammatory cytokine IL-1β, suggesting that, like DUSP1, DUSP10 negatively regulates the inflammatory response in the airway [28]. Importantly, rhinoviral infection of airway epithelial cells or monocytes causes the release of IL-1β [28,101]; and combined stimulation with rhinovirus and IL-1β leads to an even greater inflammatory response in DUSP10 knock down primary bronchial epithelial cells from both healthy and COPD donors [28]. This identifies DUSP10 as a central regulator of the inflammatory response to respiratory viruses: infection of epithelial cells induces release of IL-1β, which acts back on the epithelium to promote inflammation, which is negatively regulated by DUSP10.

The role of DUSP10 in respiratory viral infection has also been examined in vivo: DUSP10 knock out mice infected with influenza had elevated levels of IL-6 in bronchoalveolar lavage (BAL) than wild-type mice. Interestingly, DUSP10 knock out mice also had decreased viral titres and better survival in response to infection. This was associated with raised expression and phosphorylation of IRF3, and therefore increased interferon (IFN) expression. Further investigation established that DUSP10 and IRF3 directly interact, indicating IRF3 as a novel substrate for DUSP10 and highlighting the importance of DUSP10 in regulating not only the inflammatory response, but also the anti-viral response.

Sustained, uncontrolled pulmonary inflammation can lead to acute lung injury, often seen in severe influenza infection. Murine models of acute lung injury can be generated by intratracheal injection of lipopolysaccharide (LPS), a TLR4 agonist. When DUSP10 knock out mice were utilized in an acute lung injury model, they exhibited greater disease severity than wild-type mice, with increased lung injury and pulmonary edema [102]. This was associated with augmented neutrophil influx in the lungs, and inflammatory cytokines in BAL. BMDMs isolated from these mice had elevated activation of p38 and JNK, and to a lesser extent ERK, in response to LPS treatment. Adoptive transfer of these BMDMs into wild-type mice led to enhanced lung inflammation in response to intratracheal LPS injection than the transfer of wild-type BMDMs [102]. This is in keeping with the in vitro data described above, and demonstrates that DUSP10 has an anti-inflammatory role in the airway, and is important in limiting immune-mediated lung damage.

2.4.3. DUSP4/MKP2

Interestingly, one DUSP has been found to have a pro-inflammatory role in murine models of acute lung injury. In response to intratracheal LPS injection, DUSP4 (MKP2) knock out mice had decreased inflammatory cytokines in BAL and neutrophil infiltration in to the lung [103]. These data fit with an earlier study showing a pro-inflammatory role for DUSP4 in sepsis, with improved survival in DUSP4 knock out mice [104]. BMDMs taken from these mice produced lower levels of inflammatory cytokines in response to LPS injection than wild-type mice, associated with reduced activation of p38 and JNK, but increased activation of ERK. The authors suggest this was due to ERK-induced DUSP1 transcription, as has been demonstrated previously [105]. These studies indicate that different DUSPs may have pro- or anti-inflammatory roles in pulmonary inflammation. It should be noted that Al-Mutairi et al. found conflicting results, with DUSP4 knock out BMDMs releasing higher levels of inflammatory cytokines in response to LPS, although it is unclear why these studies differ [106].

3. T Cell Responses

Around 50% of asthma cases have an allergic phenotype, characterized by predominantly eosinophilic inflammation and T helper 2 (Th2) responses [19,107]. Higher levels of several Th2 cytokines have been found in BAL of asthmatics, including IL-4, IL-5, IL-13, IL-25, IL-33, and TSLP [108–110], and the levels of Th2 cytokines correlate with severity of asthma exacerbation [20]. The Th1/Th2 balance is also crucial for the immune control of respiratory viral infection. Asthmatics experimentally infected with rhinovirus had increased viral titres compared to infected healthy controls, with greater airway inflammation, bronchial hyperreactivity, and reductions in lung function associated with increased levels of IL-4, IL-5 and IL-13 in BAL [19]. The MAPKs have been implicated in induction of Th2 cytokines in the airway. Inhibition of p38 or ERK pathways in primary nasal epithelial cells or alveolar macrophages decreased release of IL-33 in response to TNF-α stimulation or RSV infection, respectively [74,75]. ERK and p38 inhibitors have also been used to confirm roles for these pathways in TSLP production in ASM cells in response to TNF-α or IL-1β stimulation [111]. Transcription of TSLP in ASM cells is also partially dependent on the AP-1 transcription factor, suggesting the involvement of JNK [112].

Early infection with RSV has been linked to the development of asthma, possibly through skewing the immune response towards a Th2 phenotype. Cytokine profiles of children infected with RSV revealed an expansion of Th2 cytokines and decreased Th1 cytokines [113], and RSV infection of mouse pups led to increased Th2 responses and impaired regulatory T (Treg) cell responses [114]. Enhanced T cell recruitment in RSV infection correlates with worsening symptoms [115], and ablation of either CD4+ cells or CD8+ cells in mouse models mitigates disease severity [116]. This Th2 skewing in response to infection may involve p38 MAPK. Infection of monocyte-derived dendritic cells with RSV induced expression of indoleamine-2,3-dioxygenase, an enzyme which favors Th2 differentiation by inducing apoptosis in Th1 cells. The expression of indoleamine-2,3-dioxygenase was reduced by treatment with the p38 inhibitor SB202190 [117].

Taken together, this work illustrates the importance of regulating the T cell response, and the T helper subset balance. In addition to affecting T cell activation through cytokine production by epithelial cells, the MAPKs have roles in T cells themselves, affecting their activation, proliferation and function [118].

The Roles of DUSPs in T Cell Responses

Several studies have implicated DUSPs in the regulation of T cells and in the differentiation of T helper subsets (Table 1). DUSP1 knock out mice have been utilized to demonstrate roles for DUSP1 in T cell priming, proliferation, and T cell subset skewing. Antigen presenting cells (APCs), such as dendritic cells, have roles in this altered response. DUSP1 knock out dendritic cells had increased activation of p38, and thus altered cytokine production. This led to impaired priming of naïve wild-type T cells, with reduced differentiation into Th1 cells and augmented Th17 and Treg cell differentiation [119]. DUSP1 knock out T cells exhibit reduced proliferation in response to activation with anti-CD3 antibodies, and lower levels of IFN-γ and IL-17, Th1 and Th17 cytokines, respectively; while the Th2 cytokine IL-4 levels remained unchanged. These studies emphasize the different roles of DUSP1 in different cell types, with knock out having differing effects on APC mediated and T cell intrinsic responses. The overall effect of DUSP1 knock out was observed in influenza infection, with a decline in Th1 and CD8$^+$ T cell numbers, leading to impaired control of the virus and greater disease severity. This altered response was associated with decreased nuclear translocation of NFATc1, a transcription factor important for T cell proliferation and function [120]. JNK was previously found to negatively regulate NFATc1 by phosphorylation [121], implying that the impaired T cell responses in DUSP1 knock out mice were due to increased JNK activation.

DUSP10 also plays a role in T cell proliferation. DUSP10 knock out mice have decreased numbers of virus specific CD4$^+$ cells and CD8$^+$ cells in the lung in response to influenza infection [100]. These mice also have diminished CD4$^+$ cell proliferation in response to anti-CD3 and anti-CD28 antibodies; however, T cell effector functions were increased, with greater levels of Th1 and Th2 cytokine release. This elevated cytokine release was also observed in response to secondary infection with lymphocytic choriomeningitis virus, leading to immune-mediated death [122].

Three other DUSPs have also been implicated in T helper subset skewing: DUSP4, DUSP5 and DUSP16. DUSP4 negatively regulates Treg cell differentiation through inactivating STAT5 [123]. STAT5 is activated by IL-2, and is required for induction of Treg cells [124]. Overexpression of DUSP4 in HEK-293T cells reduced phosphorylation of STAT5 in response to IFN-β stimulation, and DUSP4 and STAT5 were co-immunoprecipitated, indicating that DUSP4 directly dephosphorylates STAT5. DUSP4 knock out mice had increased numbers of Treg cells and fewer Th17 cells [123]. In contrast to the negative regulatory role of DUSP4 in Treg cell generation, DUSP5 (hVHR3) seems to act as a positive regulator. Mice overexpressing DUSP5 had decreased inflammation and disease severity in a collagen-induced arthritis model, due to raised numbers of Treg cells and higher STAT5 activation, and reduced Th17 cells and lower STAT3 activation [125]. DUSP16 (MKP7) knock out shows embryonic lethality; however, mice expressing a dominant negative DUSP16 protein have been generated, and have an altered Th1/Th2 balance. T cells isolated from these mice produce increased levels of IFN-γ and diminished IL-4, IL-5 and IL-13 in response to anti-CD3 and anti-CD28 antibodies or ovalbumin [126,127]. In contrast to this, mice with a DUSP16 knock out specifically in the hematopoietic compartment do not display altered Th1 or Th2 responses, but demonstrate a role for the protein in regulating IL-2 production [128]. These mice had enhanced release of IL-2 and T cell proliferation, compared to wild-type mice, in response to anti-CD3 antibodies. This was associated with increased activation of ERK, which is critical for IL-2 expression [128]. IL-2 has previously been shown to inhibit the expansion of Th17 cells [129], and these mice exhibited a decrease in numbers of IL-17 producing cells, which was reversed by treatment with the ERK inhibitor U0126 [128].

In addition to their roles in T cell proliferation and T helper subset skewing, DUSPs have also been found to regulate T cell receptor (TCR) signaling. Binding of the TCR to antigen leads to recruitment of Lck, a tyrosine kinase, which phosphorylates the ζ chain, leading to ZAP-70 recruitment and the initiation of a range of signaling pathways [130]. Lck knock out mice emphasize its importance in the induction of TCR signaling [131]. T cells isolated from DUSP22 (JKAP) knock out mice had higher activation levels of several molecules downstream of the TCR, including Lck, ZAP-70, IKK, and the MAPKs, in response to anti-CD3 antibodies. DUSP22 and Lck were co-immunoprecipitated from murine splenic T cells, and DUSP22 was found to dephosphorylate Lck on Tyr394, inactivating it [132].

Further downstream of ZAP-70, TAB1 is activated by PKC-θ. TAB1 binds to TAK1, inducing an activating conformational change, triggering the IKK, p38 and JNK pathways. When DUSP14 (MKP-L) knock out T cells were stimulated with anti-CD3 antibodies, the numbers of activated CD69$^+$ T cells were significantly higher than in wild-type T cells [133]. To investigate the reason behind this, HEK293T cells were transfected with wild-type or non-functional mutant DUSP14 and stimulated with anti-CD3 antibodies. Levels of activated Lck and ZAP-70 were unchanged between cells expressing wild-type and mutated DUSP14; however, IKK and MAPK activation was increased in cells expressing mutated DUSP14. Mass spectrometry was used to identify the binding protein and target of DUSP14 as TAB1, and further analysis revealed DUSP14 dephosphorylates TAB1 at the Ser358 residue [133].

These data illustrate that DUSPs have fundamental roles in adaptive immunity, affecting the activation, proliferation and differentiation of T helper cells. Although many of these studies have examined T cells in isolation, they identify the DUSPs as important regulators and suggest essential roles for them in airway diseases.

Table 1. Roles of DUSPs in T cells.

DUSP	Regulates Proliferation	Regulates TCR Signaling	Regulates Subset Differentiation				Reference
			Th1	Th2	Th17	Treg	
DUSP1	X		Promotes		Promotes	Inhibits	[120]
DUSP4					Promotes	Inhibits	[123]
DUSP5					Inhibits	Promotes	[125]
DUSP10	X						[122]
DUSP14	X	X					[133]
DUSP16	X		Inhibits	Promotes	Promotes		[127,128]
DUSP22	X	X					[132]

* Blank boxes = not determined.

4. The Role of DUSP1 in Steroid Treatment

Exacerbations of asthma or COPD are treated with corticosteroids to limit the inflammatory response [134]. Steroids interact with the glucocorticoid receptor (GR) in the cytosol, inducing a conformational change, which allows the GR to translocate to the nucleus where it interacts with and inhibits transcription factors, such as NF-κB and AP-1 [135,136]. More recent evidence revealed that steroids mediate many of their actions through DUSP1. Treatment with the glucocorticoid dexamethasone increased DUSP1 expression in airway epithelial cell lines [137,138] and airway smooth muscle cells [139–141]. In mouse models, dexamethasone treatment reduced the release of inflammatory cytokines, TNF-α and IL-6, in serum in response to LPS injection. This inhibitory action of dexamethasone was weakened in DUSP1 knock out mice [142]. BMDMs or peritoneal macrophages isolated from DUSP1 knock out mice show that this was due to impaired inhibition of MAPK activation, and thus cytokine release, in response to LPS when DUSP1 is not present [142,143]. Dexamethasone treatment can also promote wound healing responses, upregulating proteins such as arginase 1 and fibroblast growth factors. This was diminished in peritoneal macrophages isolated from DUSP1 knock out mice in response to alternative macrophage activators IL-4 and IL-13, indicating that DUSP1 both restricts inflammation and promotes wound healing [144]. Interestingly, bone-marrow derived mast cells from DUSP1 knock out mice did not differ from wild-type in the levels of cytokines released in response to IgE cross-linking and dexamethasone treatment. This may be due to redundancy within the DUSP family, as dexamethasone also upregulated DUSPs 2, 4 and 9 [145]. This suggests the actions of DUSPs in dexamethasone treatment differ depending on cell type and stimulus. The role of DUSP1 in steroid treatment of the airway epithelium has also been investigated. siRNA knock down of DUSP1 in the A549 cell line blocked the anti-inflammatory action of dexamethasone on MAPK activation and cytokine release in response to IL-1β [146,147].

Around 10% of asthmatics are resistant to steroid treatment [148]. Several studies have examined the different responses between asthmatics who are sensitive to steroids, and those who are resistant. Bronchial biopsies from steroid-sensitive asthmatics show a reduction in JNK activation after treatment with dexamethasone, which was not seen in bronchial biopsies from steroid-resistant asthmatics [149]. Higher activation levels of p38 were also detected in alveolar macrophages isolated from steroid-resistant asthmatics than in steroid-sensitive cells. This was associated with reduced expression of DUSP1 in response to dexamethasone [150]. Lower expression of DUSP1 was also found in peripheral blood neutrophils from steroid-resistant asthmatics in comparison to steroid-sensitive asthmatics [151].

Exacerbations of asthma and COPD caused by viral or bacterial infection are also more resistant to steroid treatment than non-viral exacerbations [152,153]. Rhinovirus infections impair the anti-inflammatory actions of steroids, partly by reducing nuclear translocation of the glucocorticoid receptor [154]. Treatment of A549 cells with dexamethasone reduced inflammatory cytokine release and upregulated DUSP1 in response to IL-1β stimulation. However, when these cells were infected with rhinovirus, this suppression by dexamethasone was abrogated, as was the upregulation of DUSP1 expression [154]. Treatment of A549 cells with the TLR2 ligand Pam_3CSK_4 also induced steroid resistance, but had no effect on DUSP1 expression. However, Pam_3CSK_4 treatment did induce oxidative stress, and a proportion of the DUSP1 present in these cells was oxidized [155]. Oxidation of the catalytic cysteine residue in the active site of DUSPs renders them inactive [156]. These findings suggest that steroid insensitivity in asthmatics, or in infected airways, may be due to a defect in DUSP1 expression or activation. Furthermore, polymorphisms in the DUSP1 gene have been associated with steroid responsiveness [157]. The roles of other DUSPs in steroid treatment remain to be investigated.

Another therapy commonly used to treat asthmatics are bronchodilators, such as $β_2$-recpetor agonists. In addition to bronchodilation, $β_2$-agonists also have some anti-inflammatory effects. Treatment of cells with $β_2$-agonists increases intracellular levels of cAMP [158]. The promoter of DUSP1 contains a cAMP-response element [159], and $β_2$-agonist treatment of airway epithelial cell lines and airway smooth muscle cells has been found to upregulate DUSP1 expression [160,161]. The role of DUSP1 in the anti-inflammatory action of the $β_2$-agonist salbutamol was investigated by inducing paw edema in wild-type and DUSP1 knock out mice. Salbutamol treatment reduced the level of inflammation by around 70% in wild-type mice, but only by around 40% in DUSP1 knock out mice [137]. This signifies that DUSP1 also has an important role in mediating the anti-inflammatory effects of $β_2$-agonists, in addition to corticosteroids.

5. Studies Linking DUSPs to Asthma and Sarcoidosis

It has been suggested that several of the DUSPs may be dysregulated in people with asthma. This is an intriguing explanation for the excessive inflammatory response seen in these patients, particularly because activation of the MAPK proteins is elevated in asthmatics. Baseline levels of phosphorylated p38 are higher in bronchial epithelial cells isolated from asthmatics than healthy controls, and phosphorylated ERK is greater in T cells isolated from asthmatics both at baseline and in response to anti-CD3 antibodies [162,163]. A study in 2008 isolated nasal epithelial cells from healthy individuals and patients with house dust mite allergy, a common allergy associated with asthma. They performed a microarray to determine any gene expression changes in response to stimulation with house dust mite. In non-allergic controls, DUSP1 mRNA expression was upregulated in response to house dust mite challenge; however, in allergic patients, DUSP1 expression did not alter [164]. Altered expression and activation of p38 and DUSP1 has also been observed in sarcoidosis. Sarcoidosis is a systemic inflammatory disorder, often characterized by granulomas within the lung. Rastogi et al. isolated and cultured leukocytes from the BAL of controls and sarcoidosis patients [165]. Activation levels of p38 and inflammatory cytokine production were higher in cells isolated from patients, both at baseline and after PRR stimulation. This coincided with much lower DUSP1 protein expression in controls than sarcoidosis patients [165]. These findings suggest impaired DUSP1 upregulation as a reason for enhanced inflammatory responses in asthma and sarcoidosis.

Cigarette smoke exposure can lead to the development of asthma and is the most common cause of COPD [166]. This is partially through the induction of inflammatory responses in the lung, with higher levels of inflammatory cytokines detected in the lungs of smokers [167]. Higher levels of p38 activation have also been detected in smokers' lungs [168], and treatment of BEAS-2B cells with p38 inhibitors reduced the release of cytokines in response to cigarette smoke [169]. DUSP1 may also have a role in this process, as ferrets exposed to cigarette smoke for six months had reduced levels of DUSP1 protein in lung tissue, although the functional effects of this have yet to be examined [170].

DUSP10 may also be differentially expressed in asthmatic patients. A transcriptional profile of Th2 cells taken from asthmatic and healthy subjects showed lower baseline mRNA expression of DUSP10 in the asthmatic Th2 cells than the healthy cells [171]. Intriguingly, a single nucleotide polymorphism in the DUSP4 gene was identified in a genetic screen for variants associated with severe asthma. However, this was not statistically significant, possibly due to the limited number of patients in the study [172]. Studies examining the relationship of other DUSPs and asthma would be of interest and are yet to be carried out.

6. Conclusions

The MAPK pathways have important roles in airway inflammation and are aberrantly activated in several inflammatory airway diseases. This may in part be due to altered expression of DUSPs, with lower baseline levels of DUSP10 and lower induction of DUSP1 expression upon allergen stimulation or steroid treatment in asthmatics. DUSPs have central roles in regulating inflammation, therefore, this aberrant expression could have important functions in the pathogenesis of lung inflammation. Underlying airway disease also leads to greater susceptibility to lower respiratory tract infections, due to impaired control of viral replication. The literature discussed here suggest a possible role for DUSPs in controlling viral-induced exacerbations of airway disease, not only in terms of regulating the MAPKs and their roles in viral life cycles, but also IFN production, T cell proliferation and Th2 skewing. Current treatments of airway inflammation are not always effective and cause significant side effects. Therefore, the development of new, more specific, treatments is of extreme importance. MAPK inhibitors have been investigated for therapeutic application with varying success [173]. An alternative method of reducing inflammation would be via upregulation of DUSPs. Mechanisms by which this may be achieved have been reviewed previously [174]. DUSPs represent potential targets for novel anti-inflammatory treatments of airway disease and future work into their roles in the airway is imperative.

Author Contributions: G.C.A.M.; writing—original draft preparation, Y.Z. and L.C.P.; writing—review and editing.

Funding: This work was supported by grants from the Singapore National Medical Research Council (NMRC/OFIRG/0059/2017), A*STAR-NHMRC bilateral grant (NHMRC2017-SG006), and the National Research Foundation, Prime Minister's Office, Singapore, under its Campus of Research Excellence and Technological Enterprise (CREATE) program.

Conflicts of Interest: The authors declare no conflict of interest.

Abbreviations

AP-1	Activator protein 1
APC	Antigen presenting cell
Arp	Actin regulated protein
ASM	Airway smooth muscle
ATP	Adenosine triphosphate
BAL	Bronchoalveolar lavage
BMDM	Bone marrow derived macrophages
cAMP	Cyclic adenosine monophosphate
CCL	C-C motif chemokine ligand
CD	Cluster of differentiation
COPD	Chronic obstructive pulmonary disease
CXCL	C-X-C motif chemokine ligand
dsRNA	Double stranded ribonucleic acid
DUSP	Dual specificity phosphatase
EGF	Epidermal growth factor

EGFR	Epidermal growth factor receptor
ERK	Extracellular signal regulated kinase
G-CSF	Granulocyte colony stimulating factor
GM-CSF	Granulocyte macrophage colony stimulating factor
GR	Glucocorticoid receptor
GRO	Growth related oncogene
Hsp	Heat shock protein
hVHR3	Human vaccinia virus related phosphatase 3
IFN	Interferon
IFNAR	Interferon-α/β receptor
Ig	Immunoglobulin
IKK	IκB kinase
IL	Interleukin
IRF	Interferon regulatory factor
ISG	Interferon stimulated gene
ISGF	Interferon stimulated gene factor
JAK	Janus kinase
JKAP	JNK pathway associated phosphatase
JNK	c-Jun N-terminal kinase
Lck	Lymphocyte specific protein tyrosine kinase
LPS	Lipopolysaccharide
MAPK	Mitogen activated protein kinase
MAVS	Mitochondrial antiviral signaling protein
MEK	MAPK/ERK kinase
MKP	MAPK phosphatase
MUC	Mucin
MyD88	Myeloid differentiation primary response gene
NF-κB	Nuclear factor of κ-light-chain-enhancer of activated B cells
NFATc	Nuclear factor of activated T cells
PKC	Protein kinase C
Poly(I:C)	Polyinosinic:polycytidylic acid
PRR	Pattern recognition receptor
RhoA	Ras homolog gene family member A
RLR	Retinoic acid inducible gene-like receptor
RSV	Respiratory syncytial virus
siRNA	Small interfering RNA
ssRNA	Single stranded RNA
STAT	Signal transducer and activator of transcription
Syk	Spleen tyrosine kinase
TAB	Transforming growth factor β-activated kinase 1-binding protein
TAK	Transforming growth factor β-activated kinase
TCR	T cell receptor
Th	T helper
TLR	Toll like receptor
TNF	Tumor necrosis factor
Treg	T regulatory
TRIF	TIR-domain containing adaptor-inducing interferon-β
TSLP	Thymic stromal lymphoprotein
Tyk	Tyrosine kinase
vRNP	Viral ribonucleoprotein
ZAP	Zeta-chain associated protein kinase

References

1. Collaborators, G.C.R.D. Global, regional, and national deaths, prevalence, disability-adjusted life years, and years lived with disability for chronic obstructive pulmonary disease and asthma, 1990–2015: A systematic analysis for the global burden of disease study 2015. *Lancet Respir. Med.* **2017**, *5*, 691–706.
2. WHO. *Global Health Estimates*; WHO: Geneva, Switzerland, 2018.
3. Barnes, P.J. Immunology of asthma and chronic obstructive pulmonary disease. *Nat. Rev. Immunol.* **2008**, *8*, 183–192. [CrossRef] [PubMed]
4. Atmar, R.L.; Guy, E.; Guntupalli, K.K.; Zimmerman, J.L.; Bandi, V.D.; Baxter, B.D.; Greenberg, S.B. Respiratory tract viral infections in inner-city asthmatic adults. *Arch. Intern. Med.* **1998**, *158*, 2453–2459. [CrossRef] [PubMed]
5. Heymann, P.W.; Carper, H.T.; Murphy, D.D.; Platts-Mills, T.A.E.; Patrie, J.; McLaughlin, A.P.; Erwin, E.A.; Shaker, M.S.; Hellems, M.; Peerzada, J.; et al. Viral infections in relation to age, atopy, and season of admission among children hospitalized for wheezing. *J. Allergy Clin. Immunol.* **2004**, *114*, 239–247. [CrossRef] [PubMed]
6. Johnston, N.W.; Johnston, S.L.; Duncan, J.M.; Greene, J.M.; Kebadze, T.; Keith, P.K.; Roy, M.; Waserman, S.; Sears, M.R. The september epidemic of asthma exacerbations in children: A search for etiology. *J. Allergy Clin. Immunol.* **2005**, *115*, 132–138. [CrossRef] [PubMed]
7. Khetsuriani, N.; Kazerouni, N.N.; Erdman, D.D.; Lu, X.Y.; Redd, S.C.; Anderson, L.J.; Teague, W.G. Prevalence of viral respiratory tract infections in children with asthma. *J. Allergy Clin. Immunol.* **2007**, *119*, 314–321. [CrossRef] [PubMed]
8. Kling, S.; Donninger, H.; Williams, Z.; Vermeulen, J.; Weinberg, E.; Latiff, K.; Ghildyal, R.; Bardin, P. Persistence of rhinovirus rna after asthma exacerbation in children. *Clin. Exp. Allergy* **2005**, *35*, 672–678. [CrossRef]
9. Miller, E.K.; Lu, X.Y.; Erdman, D.D.; Poehling, K.A.; Zhu, Y.W.; Griffin, M.R.; Hartert, T.V.; Anderson, L.J.; Weinberg, G.A.; Hall, C.B.; et al. Rhinovirus-associated hospitalizations in young children. *J. Infect. Dis.* **2007**, *195*, 773–781. [CrossRef]
10. McManus, T.E.; Marley, A.M.; Baxter, N.; Christie, S.N.; O'Neill, H.J.; Elborn, J.S.; Coyle, P.V.; Kidney, J.C. Respiratory viral infection in exacerbations of copd. *Respir. Med.* **2008**, *102*, 1575–1580. [CrossRef]
11. Rohde, G.; Wiethege, A.; Borg, I.; Kauth, M.; Bauer, T.T.; Gillissen, A.; Bufe, A.; Schultze-Werninghaus, G. Respiratory viruses in exacerbations of chronic obstructive pulmonary disease requiring hospitalisation: A case-control study. *Thorax* **2003**, *58*, 37–42. [CrossRef]
12. Newton, K.; Dixit, V.M. Signaling in innate immunity and inflammation. *Cold Spring Harb. Perspect. Biol.* **2012**, *4*, 19. [CrossRef] [PubMed]
13. Sha, Q.; Truong-Tran, A.Q.; Plitt, J.R.; Beck, L.A.; Schleimer, R.P. Activation of airway epithelial cells by toll-like receptor agonists. *Am. J. Respir. Cell Mol. Biol.* **2004**, *31*, 358–364. [CrossRef] [PubMed]
14. Vermeulen, L.; De Wilde, G.; Van Damme, P.; Vanden Berghe, W.; Haegeman, G. Transcriptional activation of the NF-κB p65 subunit by mitogen- and stress-activated protein kinase-1 (msk1). *EMBO J.* **2003**, *22*, 1313–1324. [CrossRef] [PubMed]
15. Edwards, M.R.; Regamey, N.; Vareille, M.; Kieninger, E.; Gupta, A.; Shoemark, A.; Saglani, S.; Sykes, A.; Macintyre, J.; Davies, J.; et al. Impaired innate interferon induction in severe therapy resistant atopic asthmatic children. *Mucosal Immunol.* **2013**, *6*, 797–806. [CrossRef] [PubMed]
16. Sykes, A.; Macintyre, J.; Edwards, M.R.; del Rosario, A.; Haas, J.; Gielen, V.; Kon, O.M.; McHale, M.; Johnston, S.L. Rhinovirus-induced interferon production is not deficient in well controlled asthma. *Thorax* **2014**, *69*, 240–246. [CrossRef]
17. Jakiela, B.; Brockman-Schneider, R.; Amineva, S.; Lee, W.M.; Gern, J.E. Basal cells of differentiated bronchial epithelium are more susceptible to rhinovirus infection. *Am. J. Respir. Cell Mol. Biol.* **2008**, *38*, 517–523. [CrossRef] [PubMed]
18. Lachowicz-Scroggins, M.E.; Boushey, H.A.; Finkbeiner, W.E.; Widdicombe, J.H. Interleukin-13-induced mucous metaplasia increases susceptibility of human airway epithelium to rhinovirus infection. *Am. J. Respir. Cell Mol. Biol.* **2010**, *43*, 652–661. [CrossRef]

19. Message, S.D.; Laza-Stanca, V.; Mallia, P.; Parker, H.L.; Zhu, J.; Kebadze, T.; Contoli, M.; Sanderson, G.; Kon, O.M.; Papi, A.; et al. Rhinovirus-induced lower respiratory illness is increased in asthma and related to virus load and th1/2 cytokine and il-10 production. *Proc. Natl. Acad. Sci. USA* **2008**, *105*, 13562–13567. [CrossRef]

20. Zambrano, J.C.; Carper, H.T.; Rakes, G.P.; Patrie, J.; Murphy, D.D.; Platts-Mills, T.A.E.; Hayden, F.G.; Gwaltney, J.M.; Hatley, T.K.; Owens, A.M.; et al. Experimental rhinovirus challenges in adults with mild asthma: Response to infection in relation to IGE. *J. Allergy Clin. Immunol.* **2003**, *111*, 1008–1016. [CrossRef]

21. Gern, J.E.; French, D.A.; Grindle, K.A.; Brockman-Schneider, R.A.; Konno, S.I.; Busse, W.W. Double-stranded rna induces the synthesis of specific chemokines by bronchial epithelial cells. *Am. J. Respir. Cell Mol. Biol.* **2003**, *28*, 731–737. [CrossRef]

22. Dumitru, C.A.; Dreschers, S.; Gulbins, E. Rhinoviral infections activate p38map-kinases via membrane rafts and rhoa. *Cell. Physiol. Biochem.* **2006**, *17*, 159–166. [CrossRef] [PubMed]

23. Lau, C.; Wang, X.; Song, L.; North, M.; Wiehler, S.; Proud, D.; Chow, C.W. Syk associates with clathrin and mediates phosphatidylinositol 3-kinase activation during human rhinovirus internalization. *J. Immunol.* **2008**, *180*, 870–880. [CrossRef] [PubMed]

24. Wang, X.M.; Lau, C.; Wiehler, S.; Pow, A.; Mazzulli, T.; Gutierrez, C.; Proud, D.; Chow, C.W. Syk is downstream of intercellular adhesion molecule-1 and mediates human rhinovirus activation of p38 mapk in airway epithelial cells. *J. Immunol.* **2006**, *177*, 6859–6870. [CrossRef] [PubMed]

25. Maruoka, S.; Hashimoto, S.; Gon, Y.; Nishitoh, H.; Takeshita, I.; Asai, Y.; Mizumura, K.; Shimizu, K.; Ichijo, H.; Horie, T. Ask1 regulates influenza virus infection-induced apoptotic cell death. *Biochem. Biophys. Res. Commun.* **2003**, *307*, 870–876. [CrossRef]

26. Young, P.R.; McLaughlin, M.M.; Kumar, S.; Kassis, S.; Doyle, M.L.; McNulty, D.; Gallagher, T.F.; Fisher, S.; McDonnell, P.C.; Carr, S.A.; et al. Pyridinyl imidazole inhibitors of p38 mitogen-activated protein kinase bind in the ATP site. *J. Biol. Chem.* **1997**, *272*, 12116–12121. [CrossRef] [PubMed]

27. Griego, S.D.; Weston, C.B.; Adams, J.L.; Tal-Singer, R.; Dillon, S.B. Role of p38 mitogen-activated protein kinase in rhinovirus-induced cytokine production by bronchial epithelial cells. *J. Immunol.* **2000**, *165*, 5211–5220. [CrossRef]

28. Manley, G.C.A.; Stokes, C.A.; Marsh, E.K.; Sabroe, I.; Parker, L.C. Dusp10 negatively regulates the inflammatory response to rhinovirus through il-1β signalling. *J. Virol.* **2018**, *93*, e01659-18. [CrossRef]

29. Pazdrak, K.; Olszewska-Pazdrak, B.; Liu, T.S.; Takizawa, R.; Brasier, A.R.; Garofalo, R.P.; Casola, A. Mapk activation is involved in posttranscriptional regulation of rsv-induced rantes gene expression. *Am. J. Physiol.-Lung Cell. Mol. Physiol.* **2002**, *283*, L364–L372. [CrossRef]

30. Galvan Morales, M.A.; Cabello Gutierrez, C.; Mejia Nepomuceno, F.; Valle Peralta, L.; Valencia Maqueda, E.; Manjarrez Zavala, M.E. Parainfluenza virus type 1 induces epithelial il-8 production via p38-mapk signalling. *J. Immunol. Res.* **2014**, *2014*, 515984. [CrossRef]

31. Meusel, T.R.; Imani, F. Viral induction of inflammatory cytokines in human epithelial cells follows a p38 mitogen-activated protein kinase-dependent but NF-κB-independent pathway. *J. Immunol.* **2003**, *171*, 3768–3774. [CrossRef]

32. Wei, D.; Huang, Z.H.; Zhang, R.H.; Wang, C.L.; Xu, M.J.; Liu, B.J.; Wang, G.H.; Xu, T. Roles of p38 mapk in the regulation of the inflammatory response to swine influenza virus-induced acute lung injury in mice. *Acta Virol.* **2014**, *58*, 374–379. [CrossRef] [PubMed]

33. Davies, S.P.; Reddy, H.; Caivano, M.; Cohen, P. Specificity and mechanism of action of some commonly used protein kinase inhibitors. *Biochem. J.* **2000**, *351*, 95–105. [CrossRef] [PubMed]

34. Liu, K.; Gualano, R.C.; Hibbs, M.L.; Anderson, G.P.; Bozinovski, S. Epidermal growth factor receptor signaling to erk1/2 and stats control the intensity of the epithelial inflammatory responses to rhinovirus infection. *J. Biol. Chem.* **2008**, *283*, 9977–9985. [CrossRef] [PubMed]

35. Newcomb, D.C.; Sajjan, U.S.; Nagarkar, D.R.; Goldsmith, A.M.; Bentley, J.K.; Hershenson, M.B. Cooperative effects of rhinovirus and tnf-alpha on airway epithelial cell chemokine expression. *Am. J. Physiol.-Lung Cell. Mol. Physiol.* **2007**, *293*, L1021–L1028. [CrossRef] [PubMed]

36. Chen, W.; Monick, M.M.; Carter, A.B.; Hunninghake, G.W. Activation of erk2 by respiratory syncytial virus in a549 cells is linked to the production of interleukin 8. *Exp. Lung Res.* **2000**, *26*, 13–26. [CrossRef] [PubMed]

37. Crotta, S.; Davidson, S.; Mahlakoiv, T.; Desmet, C.J.; Buckwalter, M.R.; Albert, M.L.; Staeheli, P.; Wack, A. Type i and type iii interferons drive redundant amplification loops to induce a transcriptional signature in influenza-infected airway epithelia. *PLoS Pathog.* **2013**, *9*, 12. [CrossRef] [PubMed]

38. Khaitov, M.R.; Laza-Stanca, V.; Edwards, M.R.; Walton, R.P.; Rohde, G.; Contoli, M.; Papi, A.; Stanciu, L.A.; Kotenko, S.V.; Johnston, S.L. Respiratory virus induction of alpha-, beta- and lambda-interferons in bronchial epithelial cells and peripheral blood mononuclear cells. *Allergy* **2009**, *64*, 375–386. [CrossRef] [PubMed]

39. Okabayashi, T.; Kojima, T.; Masaki, T.; Yokota, S.; Imaizumi, T.; Tsutsumi, H.; Himi, T.; Fujii, N.; Sawada, N. Type-iii interferon, not type-i, is the predominant interferon induced by respiratory viruses in nasal epithelial cells. *Virus Res.* **2011**, *160*, 360–366. [CrossRef]

40. Cakebread, J.A.; Xu, Y.H.; Grainge, C.; Kehagia, V.; Howarth, P.H.; Holgate, S.T.; Davies, D.E. Exogenous ifn-beta has antiviral and anti-inflammatory properties in primary bronchial epithelial cells from asthmatic subjects exposed to rhinovirus. *J. Allergy Clin. Immunol.* **2011**, *127*, U1148–U1416. [CrossRef]

41. Zeng, H.; Goldsmith, C.; Thawatsupha, P.; Chittaganpitch, M.; Waicharoen, S.; Zaki, S.; Tumpey, T.M.; Katz, J.M. Highly pathogenic avian influenza h5n1 viruses elicit an attenuated type i interferon response in polarized human bronchial epithelial cells. *J. Virol.* **2007**, *81*, 12439–12449. [CrossRef]

42. Mibayashi, M.; Martinez-Sobrido, L.; Loo, Y.M.; Cardenas, W.B.; Gale, M.; Garcia-Sastre, A. Inhibition of retinoic acid-inducible gene i-mediated induction of beta interferon by the ns1 protein of influenza a virus. *J. Virol.* **2007**, *81*, 514–524. [CrossRef] [PubMed]

43. Ren, J.; Liu, T.; Pang, L.; Li, K.; Garofalo, R.P.; Casola, A.; Bao, X. A novel mechanism for the inhibition of interferon regulatory factor-3-dependent gene expression by human respiratory syncytial virus ns1 protein. *J. Gen. Virol.* **2011**, *92*, 2153–2159. [CrossRef]

44. Xing, Z.; Cardona, C.J.; Anunciacion, J.; Adams, S.; Dao, N. Roles of the erk mapk in the regulation of proinflammatory and apoptotic responses in chicken macrophages infected with h9n2 avian influenza virus. *J. Gen. Virol.* **2010**, *91*, 343–351. [CrossRef] [PubMed]

45. Ludwig, S.; Ehrhardt, C.; Neumeier, E.R.; Kracht, M.; Rapp, U.R.; Pleschka, S. Influenza virus-induced ap-1-dependent gene expression requires activation of the jnk signaling pathway. *J. Biol. Chem.* **2001**, *276*, 10990–10998. [CrossRef]

46. Borgeling, Y.; Schmolke, M.; Viemann, D.; Nordhoff, C.; Roth, J.; Ludwig, S. Inhibition of p38 mitogen-activated protein kinase impairs influenza virus-induced primary and secondary host gene responses and protects mice from lethal h5n1 infection. *J. Biol. Chem.* **2014**, *289*, 13–27. [CrossRef] [PubMed]

47. Kong, X.; San Juan, H.; Behera, A.; Peeples, M.E.; Wu, J.; Lockey, R.F.; Mohapatra, S.S. Erk-1/2 activity is required for efficient rsv infection. *FEBS Lett.* **2004**, *559*, 33–38. [CrossRef]

48. Marchant, D.; Singhera, G.K.; Utokaparch, S.; Hackett, T.L.; Boyd, J.H.; Luo, Z.S.; Si, X.N.; Dorscheid, D.R.; McManus, B.M.; Hegele, R.G. Toll-like receptor 4-mediated activation of p38 mitogen-activated protein kinase is a determinant of respiratory virus entry and tropism. *J. Virol.* **2010**, *84*, 11359–11373. [CrossRef]

49. Marjuki, H.; Alam, M.I.; Ehrhardt, C.; Wagner, R.; Planz, O.; Klenk, H.D.; Ludwig, S.; Pleschka, S. Membrane accumulation of influenza a virus hemagglutinin triggers nuclear export of the viral genome via protein kinase c alpha-mediated activation of erk signaling. *J. Biol. Chem.* **2006**, *281*, 16707–16715. [CrossRef]

50. Pleschka, S.; Wolff, T.; Ehrhardt, C.; Hobom, G.; Planz, O.; Rapp, U.R.; Ludwig, S. Influenza virus propagation is impaired by inhibition of the raf/mek/erk signalling cascade. *Nat. Cell Biol.* **2001**, *3*, 301–305. [CrossRef]

51. Nencioni, L.; De Chiara, G.; Sgarbanti, R.; Amatore, D.; Aquilano, K.; Marcocci, M.E.; Serafino, A.; Torcia, M.; Cozzolino, F.; Ciriolo, M.R.; et al. Bcl-2 expression and p38mapk activity in cells infected with influenza a virus impact on virally induced apoptosis and viral replication. *J. Biol. Chem.* **2009**, *284*, 16004–16015. [CrossRef]

52. Choi, M.S.; Heo, J.; Yi, C.M.; Ban, J.; Lee, N.J.; Lee, N.R.; Kim, S.W.; Kim, N.J.; Inn, K.S. A novel p38 mitogen activated protein kinase (mapk) specific inhibitor suppresses respiratory syncytial virus and influenza a virus replication by inhibiting virus-induced p38 mapk activation. *Biochem. Biophys. Res. Commun.* **2016**, *477*, 311–316. [CrossRef]

53. Preugschas, H.F.; Hrincius, E.R.; Mewis, C.; Tran, G.V.Q.; Ludwig, S.; Ehrhardt, C. Late activation of the raf/mek/erk pathway is required for translocation of the rsv f protein to the plasma membrane and efficient viral replication. *Cell Microbiol.* **2018**, *21*, e12955. [CrossRef] [PubMed]

54. Rixon, H.W.M.; Brown, G.; Murray, J.T.; Sugrue, R.J. The respiratory syncytial virus small hydrophobic protein is phosphorylated via a mitogen-activated protein kinase p38-dependent tyrosine kinase activity during virus infection. *J. Gen. Virol.* **2005**, *86*, 375–384. [CrossRef] [PubMed]

55. Kumar, A.; Kim, J.H.; Ranjan, P.; Metcalfe, M.G.; Cao, W.; Mishina, M.; Gangappa, S.; Guo, Z.; Boyden, E.S.; Zaki, S.; et al. Influenza virus exploits tunneling nanotubes for cell-to-cell spread. *Sci. Rep.* **2017**, *7*, 40360. [CrossRef] [PubMed]

56. Harris, J.; Werling, D. Binding and entry of respiratory syncytial virus into host cells and initiation of the innate immune response. *Cell Microbiol.* **2003**, *5*, 671–680. [CrossRef] [PubMed]

57. Mehedi, M.; McCarty, T.; Martin, S.E.; Le Nouen, C.; Buehler, E.; Chen, Y.C.; Smelkinson, M.; Ganesan, S.; Fischer, E.R.; Brock, L.G.; et al. Actin-related protein 2 (arp2) and virus-induced filopodia facilitate human respiratory syncytial virus spread. *PLoS Pathog.* **2016**, *12*, e1006062. [CrossRef] [PubMed]

58. Mendoza, M.C.; Vilela, M.; Juarez, J.E.; Blenis, J.; Danuser, G. Erk reinforces actin polymerization to power persistent edge protrusion during motility. *Sci. Signal.* **2015**, *8*, ra47. [CrossRef] [PubMed]

59. Fischer, I.; Weber, M.; Kuhn, C.; Fitzgerald, J.S.; Schulze, S.; Friese, K.; Walzel, H.; Markert, U.R.; Jeschke, U. Is galectin-1 a trigger for trophoblast cell fusion?: The map-kinase pathway and syncytium formation in trophoblast tumour cells bewo. *Mol. Hum. Reprod.* **2011**, *17*, 747–757. [CrossRef]

60. Singh, D.; McCann, K.L.; Imani, F. Mapk and heat shock protein 27 activation are associated with respiratory syncytial virus induction of human bronchial epithelial monolayer disruption. *Am. J. Physiol.-Lung Cell. Mol. Physiol.* **2007**, *293*, L436–L445. [CrossRef]

61. Ilnytska, O.; Santiana, M.; Hsu, N.Y.; Du, W.L.; Chen, Y.H.; Viktorova, E.G.; Belov, G.; Brinker, A.; Storch, J.; Moore, C.; et al. Enteroviruses harness the cellular endocytic machinery to remodel the host cell cholesterol landscape for effective viral replication. *Cell Host Microbe* **2013**, *14*, 281–293. [CrossRef]

62. Cavalli, V.; Vilbois, F.; Corti, M.; Marcote, M.J.; Tamura, K.; Karin, M.; Arkinstall, S.; Gruenberg, J. The stress-induced map kinase p38 regulates endocytic trafficking via the gdi: Rab5 complex. *Mol. Cell* **2001**, *7*, 421–432. [CrossRef]

63. McCaskill, J.L.; Ressel, S.; Alber, A.; Redford, J.; Power, U.F.; Schwarze, J.; Dutia, B.M.; Buck, A.H. Broad-spectrum inhibition of respiratory virus infection by microrna mimics targeting p38 mapk signaling. *Mol. Ther. Nucleic Acids* **2017**, *7*, 256–266. [CrossRef] [PubMed]

64. Aikawa, T.; Shimura, S.; Sasaki, H.; Ebina, M.; Takishima, T. Marked goblet cell hyperplasia with mucus accumulation in the airways of patients who died of severe acute asthma attack. *Chest* **1992**, *101*, 916–921. [CrossRef] [PubMed]

65. Ordonez, C.L.; Khashayar, R.; Wong, H.H.; Ferrando, R.; Wu, R.; Hyde, D.M.; Hotchkiss, J.A.; Zhang, Y.; Novikov, A.; Dolganov, G.; et al. Mild and moderate asthma is associated with airway goblet cell hyperplasia and abnormalities in mucin gene expression. *Am. J. Respir. Crit. Care Med.* **2001**, *163*, 517–523. [CrossRef]

66. Innes, A.L.; Woodruff, P.G.; Ferrando, R.E.; Donnelly, S.; Dolganov, G.M.; Lazarus, S.C.; Fahy, J.V. Epithelial mucin stores are increased in the large airways of smokers with airflow obstruction. *Chest* **2006**, *130*, 1102–1108. [CrossRef] [PubMed]

67. Kuperman, D.; Schofield, B.; Wills-Karp, M.; Grusby, M.J. Signal transducer and activator of transcription factor 6 (stat6)-deficient mice are protected from antigen-induced airway hyperresponsiveness and mucus production. *J. Exp. Med.* **1998**, *187*, 939–948. [CrossRef] [PubMed]

68. Kuperman, D.A.; Huang, X.; Koth, L.L.; Chang, G.H.; Dolganov, G.M.; Zhu, Z.; Elias, J.A.; Sheppard, D.; Erle, D.J. Direct effects of interleukin-13 on epithelial cells cause airway hyperreactivity and mucus overproduction in asthma. *Nat. Med.* **2002**, *8*, 885–889. [CrossRef]

69. Fujisawa, T.; Ide, K.; Holtzman, M.J.; Suda, T.; Suzuki, K.; Kuroishi, S.; Chida, K.; Nakamura, H. Involvement of the p38 mapk pathway in il-13-induced mucous cell metaplasia in mouse tracheal epithelial cells. *Respirology* **2008**, *13*, 191–202. [CrossRef]

70. Atherton, H.C.; Jones, G.; Danahay, H. Il-13-induced changes in the goblet cell density of human bronchial epithelial cell cultures: Map kinase and phosphatidylinositol 3-kinase regulation. *Am. J. Physiol.-Lung Cell. Mol. Physiol.* **2003**, *285*, L730–L739. [CrossRef]

71. Kono, Y.; Nishiuma, T.; Okada, T.; Kobayashi, K.; Funada, Y.; Kotani, Y.; Jahangeer, S.; Nakamura, S.; Nishimura, Y. Sphingosine kinase 1 regulates mucin production via erk phosphorylation. *Pulm. Pharmacol. Ther.* **2010**, *23*, 36–42. [CrossRef]

72. Londhe, V.; McNamara, N.; Lemjabbar, H.; Basbaum, C. Viral dsrna activates mucin transcription in airway epithelial cells. *FEBS Lett.* **2003**, *553*, 33–38. [CrossRef]

73. Hashimoto, K.; Graham, B.S.; Ho, S.B.; Adler, K.B.; Collins, R.D.; Olson, S.J.; Zhou, W.; Suzutani, T.; Jones, P.W.; Goleniewska, K.; et al. Respiratory syncytial virus in allergic lung inflammation increases muc5ac and gob-5. *Am. J. Respir. Crit. Care Med.* **2004**, *170*, 306–312. [CrossRef] [PubMed]

74. Qi, F.; Bai, S.; Wang, D.; Xu, L.; Hu, H.; Zeng, S.; Chai, R.; Liu, B. Macrophages produce il-33 by activating mapk signaling pathway during rsv infection. *Mol. Immunol.* **2017**, *87*, 284–292. [CrossRef] [PubMed]

75. Park, I.H.; Park, J.H.; Shin, J.M.; Lee, H.M. Tumor necrosis factor-alpha regulates interleukin-33 expression through extracellular signal-regulated kinase, p38, and nuclear factor-κb pathways in airway epithelial cells. *Int. Forum Allergy Rhinol.* **2016**, *6*, 973–980. [CrossRef] [PubMed]

76. Mata, M.; Sarria, B.; Buenestado, A.; Cortijo, J.; Cerda, M.; Morcillo, E.J. Phosphodiesterase 4 inhibition decreases muc5ac expression induced by epidermal growth factor in human airway epithelial cells. *Thorax* **2005**, *60*, 144–152. [CrossRef]

77. Kim, S.W.; Hong, J.S.; Ryu, S.H.; Chung, W.C.; Yoon, J.H.; Koo, J.S. Regulation of mucin gene expression by creb via a nonclassical retinoic acid signaling pathway. *Mol. Cell. Biol.* **2007**, *27*, 6933–6947. [CrossRef]

78. Inoue, D.; Yamaya, M.; Kubo, H.; Sasaki, T.; Hosoda, M.; Numasaki, M.; Tomioka, Y.; Yasuda, H.; Sekizawa, K.; Nishimura, H.; et al. Mechanisms of mucin production by rhinovirus infection in cultured human airway epithelial cells. *Respir. Physiol. Neurobiol.* **2006**, *154*, 484–499. [CrossRef]

79. Zhu, L.; Lee, P.K.; Lee, W.M.; Zhao, Y.; Yu, D.; Chen, Y. Rhinovirus-induced major airway mucin production involves a novel tlr3-egfr-dependent pathway. *Am. J. Respir. Cell Mol. Biol.* **2009**, *40*, 610–619. [CrossRef]

80. Bain, J.; Plater, L.; Elliott, M.; Shpiro, N.; Hastie, C.J.; McLauchlan, H.; Klevernic, I.; Arthur, J.S.C.; Alessi, D.R.; Cohen, P. The selectivity of protein kinase inhibitors: A further update. *Biochem. J.* **2007**, *408*, 297–315. [CrossRef]

81. Denu, J.M.; Dixon, J.E. A catalytic mechanism for the dual-specific phosphatases. *Proc. Natl. Acad. Sci. USA* **1995**, *92*, 5910–5914. [CrossRef]

82. Franklin, C.C.; Kraft, A.S. Conditional expression of the mitogen-activated protein kinase (mapk) phosphatase mkp-1 preferentially inhibits p38 mapk and stress-activated protein kinase in u937 cells. *J. Biol. Chem.* **1997**, *272*, 16917–16923. [CrossRef] [PubMed]

83. Zhao, Q.; Shepherd, E.G.; Manson, M.E.; Nelin, L.D.; Sorokin, A.; Liu, Y.S. The role of mitogen-activated protein kinase phosphatase-1 in the response of alveolar macrophages to lipopolysaccharide - attenuation of proinflammatory cytokine biosynthesis via feedback control of p38. *J. Biol. Chem.* **2005**, *280*, 8101–8108. [CrossRef] [PubMed]

84. Chi, H.B.; Barry, S.P.; Roth, R.J.; Wu, J.J.; Jones, E.A.; Bennettt, A.M.; Flavell, R.A. Dynamic regulation of pro- and anti-inflammatory cytokines by mapk phosphatase 1 (mkp-1) in innate immune responses. *Proc. Natl. Acad. Sci. USA* **2006**, *103*, 2274–2279. [CrossRef] [PubMed]

85. Golebski, K.; van Egmond, D.; de Groot, E.; Roschmann, K.I.L.; Fokkens, W.J.; van Drunen, C.M. High degree of overlap between responses to a virus and to the house dust mite allergen in airway epithelial cells. *PLoS ONE* **2014**, *9*, e87768. [CrossRef] [PubMed]

86. Golebski, K.; van Egmond, D.; de Groot, E.J.; Roschmann, K.I.L.; Fokkens, W.J.; van Drunen, C.M. Egr-1 and dusp-1 are important negative regulators of pro-allergic responses in airway epithelium. *Mol. Immunol.* **2015**, *65*, 43–50. [CrossRef] [PubMed]

87. Liao, Y.; Wang, X.X.; Huang, M.; Tam, J.P.; Liu, D.X. Regulation of the p38 mitogen-activated protein kinase and dual-specificity phosphatase 1 feedback loop modulates the induction of interleukin 6 and 8 in cells infected with coronavirus infectious bronchitis virus. *Virology* **2011**, *420*, 106–116. [CrossRef] [PubMed]

88. Robitaille, A.C.; Caron, E.; Zucchini, N.; Mukawera, E.; Adam, D.; Mariani, M.K.; Gelinas, A.; Fortin, A.; Brochiero, E.; Grandvaux, N. Dusp1 regulates apoptosis and cell migration, but not the jip1-protected cytokine response, during respiratory syncytial virus and sendai virus infection. *Sci. Rep.* **2017**, *7*, 17388. [CrossRef] [PubMed]

89. Choi, J.E.; Kwon, J.H.; Kim, J.H.; Hur, W.; Sung, P.S.; Choi, S.W.; Yoon, S.K. Suppression of dual specificity phosphatase i expression inhibits hepatitis c virus replication. *PLoS ONE* **2015**, *10*, 11. [CrossRef] [PubMed]

90. Liu, D.; Scafidi, J.; Prada, A.E.; Zahedi, K.; Davis, A.E., 3rd. Nuclear phosphatases and the proteasome in suppression of stat1 activity in hepatocytes. *Biochem. Biophys. Res. Commun.* **2002**, *299*, 574–580. [CrossRef]

91. Slack, D.N.; Seternes, O.M.; Gabrielsen, M.; Keyse, S.M. Distinct binding determinants for erk2/p38alpha and jnk map kinases mediate catalytic activation and substrate selectivity of map kinase phosphatase-1. *J. Biol. Chem.* **2001**, *276*, 16491–16500. [CrossRef] [PubMed]

92. Bochkov, Y.A.; Hanson, K.M.; Keles, S.; Brockman-Schneider, R.A.; Jarjour, N.N.; Gern, J.E. Rhinovirus-induced modulation of gene expression in bronchial epithelial cells from subjects with asthma. *Mucosal Immunol.* **2010**, *3*, 69–80. [CrossRef] [PubMed]

93. Choi, J.; Callaway, Z.; Kim, H.B.; Fujisawa, T.; Kim, C.K. The role of tnf-alpha in eosinophilic inflammation associated with rsv bronchiolitis. *Pediatr. Allergy Immunol.* **2010**, *21*, 474–479. [CrossRef] [PubMed]

94. Broide, D.H.; Lotz, M.; Cuomo, A.J.; Coburn, D.A.; Federman, E.C.; Wasserman, S.I. Cytokines in symptomatic asthma airways. *J. Allergy Clin. Immunol.* **1992**, *89*, 958–967. [CrossRef]

95. Manetsch, M.; Che, W.C.; Seidel, P.; Chen, Y.; Ammit, A.J. Mkp-1: A negative feedback effector that represses mapk-mediated pro-inflammatory signaling pathways and cytokine secretion in human airway smooth muscle cells. *Cell. Signal.* **2012**, *24*, 907–913. [CrossRef] [PubMed]

96. Ordonez, C.L.; Shaughnessy, T.E.; Matthay, M.A.; Fahy, J.V. Increased neutrophil numbers and il-8 levels in airway secretions in acute severe asthma clinical and biologic significance. *Am. J. Respir. Crit. Care Med.* **2000**, *161*, 1185–1190. [CrossRef] [PubMed]

97. Wark, P.A.B.; Johnston, S.L.; Moric, I.; Simpson, J.L.; Hensley, M.J.; Gibson, P.G. Neutrophil degranulation and cell lysis is associated with clinical severity in virus-induced asthma. *Eur. Respir. J.* **2002**, *19*, 68–75. [CrossRef]

98. Shah, S.A.; Ishinaga, H.; Takeuchi, K. Clarithromycin inhibits tnf-alpha-induced muc5ac mucin gene expression via the mkp-1-p38mapk-dependent pathway. *Int. Immunopharmacol.* **2017**, *49*, 60–66. [CrossRef]

99. Tanoue, T.; Moriguchi, T.; Nishida, E. Molecular cloning and characterization of a novel dual specificity phosphatase, mkp-5. *J. Biol. Chem.* **1999**, *274*, 19949–19956. [CrossRef]

100. James, S.J.; Jiao, H.P.; Teh, H.Y.; Takahashi, H.; Png, C.W.; Phoon, M.C.; Suzuki, Y.; Sawasaki, T.; Xiao, H.; Chow, V.T.K.; et al. Mapk phosphatase 5 expression induced by influenza and other rna virus infection negatively regulates irf3 activation and type i interferon response. *Cell Rep.* **2015**, *10*, 1722–1734. [CrossRef]

101. Ganesan, S.; Pham, D.; Jing, Y.X.; Farazuddin, M.; Hudy, M.H.; Unger, B.; Comstock, A.T.; Proud, D.; Lauring, A.S.; Sajjan, U.S. Tlr2 activation limits rhinovirus-stimulated cxcl-10 by attenuating irak-1 dependent il-33 receptor signaling in human bronchial epithelial cells. *J. Immunol.* **2016**, *197*, 2409–2420. [CrossRef]

102. Qian, F.; Deng, J.; Gantner, B.; Dong, C.; Christman, J.; Ye, R. Map kinase phosphatase 5 protects against sepsis-induced acute lung injury. *J. Immunol.* **2012**, *188*, 1. [CrossRef] [PubMed]

103. Cornell, T.T.; Fleszar, A.; McHugh, W.; Blatt, N.B.; LeVine, A.M.; Shanley, T.P. Mitogen-activated protein kinase phosphatase 2, mkp-2, regulates early inflammation in acute lung injury. *Am. J. Physiol.-Lung Cell. Mol. Physiol.* **2012**, *303*, L251–L258. [CrossRef] [PubMed]

104. Cornell, T.T.; Rodenhouse, P.; Cai, Q.; Sun, L.; Shanley, T.P. Mitogen-activated protein kinase phosphatase 2 regulates the inflammatory response in sepsis. *Infect. Immunity* **2010**, *78*, 2868–2876. [CrossRef] [PubMed]

105. Ananieva, O.; Darragh, J.; Johansen, C.; Carr, J.M.; McIlrath, J.; Park, J.M.; Wingate, A.; Monk, C.E.; Toth, R.; Santos, S.G.; et al. The kinases msk1 and msk2 act as negative regulators of toll-like receptor signaling. *Nat. Immunol.* **2008**, *9*, 1028–1036. [CrossRef] [PubMed]

106. Al-Mutairi, M.S.; Cadalbert, L.C.; McGachy, H.A.; Shweash, M.; Schroeder, J.; Kurnik, M.; Sloss, C.M.; Bryant, C.E.; Alexander, J.; Plevin, R. Map kinase phosphatase-2 plays a critical role in response to infection by leishmania mexicana. *PLoS Pathog.* **2010**, *6*, 11. [CrossRef] [PubMed]

107. Woodruff, P.G.; Modrek, B.; Choy, D.F.; Jia, G.Q.; Abbas, A.R.; Ellwanger, A.; Arron, J.R.; Koth, L.L.; Fahy, J.V. T-helper type 2-driven inflammation defines major subphenotypes of asthma. *Am. J. Respir. Crit. Care Med.* **2009**, *180*, 388–395. [CrossRef] [PubMed]

108. Jackson, D.J.; Makrinioti, H.; Rana, B.M.J.; Shamji, B.W.H.; Trujillo-Torralbo, M.B.; Footitt, J.; Del-Rosario, J.; Telcian, A.G.; Nikonova, A.; Zhu, J.; et al. Il-33-dependent type 2 inflammation during rhinovirus-induced asthma exacerbations in vivo. *Am. J. Respir. Crit. Care Med.* **2014**, *190*, 1373–1382. [CrossRef] [PubMed]

109. Nino, G.; Huseni, S.; Perez, G.F.; Pancham, K.; Mubeen, H.; Abbasi, A.; Wang, J.; Eng, S.; Colberg-Poley, A.M.; Pillai, D.K.; et al. Directional secretory response of double stranded rna-induced thymic stromal lymphopoetin (tslp) and ccl11/eotaxin-1 in human asthmatic airways. *PLoS ONE* **2014**, *9*, 19. [CrossRef]

110. Beale, J.; Jayaraman, A.; Jackson, D.J.; Macintyre, J.D.R.; Edwards, M.R.; Walton, R.P.; Zhu, J.; Ching, Y.M.; Shamji, B.; Edwards, M.; et al. Rhinovirus-induced il-25 in asthma exacerbation drives type 2 immunity and allergic pulmonary inflammation. *Sci. Transl. Med.* **2014**, *6*, 11. [CrossRef]

111. Zhang, K.; Shan, L.; Rahman, M.S.; Unruh, H.; Halayko, A.J.; Gounni, A.S. Constitutive and inducible thymic stromal lymphopoietin expression in human airway smooth muscle cells: Role in chronic obstructive pulmonary disease. *Am. J. Physiol.-Lung Cell. Mol. Physiol.* **2007**, *293*, L375–L382. [CrossRef]

112. Redhu, N.S.; Saleh, A.; Halayko, A.J.; Ali, A.S.; Gounni, A.S. Essential role of NF-κB and ap-1 transcription factors in tnf-alpha-induced tslp expression in human airway smooth muscle cells. *Am. J. Physiol.-Lung Cell. Mol. Physiol.* **2011**, *300*, L479–L485. [CrossRef] [PubMed]

113. Becker, Y. Respiratory syncytial virus (rsv) evades the human adaptive immune system by skewing the th1/th2 cytokine balance toward increased levels of th2 cytokines and ige, markers of allergy—A review. *Virus Genes* **2006**, *33*, 235–252. [PubMed]

114. Krishnamoorthy, N.; Khare, A.; Oriss, T.B.; Raundhal, M.; Morse, C.; Yarlagadda, M.; Wenzel, S.E.; Moore, M.L.; Peebles, R.S.; Ray, A.; et al. Early infection with respiratory syncytial virus impairs regulatory t cell function and increases susceptibility to allergic asthma. *Nat. Med.* **2012**, *18*, 1525–1530. [CrossRef] [PubMed]

115. Lee, D.C.P.; Harker, J.A.E.; Tregoning, J.S.; Atabani, S.F.; Johansson, C.; Schwarze, J.; Openshaw, P.J.M. Cd25(+) natural regulatory t cells are critical in limiting innate and adaptive immunity and resolving disease following respiratory syncytial virus infection. *J. Virol.* **2010**, *84*, 8790–8798. [CrossRef] [PubMed]

116. Welliver, T.P.; Garofalo, R.P.; Hosakote, Y.; Hintz, K.H.; Avendano, L.; Sanchez, K.; Velozo, L.; Jafri, H.; Chavez-Bueno, S.; Ogra, P.L.; et al. Severe human lower respiratory tract illness caused by respiratory syncytial virus and influenza virus is characterized by the absence of pulmonary cytotoxic lymphocyte responses. *J. Infect. Dis.* **2007**, *195*, 1126–1136. [CrossRef] [PubMed]

117. Ajamian, F.; Wu, Y.; Ebeling, C.; Ilarraza, R.; Odemuyiwa, S.O.; Moqbel, R.; Adamko, D.J. Respiratory syncytial virus induces indoleamine 2,3-dioxygenase activity: A potential novel role in the development of allergic disease. *Clin. Exp. Allergy* **2015**, *45*, 644–659. [CrossRef] [PubMed]

118. Dong, C.; Davis, R.J.; Flavell, R.A. Map kinases in the immune response. *Annu. Rev. Immunol.* **2002**, *20*, 55–72. [CrossRef] [PubMed]

119. Huang, G.; Wang, Y.; Shi, L.Z.; Kanneganti, T.D.; Chi, H. Signaling by the phosphatase mkp-1 in dendritic cells imprints distinct effector and regulatory t cell fates. *Immunity* **2011**, *35*, 45–58. [CrossRef]

120. Zhang, Y.; Reynolds, J.M.; Chang, S.H.; Martin-Orozco, N.; Chung, Y.; Nurieva, R.I.; Dong, C. Mkp-1 is necessary for t cell activation and function. *J. Biol. Chem.* **2009**, *284*, 30815–30824. [CrossRef]

121. Chow, C.W.; Dong, C.; Flavell, R.A.; Davis, R.J. C-jun nh(2)-terminal kinase inhibits targeting of the protein phosphatase calcineurin to nfatc1. *Mol. Cell. Biol.* **2000**, *20*, 5227–5234. [CrossRef]

122. Zhang, Y.L.; Blattman, J.N.; Kennedy, N.J.; Duong, J.; Nguyen, T.; Wang, Y.; Davis, R.J.; Greenberg, P.D.; Flavell, R.A.; Dong, C. Regulation of innate and adaptive immune responses by map kinase phosphatase 5. *Nature* **2004**, *430*, 793–797. [CrossRef] [PubMed]

123. Hsiao, W.Y.; Lin, Y.C.; Liao, F.H.; Chan, Y.C.; Huang, C.Y. Dual-specificity phosphatase 4 regulates stat5 protein stability and helper t cell polarization. *PLoS ONE* **2015**, *10*, e0145880. [CrossRef] [PubMed]

124. Burchill, M.A.; Yang, J.; Vogtenhuber, C.; Blazar, B.R.; Farrar, M.A. Il-2 receptor beta-dependent stat5 activation is required for the development of foxp3+ regulatory t cells. *J. Immunol.* **2007**, *178*, 280–290. [CrossRef] [PubMed]

125. Moon, S.J.; Lim, M.A.; Park, J.S.; Byun, J.K.; Kim, S.M.; Park, M.K.; Kim, E.K.; Moon, Y.M.; Min, J.K.; Ahn, S.M.; et al. Dual-specificity phosphatase 5 attenuates autoimmune arthritis in mice via reciprocal regulation of the th17/treg cell balance and inhibition of osteoclastogenesis. *Arthritis Rheumatol.* **2014**, *66*, 3083–3095. [CrossRef] [PubMed]

126. Kumabe, S.; Itsumi, M.; Yamada, H.; Yajima, T.; Matsuguchi, T.; Yoshikai, Y. Dual specificity phosphatase16 is a negative regulator of c-jun nh2-terminal kinase activity in t cells. *Microbiol. Immunol.* **2010**, *54*, 105–111. [CrossRef] [PubMed]

127. Musikacharoen, T.; Bandow, K.; Kakimoto, K.; Kusuyama, J.; Onishi, T.; Yoshikai, Y.; Matsuguchi, T. Functional involvement of dual specificity phosphatase 16 (dusp16), a c-jun n-terminal kinase-specific phosphatase, in the regulation of t helper cell differentiation. *J. Biol. Chem.* **2011**, *286*, 24896–24905. [CrossRef] [PubMed]

128. Zhang, Y.L.; Nallaparaju, K.C.; Liu, X.; Jiao, H.P.; Reynolds, J.M.; Wang, Z.X.; Dong, C. Mapk phosphatase 7 regulates t cell differentiation via inhibiting erk-mediated il-2 expression. *J. Immunol.* **2015**, *194*, 3088–3095. [CrossRef]

129. Laurence, A.; Tato, C.M.; Davidson, T.S.; Kanno, Y.; Chen, Z.; Yao, Z.; Blank, R.B.; Meylan, F.; Siegel, R.; Hennighausen, L.; et al. Interleukin-2 signaling via stat5 constrains t helper 17 cell generation. *Immunity* **2007**, *26*, 371–381. [CrossRef]

130. Smith-Garvin, J.E.; Koretzky, G.A.; Jordan, M.S. T cell activation. *Annu. Rev. Immunol.* **2009**, *27*, 591–619. [CrossRef]

131. Molina, T.J.; Kishihara, K.; Siderovski, D.P.; van Ewijk, W.; Narendran, A.; Timms, E.; Wakeham, A.; Paige, C.J.; Hartmann, K.U.; Veillette, A.; et al. Profound block in thymocyte development in mice lacking p56lck. *Nature* **1992**, *357*, 161–164. [CrossRef]

132. Li, J.P.; Yang, C.Y.; Chuang, H.C.; Lan, J.L.; Chen, D.Y.; Chen, Y.M.; Wang, X.; Chen, A.J.; Belmont, J.W.; Tan, T.H. The phosphatase jkap/dusp22 inhibits t-cell receptor signalling and autoimmunity by inactivating lck. *Nat. Commun.* **2014**, *5*, 3618. [CrossRef] [PubMed]

133. Yang, C.Y.; Li, J.P.; Chiu, L.L.; Lan, J.L.; Chen, D.Y.; Chuang, H.C.; Huang, C.Y.; Tan, T.H. Dual-specificity phosphatase 14 (dusp14/mkp6) negatively regulates tcr signaling by inhibiting tab1 activation. *J. Immunol.* **2014**, *192*, 1547–1557. [CrossRef] [PubMed]

134. Cutrera, R.; Baraldi, E.; Indinnimeo, L.; Del Giudice, M.M.; Piacentini, G.; Scaglione, F.; Ullmann, N.; Moschino, L.; Galdo, F.; Duse, M. Management of acute respiratory diseases in the pediatric population: The role of oral corticosteroids. *Ital. J. Pediatr.* **2017**, *43*, 21. [CrossRef] [PubMed]

135. Heck, S.; Kullmann, M.; Gast, A.; Ponta, H.; Rahmsdorf, H.J.; Herrlich, P.; Cato, A.C.B. A distinct modulating domain in glucocorticoid receptor monomers in the repression of activity of the transcription factor ap-1. *EMBO J.* **1994**, *13*, 4087–4095. [CrossRef] [PubMed]

136. Ray, A.; Prefontaine, K.E. Physical association and functional antagonism between the p65 subunit of transcription factor NF-κB and the glucocorticoid receptor. *Proc. Natl. Acad. Sci. USA* **1994**, *91*, 752–756. [CrossRef] [PubMed]

137. Keranen, T.; Moilanen, E.; Korhonen, R. Suppression of cytokine production by glucocorticoids is mediated by mkp-1 in human lung epithelial cells. *Inflamm. Res.* **2017**, *66*, 441–449. [CrossRef] [PubMed]

138. King, E.M.; Holden, N.S.; Gong, W.; Rider, C.F.; Newton, R. Inhibition of NF-κB-dependent transcription by mkp-1 transcriptional repression by glucocorticoids occurring via p38 mapk. *J. Biol. Chem.* **2009**, *284*, 26803–26815. [CrossRef] [PubMed]

139. Prabhala, P.; Bunge, K.; Ge, Q.; Ammit, A.J. Corticosteroid-induced mkp-1 represses pro-inflammatory cytokine secretion by enhancing activity of tristetraprolin (ttp) in asm cells. *J. Cell. Physiol.* **2016**, *231*, 2153–2158. [CrossRef] [PubMed]

140. Issa, R.; Xie, S.P.; Khorasani, N.; Sukkar, M.; Adcock, I.M.; Lee, K.Y.; Chung, K.F. Corticosteroid inhibition of growth-related oncogene protein-alpha via mitogen-activated kinase phosphatase-1 in airway smooth muscle cells. *J. Immunol.* **2007**, *178*, 7366–7375. [CrossRef]

141. Quante, T.; Ng, Y.C.; Ramsay, E.E.; Henness, S.; Allen, J.C.; Parmentier, J.; Ge, Q.; Ammit, A.J. Corticosteroids reduce il-6 in asm cells via up-regulation of mkp-1. *Am. J. Respir. Cell Mol. Biol.* **2008**, *39*, 208–217. [CrossRef]

142. Wang, X.; Nelin, L.D.; Kuhlman, J.R.; Meng, X.; Welty, S.E.; Liu, Y. The role of map kinase phosphatase-1 in the protective mechanism of dexamethasone against endotoxemia. *Life Sci.* **2008**, *83*, 671–680. [CrossRef] [PubMed]

143. Abraham, S.M.; Lawrence, T.; Kleiman, A.; Warden, P.; Medghalchi, M.; Tuckermann, J.; Saklatvala, J.; Clark, A.R. Antiinflammatory effects of dexamethasone are partly dependent on induction of dual specificity phosphatase 1. *J. Exp. Med.* **2006**, *203*, 1883–1889. [CrossRef] [PubMed]

144. Pemmari, A.; Paukkeri, E.-L.; Hämäläinen, M.; Leppänen, T.; Korhonen, R.; Moilanen, E. Mkp-1 promotes anti-inflammatory m(il-4/il-13) macrophage phenotype and mediates the anti-inflammatory effects of glucocorticoids. *Basic Clin. Pharmacol. Toxicol.* **2018**. [CrossRef] [PubMed]

145. Maier, J.V.; Brema, S.; Tuckermann, J.; Herzer, U.; Klein, M.; Stassen, M.; Moorthy, A.; Cato, A.C.B. Dual specificity phosphatase 1 knockout mice show enhanced susceptibility to anaphylaxis but are sensitive to glucocorticoids. *Mol. Endocrinol.* **2007**, *21*, 2663–2671. [CrossRef] [PubMed]

146. Newton, R.; King, E.M.; Gong, W.; Rider, C.F.; Staples, K.J.; Holden, N.S.; Bergmann, M.W. Glucocorticoids inhibit il-1 beta-induced gm-csf expression at multiple levels: Roles for the erk pathway and repression by mkp-1. *Biochem. J.* **2010**, *427*, 113–124. [CrossRef] [PubMed]

147. Shah, S.; King, E.M.; Chandrasekhar, A.; Newton, R. Roles for the mitogen-activated protein kinase (mapk) phosphatase, dusp1, in feedback control of inflammatory gene expression and repression by dexamethasone. *J. Biol. Chem.* **2014**, *289*, 13667–13679. [CrossRef] [PubMed]

148. Trevor, J.L.; Deshane, J.S. Refractory asthma: Mechanisms, targets, and therapy. *Allergy* **2014**, *69*, 817–827. [CrossRef]

149. Loke, T.K.; Mallett, K.H.; Ratoff, J.; O'Connor, B.J.; Ying, S.; Meng, Q.; Soh, C.; Lee, T.H.; Corrigan, C.J. Systemic glucocorticoid reduces bronchial mucosal activation of activator protein 1 components in glucocorticoid-sensitive but not glucocorticoid-resistant asthmatic patients. *J. Allergy Clin. Immunol.* **2006**, *118*, 368–375. [CrossRef]

150. Bhavsar, P.; Hew, M.; Khorasani, N.; Torrego, A.; Barnes, P.J.; Adcock, I.; Chung, K.F. Relative corticosteroid insensitivity of alveolar macrophages in severe asthma compared with non-severe asthma. *Thorax* **2008**, *63*, 784–790. [CrossRef]

151. Wang, M.J.; Gao, P.F.; Wu, X.J.; Chen, Y.T.; Feng, Y.K.; Yang, Q.; Xu, Y.J.; Zhao, J.P.; Xie, J.G. Impaired anti-inflammatory action of glucocorticoid in neutrophil from patients with steroid-resistant asthma. *Respir. Res.* **2016**, *17*, 9. [CrossRef]

152. Xia, Y.X.C.; Radwan, A.; Keenan, C.R.; Langenbach, S.Y.; Li, M.N.; Radojicic, D.; Londrigan, S.L.; Gualano, R.C.; Stewart, A.G. Glucocorticoid insensitivity in virally infected airway epithelial cells is dependent on transforming growth factor-beta activity. *PLoS Pathog.* **2017**, *13*, 25. [CrossRef] [PubMed]

153. Goleva, E.; Jackson, L.P.; Harris, J.K.; Robertson, C.E.; Sutherland, E.R.; Hall, C.F.; Good, J.T.; Gelfand, E.W.; Martin, R.J.; Leung, D.Y.M. The effects of airway microbiome on corticosteroid responsiveness in asthma. *Am. J. Respir. Crit. Care Med.* **2013**, *188*, 1193–1201. [CrossRef] [PubMed]

154. Papi, A.; Contoli, M.; Adcock, I.M.; Bellettato, C.; Padovani, A.; Casolari, P.; Stanciu, L.A.; Barnes, P.J.; Johnston, S.L.; Ito, K.; et al. Rhinovirus infection causes steroid resistance in airway epithelium through nuclear factor κb and c-jun n-terminal kinase activation. *J. Allergy Clin. Immunol.* **2013**, *132*, 1075–1085. [CrossRef] [PubMed]

155. Rahman, M.M.; Prabhala, P.; Rumzhum, N.N.; Patel, B.S.; Wickop, T.; Hansbro, P.M.; Verrills, N.M.; Ammit, A.J. Tlr2 ligation induces corticosteroid insensitivity in a549 lung epithelial cells: Anti-inflammatory impact of pp2a activators. *Int. J. Biochem. Cell Biol.* **2016**, *78*, 279–287. [CrossRef] [PubMed]

156. Tephly, L.A.; Carter, A.B. Differential expression and oxidation of mkp-1 modulates tnf-alpha gene expression. *Am. J. Respir. Cell Mol. Biol.* **2007**, *37*, 366–374. [CrossRef] [PubMed]

157. Jin, Y.; Hu, D.L.; Peterson, E.L.; Eng, C.; Levin, A.M.; Wells, K.; Beckman, K.; Kumar, R.; Seibold, M.A.; Karungi, G.; et al. Dual-specificity phosphatase 1 as a pharmacogenetic modifier of inhaled steroid response among asthmatic patients. *J. Allergy Clin. Immunol.* **2010**, *126*, 618–625. [CrossRef] [PubMed]

158. Giembycz, M.A.; Newton, R. Beyond the dogma: Novel beta2-adrenoceptor signalling in the airways. *Eur. Respir. J.* **2006**, *27*, 1286–1306. [CrossRef] [PubMed]

159. Kwak, S.P.; Hakes, D.J.; Martell, K.J.; Dixon, J.E. Isolation and characterization of a human dual-specificity protein-tyrosine-phosphatase gene. *J. Biol. Chem.* **1994**, *269*, 3596–3604. [PubMed]

160. Manetsch, M.; Rahman, M.M.; Patel, B.S.; Ramsay, E.E.; Rumzhum, N.N.; Alkhouri, H.; Ge, Q.; Ammit, A.J. Long-acting beta2-agonists increase fluticasone propionate-induced mitogen-activated protein kinase phosphatase 1 (mkp-1) in airway smooth muscle cells. *PLoS ONE* **2013**, *8*, e59635. [CrossRef]

161. Kaur, M.; Chivers, J.E.; Giembycz, M.A.; Newton, R. Long-acting beta2-adrenoceptor agonists synergistically enhance glucocorticoid-dependent transcription in human airway epithelial and smooth muscle cells. *Mol. Pharmacol.* **2008**, *73*, 203–214. [CrossRef]

162. Hackett, T.L.; Singhera, G.K.; Shaheen, F.; Hayden, P.; Jackson, G.R.; Hegele, R.G.; Van Eeden, S.; Bai, T.R.; Dorscheid, D.R.; Knight, D.A. Intrinsic phenotypic differences of asthmatic epithelium and its inflammatory responses to respiratory syncytial virus and air pollution. *Am. J. Respir. Cell Mol. Biol.* **2011**, *45*, 1090–1100. [CrossRef] [PubMed]

163. Liang, Q.; Guo, L.; Gogate, S.; Karim, Z.; Hanifi, A.; Leung, D.Y.; Gorska, M.M.; Alam, R. Il-2 and il-4 stimulate mek1 expression and contribute to t cell resistance against suppression by tgf-beta and il-10 in asthma. *J. Immunol.* **2010**, *185*, 5704–5713. [CrossRef] [PubMed]

164. Vroling, A.B.; Jonker, M.J.; Luiten, S.; Breit, T.M.; Fokkens, W.J.; van Drunen, C.M. Primary nasal epithelium exposed to house dust mite extract shows activated expression in allergic individuals. *Am. J. Respir. Cell Mol. Biol.* **2008**, *38*, 293–299. [CrossRef] [PubMed]

165. Rastogi, R.; Du, W.; Ju, D.; Pirockinaite, G.; Liu, Y.; Nunez, G.; Samavati, L. Dysregulation of p38 and mkp-1 in response to nod1/tlr4 stimulation in sarcoid bronchoalveolar cells. *Am. J. Respir. Crit. Care Med.* **2011**, *183*, 500–510. [CrossRef] [PubMed]

166. Jaakkola, M.S.; Jaakkola, J.J. Effects of environmental tobacco smoke on the respiratory health of adults. *Scand. J. Work Environ. Health* **2002**, *28* (Suppl. 2), 52–70.

167. Crotty Alexander, L.E.; Shin, S.; Hwang, J.H. Inflammatory diseases of the lung induced by conventional cigarette smoke: A review. *Chest* **2015**, *148*, 1307–1322. [CrossRef]

168. Gaffey, K.; Reynolds, S.; Plumb, J.; Kaur, M.; Singh, D. Increased phosphorylated p38 mitogen-activated protein kinase in copd lungs. *Eur. Respir. J.* **2013**, *42*, 28–41. [CrossRef]

169. Li, D.; Hu, J.; Wang, T.; Zhang, X.; Liu, L.; Wang, H.; Wu, Y.; Xu, D.; Wen, F. Silymarin attenuates cigarette smoke extract-induced inflammation via simultaneous inhibition of autophagy and erk/p38 mapk pathway in human bronchial epithelial cells. *Sci. Rep.* **2016**, *6*, 37751. [CrossRef]

170. Liu, C.; Russell, R.M.; Wang, X.D. Low dose beta-carotene supplementation of ferrets attenuates smoke-induced lung phosphorylation of jnk, p38 mapk, and p53 proteins. *J. Nutr.* **2004**, *134*, 2705–2710. [CrossRef]

171. Seumois, G.; Zapardiel-Gonzalo, J.; White, B.; Singh, D.; Schulten, V.; Dillon, M.; Hinz, D.; Broide, D.H.; Sette, A.; Peters, B.; et al. Transcriptional profiling of th2 cells identifies pathogenic features associated with asthma. *J. Immunol.* **2016**, *197*, 655–664. [CrossRef]

172. Wan, Y.I.; Shrine, N.R.G.; Artigas, M.S.; Wain, L.V.; Blakey, J.D.; Moffatt, M.F.; Bush, A.; Chung, K.F.; Cookson, W.; Strachan, D.P.; et al. Genome-wide association study to identify genetic determinants of severe asthma. *Thorax* **2012**, *67*, 762–768. [CrossRef] [PubMed]

173. Arthur, J.S.C.; Ley, S.C. Mitogen-activated protein kinases in innate immunity. *Nat. Rev. Immunol.* **2013**, *13*, 679–692. [CrossRef] [PubMed]

174. Doddareddy, M.R.; Rawling, T.; Ammit, A.J. Targeting mitogen-activated protein kinase phosphatase-1 (mkp-1): Structure-based design of mkp-1 inhibitors and upregulators. *Curr. Med. Chem.* **2012**, *19*, 163–173. [CrossRef] [PubMed]

International Journal of
Molecular Sciences

MDPI

Article

Dynamics of Dual Specificity Phosphatases and Their Interplay with Protein Kinases in Immune Signaling

Yashwanth Subbannayya [1,2], Sneha M. Pinto [1,2], Korbinian Bösl [1], T. S. Keshava Prasad [2] and Richard K. Kandasamy [1,3,*]

1   Centre of Molecular Inflammation Research (CEMIR), Department of Clinical and Molecular Medicine (IKOM), Norwegian University of Science and Technology, N-7491 Trondheim, Norway; yashwanth.subbannayya@ntnu.no (Y.S.); sneha.pinto@ntnu.no (S.M.P.); korbinian.bosl@ntnu.no (K.B.)
2   Center for Systems Biology and Molecular Medicine, Yenepoya (Deemed to be University), Mangalore 575018, India; keshav@yenepoya.edu.in
3   Centre for Molecular Medicine Norway (NCMM), Nordic EMBL Partnership, University of Oslo and Oslo University Hospital, N-0349 Oslo, Norway
*   Correspondence: richard.k.kandasamy@ntnu.no; Tel.: +47-7282-4511

Received: 5 March 2019; Accepted: 25 April 2019; Published: 27 April 2019

Abstract: Dual specificity phosphatases (DUSPs) have a well-known role as regulators of the immune response through the modulation of mitogen-activated protein kinases (MAPKs). Yet the precise interplay between the various members of the DUSP family with protein kinases is not well understood. Recent multi-omics studies characterizing the transcriptomes and proteomes of immune cells have provided snapshots of molecular mechanisms underlying innate immune response in unprecedented detail. In this study, we focus on deciphering the interplay between members of the DUSP family with protein kinases in immune cells using publicly available omics datasets. Our analysis resulted in the identification of potential DUSP-mediated hub proteins including MAPK7, MAPK8, AURKA, and IGF1R. Furthermore, we analyzed the association of DUSP expression with TLR4 signaling and identified VEGF, FGFR, and SCF-KIT pathway modules to be regulated by the activation of TLR4 signaling. Finally, we identified several important kinases including LRRK2, MAPK8, and cyclin-dependent kinases as potential DUSP-mediated hubs in TLR4 signaling. The findings from this study have the potential to aid in the understanding of DUSP signaling in the context of innate immunity. Further, this will promote the development of therapeutic modalities for disorders with aberrant DUSP signaling.

Keywords: TLR signaling; hematopoietic cells; integrated omics analysis

---

1. Introduction

Reversible phosphorylation and dephosphorylation events serve as regulatory switches that control the structure, activity as well as the localization of the proteins in subcellular space thereby influencing vital biological processes [1,2]. A coordinated interplay between protein kinases (PKs) and protein phosphatases is crucial to regulate these intracellular signaling events as perturbation events in the basal phosphorylation levels of proteins can lead to undesirable consequences including the development of diseases such as cancers [3]. Over the years, more than 500 PKs have been reported [4], a majority of which are druggable [5]. On the contrary, protein phosphatases although being essential regulators of signaling, have drawn less attention. Among the protein phosphatases, the dual-specificity phosphatase (DUSP) family of phosphatases are the most diverse group with a wide-ranging preference for substrates. A unique feature that distinguishes DUSPs from other protein phosphatases is their ability to dephosphorylate both serine/threonine and tyrosine residues within the same substrates [6]. Recent studies have proposed that there are about 40 members of the DUSP family and nine subfamilies [7]. These DUSPs have been implicated as critical modulators of several important signaling pathways that are dysregulated in various diseases.

DUSPs include the Mitogen-activated protein kinase phosphatases (MKPs) and their role as regulators of MAPK signaling mediated cellular processes in both innate and adaptive immunity have been widely discussed [8–10]. For instance, DUSP1 also known as MKP-1 has been found to be a primary regulator of innate immunity [11] and was identified as an important regulator of T cell activation [12]. Additionally, it was also shown to regulate IL12-mediated Th1 immune response through enhanced expression of IRF1 [13]. Upon LPS treatment, DUSP1-deficient mouse macrophages showed increased expression and activation of p38MAPK leading to increased production of chemokines such as CCL3, CCL4, and CXCL2 thereby increasing the susceptibility to lethal LPS shock [14]. In the same study, DUSP1-deficient murine macrophages primed with LPS showed transient increase in JNK activity, elevated levels of pro-inflammatory cytokines and increased p38MAPK activation. Further, DUSP10-deficient mice induced with autoimmune encephalomyelitis showed reduced incidence and severity and prevented LPS-induced vascular damage by regulation of superoxide production in neutrophils [15] indicating its key role in innate and adaptive immune responses. Similarly, other DUSPs such as DUSP2 and DUSP5 have been found to participate in the positive regulation of inflammatory processes [16] and are required for normal T cell development and function [17].

In addition to their potential role in immune regulation, studies have discussed the association of DUSPs namely DUSP1, DUSP4 and DUSP6 in oncogenesis especially in the epithelial-to-mesenchymal transition of breast cancer cells and the maintenance of cancer stem cells [18]. Inhibition of DUSP1 and DUSP6 induces apoptosis of highly aggressive breast cancer cells through the increased activation of MAPK signaling [19]. Furthermore, DUSP1-deficient mice form rapidly growing head and neck tumors causing increased tumor-associated inflammation [20]. In addition to members of MKP subfamily, members of the Protein tyrosine phosphatase type IV subfamily (PTPIV, also known as PRLs) have also been suggested to be potential anti-tumor immunotherapy targets due to their role in carcinogenesis with antibody therapy against PRL proteins inhibiting metastasis in PRL-expressing tumors [21,22]. Additionally, PTP4A3 (PRL-3) has been reported to trigger tumor angiogenesis through the recruitment of endothelial cells [23]. Owing to their regulatory roles in cancer and immunological disorders, DUSPs have been identified as promising therapeutic targets of these diseases [24].

Although the role of certain members of DUSPs have been well characterized, the mechanism by which other members, especially atypical DUSPs, modulate immune response is still largely unknown. Furthermore, the interplay between members of the DUSP family and PKs and their reciprocal actions is minimally understood. Systems biology and integrative biology offer several approaches to identify molecular mechanisms operating behind biological processes in unprecedented detail. Integrated approaches such as Proteogenomics can provide macro-resolution snapshots to facilitate understanding of intricate molecular mechanisms in cancers [25,26] and infectious diseases [27,28]. Applying integrated approaches in the context of immunology can, therefore, offer unique insights into mechanisms of innate and adaptive immunity. In the past few years, several high-throughput datasets were published on naïve and activated immune and hematopoietic cells [29–38]. In this study, an integrated meta-analysis of high-throughput omics datasets related to innate immunity was carried out to delineate the expression dynamics of DUSPs in hematopoietic cells. In addition, the signaling crosstalk between DUSPs with the members of the protein kinase families in immune cells was deciphered. Finally, we analyzed the association of DUSP signaling pathways downstream of TLR4 signaling. Collectively, this study provides potential DUSP-mediated signaling pathways and hubs thus facilitating a better understanding of DUSP signaling in innate immunity.

2. Results

2.1. DUSP Classification into Subfamilies and Evolutionary Conservation

We compiled a list of DUSPs from previously published studies and used it to for further analysis. We also aligned protein sequences for known DUSPs, classified them into subfamilies according to their clustering patterns and validated the classification using domain analysis. The classification of DUSPs based on sequence similarity was performed by multiple sequence alignment analysis of DUSP protein sequences (Figure 1a). For the analysis, the list of DUSPs and their subfamilies was compiled from Chen et al. [7], and the results were found to be concordant with this classification system. The classification includes members belonging to CDC14, DSP1, DSP14, DSP15, DSP3, DSP6, DSP8, PRL, and Slingshot subfamilies. Protein domain analysis of all DUSP members also validated the sub-classification of DUSPs (Figure 1b). Most DUSP subfamily members exhibited similar architectures with a common DSPc domain. However, members of a few subfamilies contained additional domains besides DSPc namely CDC14 subfamily (N-terminal DSP domain), DSP1, DSP6, and DSP8 (Rhodanese-like domain) subfamilies and members of Slingshot subfamily (DEK domain at the carboxy terminus). Next, we aimed to determine the evolutionary conservation of dual specificity phosphatases across eukaryotic species by calculating the number of orthologs for all human DUSPs obtained from Homologene database [39]. Our analysis revealed the distribution of DUSPs ranging from a minimum, of six orthologs for DUSP2 to 20 for DUSP12 (Figure 1c). The distribution of the entire human proteome was similar with a minimum of one and a maximum of 21 ortholog counts. Most DUSPs were found to be conserved in mammals, and particularly among primates suggesting evolutionary conservation across eukaryotic species.

2.2. Expression of Dual Specificity Phosphatases and Protein Kinases in Hematopoietic Cells, Primary and Secondary Lymphoid Organs

Earlier reports suggest that the expression of DUSPs are regulated during development in a cell type-specific manner or upon cell activation in contrast to their ubiquitous substrate expression [8]. In addition, PKs play important roles in immunity [40,41] and are widely known to be modulated by DUSPs. In order to determine the extent of expression of DUSPs and PKs across human hematopoietic cells, we analyzed the proteomic data from Rieckemann et al. [29] as it is currently the largest high-resolution dataset containing expression data pertaining to 28 different hematopoietic cell types analyzed on a single platform (Figure S1, Table S1). On an average, 15 DUSPs and 240 PKs were found to be expressed across hematopoietic cells (Figure 2a) at the protein level. Among the various cell lineages, T8 TEMRA (terminally differentiated effector memory T cells which express CD45RA, as opposed to TEM cells which are CDC45RA-negative) cells expressed the highest number of DUSPs (19), while we did not identify expression of DUSPs from the erythrocyte data. On the contrary, 264 PKs were found to be expressed in NK CD56bright cells, whereas only 15 were found to be expressed in erythrocytes. NK CD56bright cell types have been previously described to be regulatory in nature influencing innate immunity through cytokine production as opposed to NK CD56dim cells, which have cytotoxic activity [42].

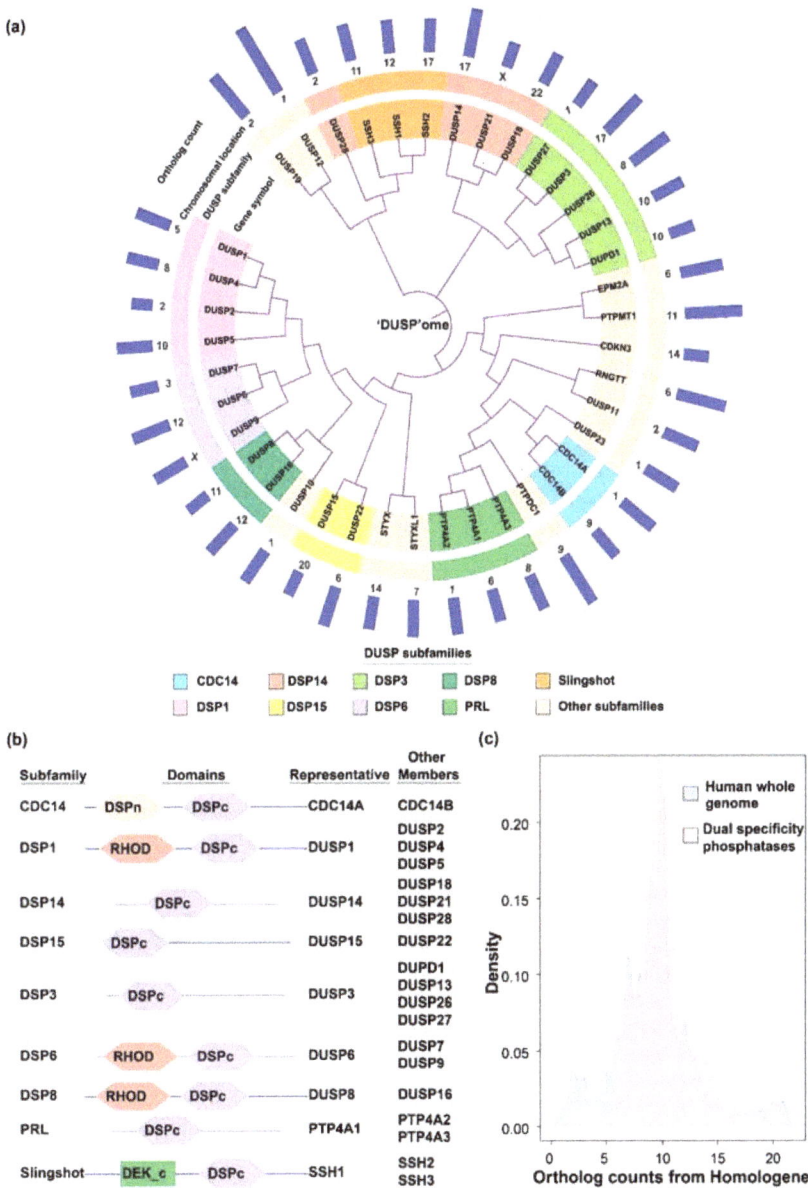

Figure 1. (**a**) Dendrogram describing the sequence similarity of members of the dual specificity phosphatase (DUSP) family. Protein sequences for various members of the dual specificity family obtained from RefSeq were aligned using Clustal Omega. The DUSPs were classified into subfamilies and their chromosomal location mapped using data and classification system from Chen et al. [7]. The tree shows largely distinct clustering of distinct DUSP subfamilies. (**b**) Domain architecture of DUSP subfamilies. Members of the DUSP family were subjected to domain analysis using SMART domain prediction. (**c**) Conservation of dual specificity phosphatases across species. Ortholog counts were obtained for all human genes from Homologene and the density of ortholog counts for DUSP family members was plotted against the density of ortholog counts for all human genes in the background. The graph largely indicates conservation of DUSPs across various species.

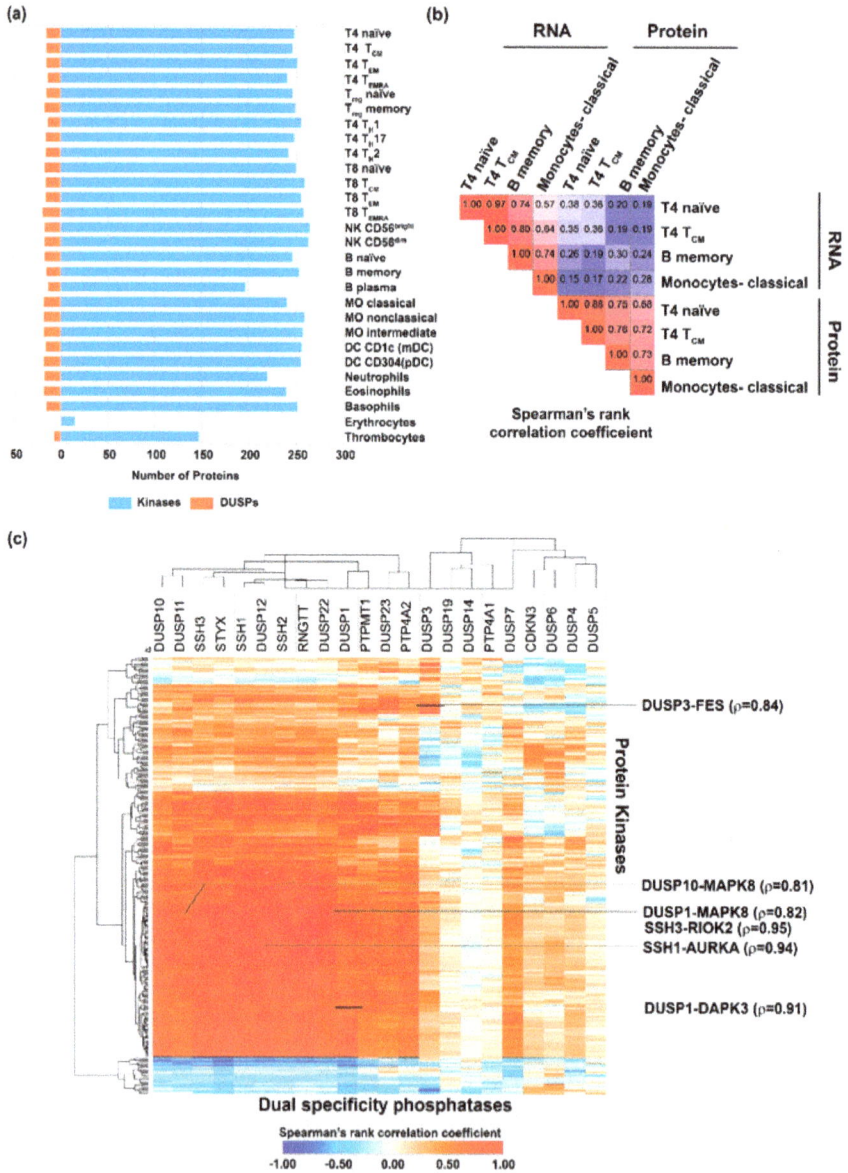

Figure 2. (**a**) Protein kinases and dual specificity phosphatase expression in hematopoietic cells. Hematopoietic cell expression data for protein kinase and dual specificity phosphatases (DUSPs) were obtained from Rieckmann et al., Nat Immunol (2017) [29]. All hematopoietic cells except erythrocytes and thrombocytes expressed similar number of kinases and DUSPs. (**b**) Correlation of transcriptomic and proteomic data of cells. Transcriptomic and proteomic data for T4 naïve, T4 TCM, B memory and classical monocytes from Rieckmann et al., Nat Immunol (2017) [29] showed poor correlation. (**c**) Correlation of protein kinase and DUSP expression patterns in hematopoietic cells. The correlation was carried out using Spearman's rank correlation to identify kinase-DUSP pairs that may have reciprocal activities. The kinase-DUSP pairs with high correlation coefficients are shown on the right-hand side.

Among the DUSP family members, 11 DUSPs were found to be expressed in a large majority of the hematopoietic cells analyzed. These include DUSP12, DUSP23, DUSP3, SSH3 and PTP4A2, which were found to be expressed in 27 of 28 hematopoietic cells and DUSP11, DUSP22, PTPMT1, RNGTT, SSH1 and SSH2 found to be expressed in 26 of the 28 cells. 18 members of the DUSP family were not identified in any of the cell types. Furthermore, we did not observe restricted expression of DUSP members to any one cell type. Among the PKs, only 287 were found to be identified in at least one hematopoietic cell type with the rest not being identified in any cell type. While 13 members of the PKs including MAPK1, MAP2K2, ILK, and ROCK2 were found to have ubiquitous expression, four kinases including PRKCD, KALRN, MYLK and PTK2 were found to have restricted expression in thrombocytes. At least five PKs including MKNK2, STK33, CSNK1E, TAOK2 and BLK were found to be restricted to 2 of the 28 cell types. Of these, MKNK2, STK33 and BLK were expressed commonly in B memory cells. STK33 was expressed in B naïve and B memory cells and seemed to be linked with B cell types. We compared proteomics and transcriptomics datasets to see if changes at the transcript level could be equated with changes in protein expression (Figure 2b, Table S2).

Overall, we observed the expression of kinases and DUSPs to be relatively lower compared to housekeeping proteins such as tubulin (TUBB). The notable exceptions were PRKCB with abundant expression in B memory cells and plasmacytoid dendritic cells and PRKDC with nearly ubiquitous high expression across all hematopoietic cell types. Cell type enriched expression patterns were observed for several kinases including MAPKAPK3 (high in eosinophils), BTK (basophils), ILK (thrombocytes) and HCK with highly abundant expression in classical, non-classical and intermediate monocytes. While ILK in thrombocytes has been implicated in the regulation of integrin signaling [43], HCK has been described as a key regulator of gene expression in activated monocytes [44]. The importance of these expression patterns need to be further investigated and may be important in the regulation of cell-specific processes.

We analyzed tissue expression levels of DUSPs and PKs in immune-related tissues. We chose to study expression data pertaining to lymphoid organs that are the major components of the immune system involved in producing B- and T-cells (primary) and are responsible for the coordinating the cell-mediated immune response (secondary) [45]. The primary lymphoid organs consist of the thymus and the bone marrow while the secondary lymphoid organs consist of lymph nodes, spleen, tonsils, and the mucosa-associated lymphoid tissues such as Peyer's patches [46]. More recently, the appendix has been deemed to be a lymphoid organ capable of carrying out immunological functions [47,48]. For the analysis, we compiled gene expression data (Table S3) from multiple projects including FANTOM5 [49], Genotype-Tissue Expression (GTEx) Project [50], The Human Protein Atlas [51,52], Illumina Body Map [53], NIH Roadmap Epigenomics Mapping Consortium [54] and the ENCODE project [55]. The PK and DUSP expression in these various tissue expression datasets largely correlated (Figure 3a,b). A majority of the 505 PKs were found to be expressed in at least one lymphoid organ and expression of only a mere 7 kinases were not observed. A significant observation was that all the members of the DUSP family were present in at least one lymphoid organ. 460 of the 505 PKs were found to be expressed in all 7 lymphoid organs while 36 of the 40 DUSPs were expressed in as many lymphoid organs. Two kinases-EPHA8 and PAK5 were found to be restricted to only lymphoid organ -spleen and thymus respectively. DUSP21 was found to be restricted in the bone marrow while DUPD1 was found to be restricted to only the thymus and the tonsil. Similarity matrices indicated a similar distribution of kinases and DUSP expression across primary and secondary lymphoid organs suggesting potential reciprocal pairs (Figure 3a,b). In conclusion, members of the DUSP family were found to be expressed in every lymphoid tissue, thereby suggesting DUSP mediated control of protein kinase activity.

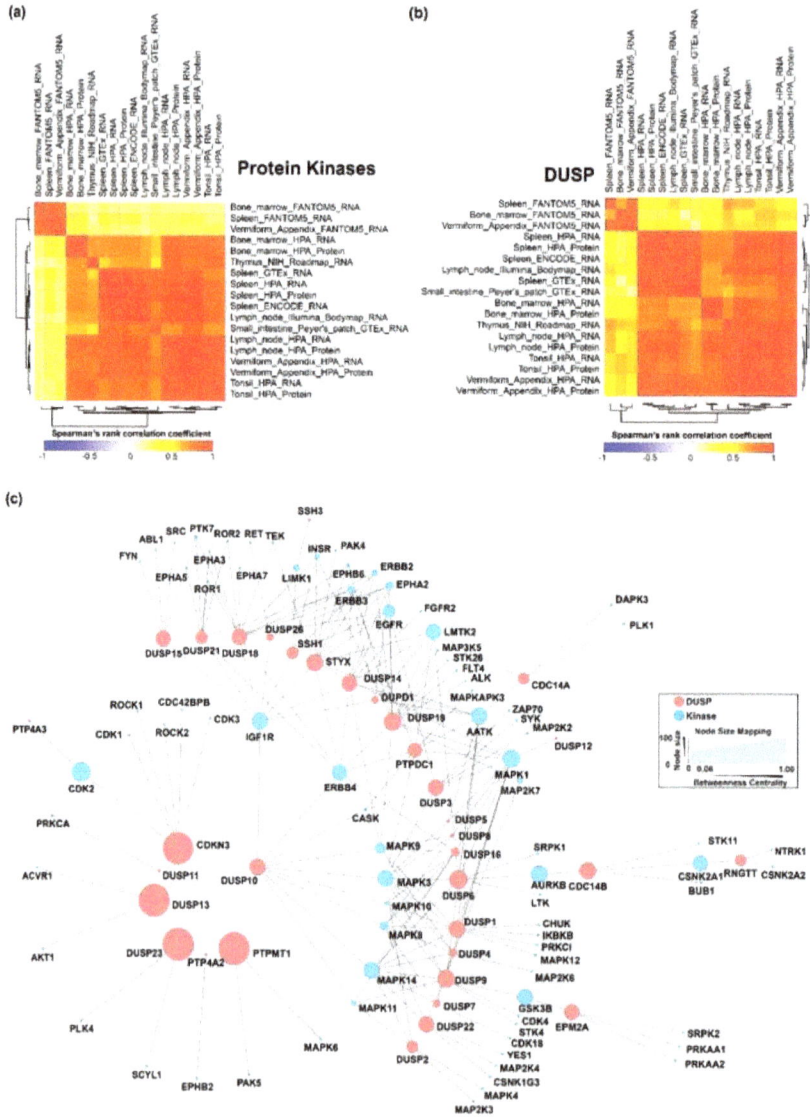

Figure 3. (**a**) Similarity matrix for protein kinase expression in primary and secondary lymphoid organs. (**b**) Similarity matrix for DUSP expression in primary and secondary lymphoid organs. The expression data Bone marrow, spleen, thymus, lymph node, tonsil, appendix and Peyer's patches were obtained from various studies including FANTOM5, HPA, GTEx, ENCODE, Illumina Bodymap and NIH Roadmap project consortia. The PK and DUSP expression in these various tissue expression datasets largely correlated (**c**) DUSP-Protein kinase interaction network. Protein-protein interaction data between DUSP and protein kinases obtained from Compartmentalized Protein-Protein Interaction (comPPI) Database were analyzed in Cytoscape using Network Analyzer to obtain network properties including Betweenness Centrality. Protein kinases with high Betweenness Centrality indicate primary regulatory proteins associated with DUSPs.

2.3. Correlation of Dual Specificity Phosphatase Activity with Protein Kinase Activity in Immune Cells

Since it is widely known that DUSPs regulate protein kinases, identifying DUSPs and protein kinases pairs with reciprocal activities could identify regulatory mechanisms that can be potentially exploited to develop strategies for therapeutic interventions in conditions such as systemic inflammatory and autoimmune disorders. To identify potentially reciprocal DUSP-kinase pairs and to determine its probable role in immune cells, we correlated the expression profiles of DUSPs and PKs expressed in hematopoietic cells from Rieckemann et al. [29] (Figure 2c, Table 1, Table S4).

Table 1. A table summarizing representative known and potentially novel human and murine DUSP-kinase pairs with reciprocal activities.

DUSP-Protein Kinase Pair (Spearman's Rank Correlation Coefficient ϱ)	Relationship	Known/Novel
SSH1-AURKA (0.94)	Aurora Kinase A (AURKA) overexpression increased slingshot kinase 1 (SSH1) expression in breast cancer cells [56]	Known
DUSP1-MAPK7 (0.79)	DUSP1 gene silencing increased MAPK7 expression in osteosarcoma cells [57]	Known
DUSP1-MAPK8 (0.82)	DUSP1 gene silencing increased MAPK8 expression in osteosarcoma cells [57]	Known
DUSP10-MAPK8 (0.81)	DUSP10 (MKP-5) dephosphorylates MAPK8 (JNK) [58]	Known
Dusp16-Mapk8 (0.83)	DUSP-16 (MKP-7) regulates MAPK8 (JNK) in LPS-activated macrophages [59]	Known
SSH1-CDK17(0.94)		Novel
DUSP12-CLK3(0.97)		Novel
SSH1-LMTK2 (0.97)		Novel
DUSP12-AURKA (0.96)		Novel
Ptpmt1-Mast3 (0.97)		Novel
Dusp1-Egfr (0.73)		Novel
Dusp16-Lrrk2 (0.86)		Novel
Dusp10-Igf1r (0.99)		Novel

We identified 231 DUSP-kinase pairs with similar coexpression patterns in immune cells with a Spearman's rank correlation coefficient of 0.9 or more indicating high confident reciprocal pairs. Similarly, 701 DUSP-kinase pairs with similar coexpression patterns in immune cells had a Spearman's rank correlation coefficient of 0.8 or more. Some of the pairs identified include SSH1-AURKA (ϱ = 0.94), DUSP1- MAPK7 (ϱ = 0.79), DUSP1-MAPK8 (ϱ= 0.82), DUSP10-MAPK8 (ϱ = 0.81). Among these, the role of DUSP1 and DUSP10 as negative regulators of MAPK8 is well known. The significance of SSH1-AURKA pair in immune cells is currently not well described at this time.

We also used an interactome-based approach to identify co-expressed and co-interacting proteins to identify potential DUSP-kinase regulatory mechanisms. Interactome analysis to identify reciprocal DUSP-protein kinase pairs from baseline protein-protein interaction (PPI) data from the comPPI database resulted in the generation of an interaction network containing 1715 DUSP-specific interactions with 1276 nodes (Figure S2). Since the network was too complex to comprehend the interplay between DUSPs and PKs, we separated interactions between DUSPs and kinases and generated an additional kinase and DUSP-specific network. This network contained 195 protein-protein interactions between 35 DUSP members and 82 PKs (Figure 3c, Table S5). Several potential hub proteins that communicate with the dual specificity phosphatase family members were deduced from the interaction network. Among the PKs, MAPK1 and MAPK3 had 12 and 10 interactions (directed edges) respectively with DUSPs. Other PKs with a high number of interactions included MAPK14 (9), IGF1R (9), AATK (9) and ERBB4 (8). Among the DUSPs, DUSP18 and DUSP19 had the most number of directed edges -14 each. Other DUSP family members with several directed edges included STYX (13), DUSP 1 (11), DUSP14 (10), DUSP9 (9) and DUSP 10 (9). It is interesting to note that DUSP19, and STYX are some of the poorly characterized members. The PKs with the highest betweenness centrality included MAPK1, IGF1R, AURKB, and AATK. Combining the directed edge and the betweenness centrality data revealed several kinases belonging to the MAPK family and receptor tyrosine kinases such as MAPK1, MAPK3, IGF1R, AATK, MAPK14, ERBB4, LMTK2, MAPK9 and MAPK8 to be strongly associated with DUSPs.

2.4. Expression Landscape and Signaling Dynamics of DUSPs and Kinases in Activated Immune Cells

The data provided by Rieckemann et al. also includes steady-state and activated protein expression profiles of 17 cell types which were analyzed to determine the effect of various activating ligands on the expression of DUSPs and PKs (Table S6). Our analysis resulted in the identification of 152 events of differential expression of 18 DUSPs across 17 cells types (Figure S3a) Of these, 57 and 95 were found to be overexpressed and downregulated respectively across the activated cell types. Similarly, we identified 2311 events of differential expression of 269 PKs across 17 cells types (Figure S3b). Of these, 1058 and 1253 were found to be overexpressed and downregulated respectively across the activated cell types. Several kinases and DUSPs were found to be differentially expressed in multiple activated cell types. The most overexpressed DUSPs included DUSP12 (9 cell types), DUSP23 (7 cell types), DUSP1, DUSP10 (6 cell types each) and STYX (4 cell types), while the most downregulated DUSPs included SSH3 (13 cell types), PTPMT1 (12 cell types), DUSP3 (10 cell types), SSH2 (9 cell types), DUSP1 (8 cell types) and SSH1 (7 cell types). Similarly, among the kinases the most overexpressed included CDK1 (14 cell types), PLK1 (14 cell types), TGFRB1 (13 cell types), RIOK3 (13 cell types) and RIOK2 (13 cell types) while the most downregulated included ATM (15 cell types), MAST3 (14 cell types), PRKACB (13 cell types), MAP3K5, LMTK2 and SYK (12 cell types each).

There have been several studies focused on the genome-wide effects of TLR ligands on hematopoietic cells such as monocytes [60]. In the current study, we aimed to determine the specific roles of DUSPs and PKs in TLR4 signaling. An integrated analysis of RNA and protein expression datasets [29,33,61] pertaining to dendritic cells (DCs) and monocytes (MOs) activated by LPS was performed (Table S17). The data were categorized into murine DCs (mDCs), human DCs (hDCs) and human MOs (hMOs) and a list of molecules differentially expressed in response to LPS was generated (Table S8). 57 proteins including 4 DUSPs and 53 PKs were found to be overexpressed in mDCs while 80 were downregulated (2 DUSPs, 78 PKs) in response to LPS. In hDCs, 50 proteins (4 DUSPs, 46 PKs) and 154 (10 DUSPs, 144 PKs) proteins were found to were found to be overexpressed and downregulated respectively. In the case of hMOs, 35 (1 DUSP, 34 PKs) were overexpressed and 80 (6 DUSPs, 74 PKs) were downregulated.

DUSPs overexpressed in dendritic cells included Dusp1/DUSP1 (mDCs and hDCs), Dusp14, Dusp16, Ptp4a2 (all in mDCs), DUSP5, DUSP7, and DUSP 10 (all in hDCs). Downregulated DUSPs included Dusp3, Dusp19 (mDCs), DUSP4, DUSP11, DUSP12, DUSP23, PTP4A2, PTPMT1, SSH1, SSH2, SSH3, and STYX (hDCs) (Figure 4a and Figure S5a–c). In hMOs stimulated with LPS, DUSP 10 was overexpressed while DUSP1, DUSP11, PTP4A2, PTPMT1, RNGTT, and SSH3 were found to be downregulated. DUSP1 (Dusp1) seems to be important in both hDCs and mDCs signaling as it was found to be upregulated in both species in response to LPS and is in concordance with previously published studies on dendritic cells stimulated with LPS [14]. However, it was found to be downregulated in hMOs. The overexpression of DUSP1 in dendritic cells identified from our analysis is in concordance with previously published studies on dendritic cells stimulated with LPS. DUSP10 was found to be upregulated in both hDCs and hMOs while in mDCs, it was not found to be differentially expressed. Among differentially expressed genes in dendritic cells, Dusp14 and Dusp16 were exclusively overexpressed in the dendritic cells from mice while DUSP5, DUSP7, and DUSP10 seemed to be exclusive to humans. Taken together, our analysis suggests probable existence of species-specific differential expression of DUSPs in TLR4 signaling. In a recent paper, human and murine macrophages were found to have varying mechanisms of immunometabolism [62]. Functional analysis of differentially expressed PKs and DUSPs in murine dendritic cells stimulated with LPS showed enrichment of several processes including MAPK cascade, response to reactive oxygen species, cellular senescence, and cell migration pathways among others (Figure 4b).

(a)

(b)

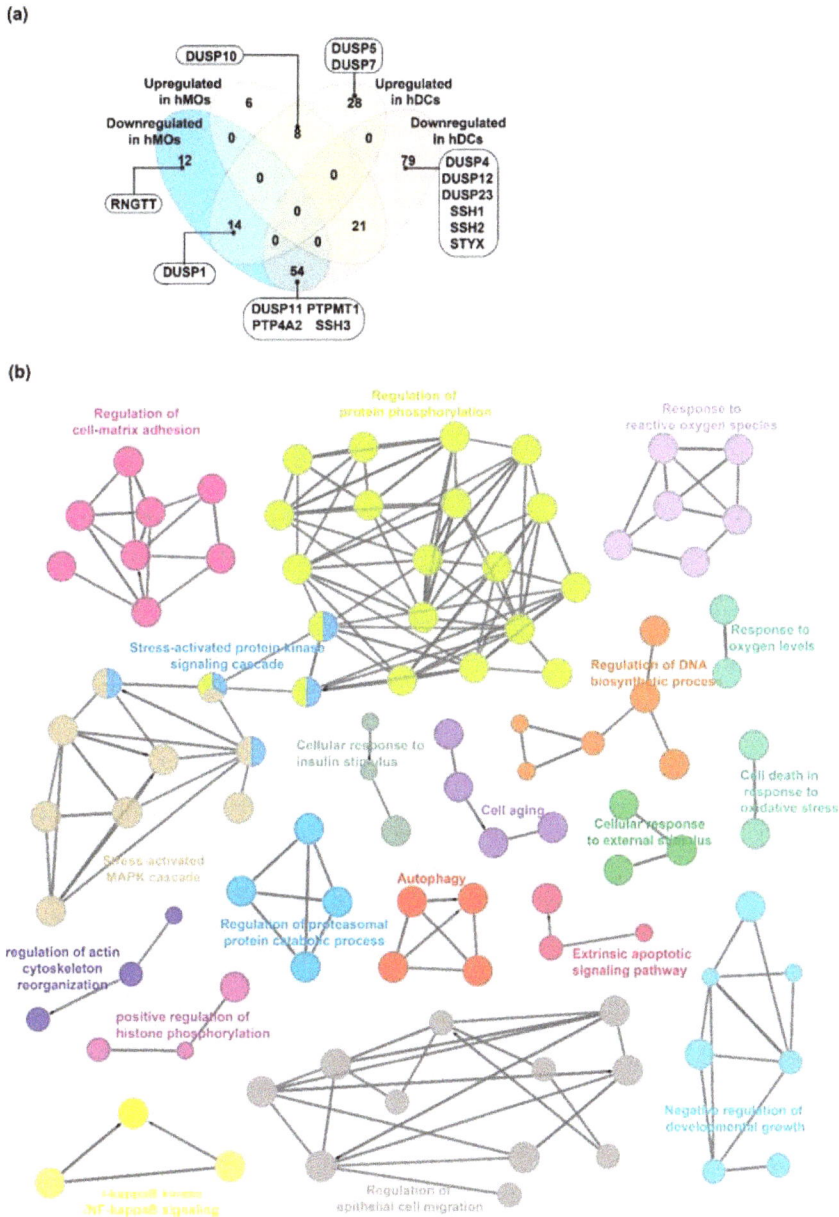

Figure 4. DUSP and kinase dynamics in response to Toll-like receptor 4 ligand-LPS. (**a**) Venn diagram showing differentially expressed protein/transcripts in human monocytes and dendritic cells stimulated with LPS. Members of the DUSP family are indicated within insets. (**b**) Enriched biological processes in murine dendritic cells stimulated with LPS. Differentially expressed protein kinases and DUSPs in response to LPS were analyzed using ClueGO in Cytoscape. Different colors indicate clusters of similar processes.

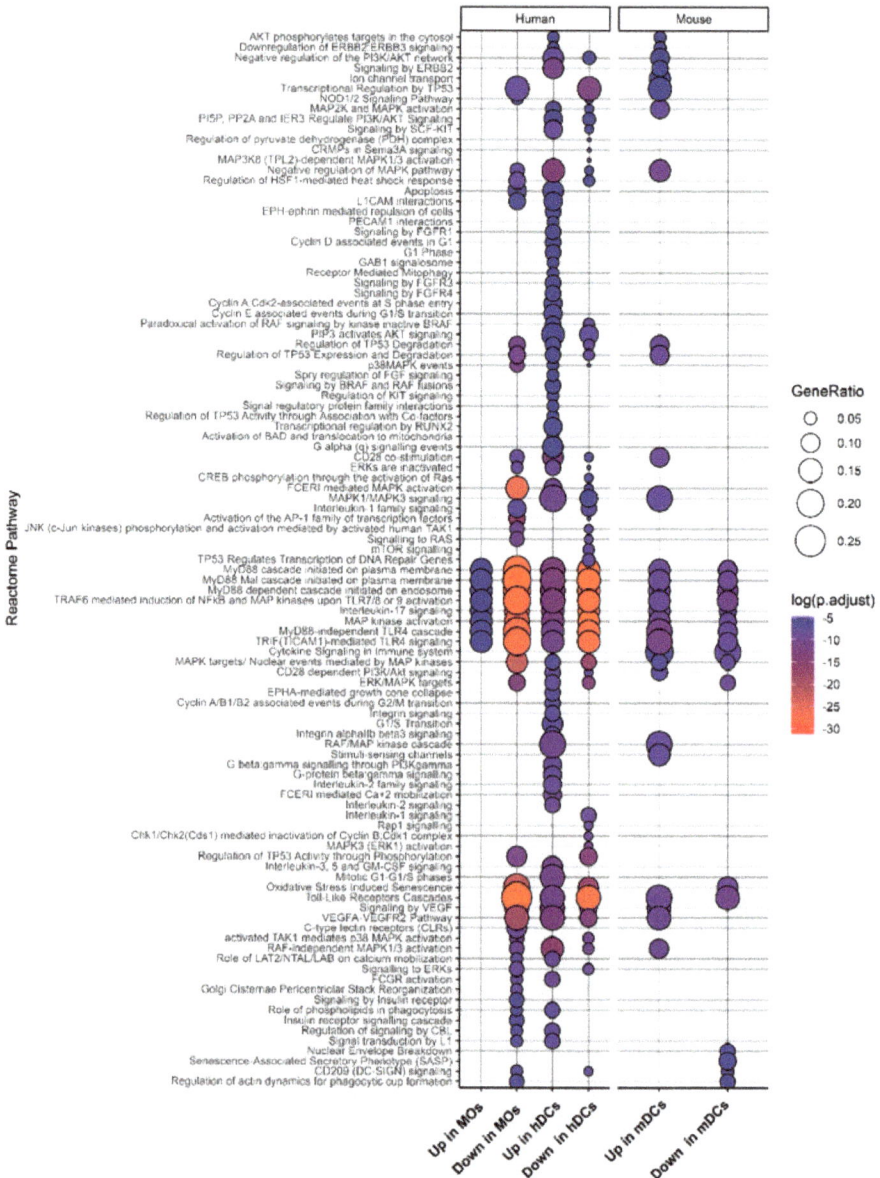

Figure 5. Pathway analysis of DUSPs and kinases differentially expressed in response to LPS. Differentially expressed genes were tested for hypergeometric enrichment of Reactome Pathways. Genesets with less than 10 genes were excluded from the analysis and *p*-values were adjusted by Benjamini–Hochberg (FDR) correction.

We next performed pathway enrichment analysis using pathway data from Reactome database for DUSP and PKs showing differential expression upon LPS stimulation (Figure 5, Table S9). We found several pathways including TLR signaling, MyD88 cascade, and MAPK signaling pathways to be enriched across LPS-activated MOs and DCs. Oxidative stress-induced senescence pathways were downregulated in mDCs, hDCs, and hMOs. VEGF signaling also seemed to be affected through

DUSP and PKs after stimulation with LPS. While induction of TLR signaling by LPS is widely known, there are a few reports on the induction of VEGF signaling by LPS stimulation [63]. FGFR signaling and SCF-KIT signaling pathways were up in hDCs, while C-lectin receptor signaling pathway was downregulated in hMOs upon LPS stimulation. Apoptotic pathways was found to be downregulated in hMOs while being upregulated in hDCs suggesting opposing cell death phenotypes in hMOs and hDCs in response to LPS.

To identify DUSP-kinase pairs with reciprocal activities, correlation analysis between DUSP and kinase expression in mDCs and mMOs treated with LPS was carried out (Figure S5d, Table S10). Our analysis resulted in the identification of several important pairs including Ptpmt1-Mast3 (ϱ = 0.97), Dusp1-Egfr (ϱ = 0.73), Dusp1-Mapk8 (ϱ = 0.66), Dusp16-Mapk8 (ϱ = 0.83), Dusp16-Lrrk2 ϱ = (0.86), Dusp10-Igf1r (ρ = 0.99).

In order to identify the interacting partners of DUSPs in TLR4 signaling, we carried out interactome analysis of members of the DUSP and protein kinase families that were found to be differentially expressed in dendritic cells and monocytes (Figures S6–S8). Analysis of network properties identified several proteins that seemed to be regulated by DUSP signaling. In the networks of differentially expressed proteins identified in activated mDCs, proteins including Akt1, Lrrk2, Pim1, Dusp1, Dusp16 (overexpressed), Mapk1/3, Pik3cg, Cdk1, Dusp3 and Dusp19 (downregulated) had a high number of edges and high betweenness centrality suggesting their importance in innate immunity. In activated hDCs, proteins including MAPK1, CDK2, PIK3CA, MAPK13, DUSP1, DUSP5, DUSP7 and DUSP10 (upregulated), PRKACA, LRRK2, CHUK and DUSP13 (downregulated) had a high betweenness centrality and were identified to be relevant in LPS-induced signaling. Similarly, in activated hMOs, PRKCZ, CDK6, DUSP10, TGFBR1 (upregulated), LRRK2, ATM, ATR, MAPK1, DUSP1, MAP3K1 (downregulated) among others had a high betweenness centrality.

3. Discussion

Dual-specificity phosphatases (DUSPs) are a family of phosphatases that can act on both serine/threonine and tyrosine residues of several protein substrates leading to wide-ranging effects on cellular signaling and biological processes. With the exact number of DUSP members still being controversial [7,64–68], we chose to consider the latest classification described by Chen et al. [7], consisting of 40 DUSPs with 9 subfamilies containing more than one member. We validated the subfamily-based classification of DUSPs using two approaches namely-sequence alignment and SMART-based domain analysis. Evolutionary conservation analysis of DUSP family members revealed high sequence conservation of all DUSP members, especially in higher mammals.

So as to determine the extent of expression of dual specificity phosphatases and protein kinases across the human hematopoietic cells and lymphoid organs, publicly available datasets were mined. Since hematopoietic cells are derived from lymphoid tissues, we chose to analyze expression profiles of these. To date, Rieckemann et al. have provided the largest expression dataset pertaining to 28 different hematopoietic cell types consisting of high throughput omics data acquired under a single platform containing. Our analysis of this dataset revealed new insights into the expression dynamics of several understudied DUSPs such as slingshot family of phosphatases (SSH1, SSH2 and SSH3) and STYX in various hematopoietic cell types. Additionally, our analysis also resulted in the evaluation of the expression patterns of several understudied protein kinases described by Huang et al. [69] including STK17A, SCYL3, MAST3, CSNK1G2, and the RIO family of kinases (RIOK1, RIOK2 and RIOK3). Furthermore, among the hematopoietic cells, erythrocytes and thrombocytes (platelets) expressed the least number of DUSPs and PKs The restricted expression patterns of these proteins could be potentially exploited for therapeutic modalities in erythrocyte and platelet disorders. In fact, four different PKs (PRKCD, KALRN, MYLK, and PTK2) showed restricted expression in thrombocytes. PRKCD has been previously reported to modulate collagen-induced platelet aggregation [70] while MYLK and PTK2 have been reported to be important in megakaryopoiesis [71,72]. Further, the different patterns of DUSP and PK expression in erythrocytes and thrombocytes compared to the rest of the hematopoietic cells may be attributed to their structures, diverse biological functions performed by these cells, their numbers in the human body and the differing lineages that form erythrocytes and thrombocytes during hematopoiesis.

Among the 701 DUSP-kinase pairs with potential reciprocal activities identified from the unstimulated hematopoietic cell expression data, several pairs with previously known reciprocal actions were also identified thereby proving our analysis methods. Among these, SSH1-AURKA pair was identified with a high Spearman's rank correlation coefficient ($\varrho = 0.94$). Aurora Kinase A (AURKA) overexpression has been previously reported to increase the expression of slingshot kinase 1 (SSH1) resulting in increased cofilin activation and migration of breast cancer cells [56]. Other notable pairs that we identified from the correlation and that were previously reported in the literature included DUSP1-MAPK7 ($\varrho = 0.79$) and DUSP1-MAPK8 ($\varrho = 0.82$). DUSP1 gene silencing has been shown to increase the expression of MAPK7 and MAPK8 transcripts in osteosarcoma cells suggesting reciprocal actions between them [57]. DUSP10 (MKP-5) has been well known to dephosphorylate MAPK8 (JNK) [58] and has also been implicated in immune function. Knocking down DUSP10 expression increased JNK activity and inflammation in murine mesangial cells while its overexpression led to decreased JNK activation [73]. A study by Zhang et al. further found that Dusp10-deficient murine cells exhibited increased JNK (MAPK8) activity, elevated levels of proinflammatory cytokines and increased T cell activation [74]. In our correlation analysis, the DUSP10-MAPK8 pair was identified with a Spearman's rank correlation coefficient of 0.81.

We used a subset of omics datasets for investigating the expression of DUSPs and their interplay with PKs upon LPS stimulation in activated dendritic cells and monocytes. Though a previous study on the meta-analysis of TLR4 signaling datasets exists, its focus was on activated macrophages [75]. Biological process-based enrichment upon LPS stimulation showed the enrichment of several processes including MAPK cascade, response to reactive oxygen species, cellular senescence and cell migration, autophagy pathways among others. While these processes are widely known in the context of macrophages, the effects of LPS stimulation seem to be similar in dendritic cells as well. We also correlated DUSP and PK expression in these cells and carried out interactome analysis to identify key molecules that are potential regulators of LPS-induced signaling. Correlation analysis indicated the presence of DUSP-PK pairs with reciprocal activities in response to activation. Some of these pairs including DUSP1-MAPK8, DUSP1-MAPK8 have already been described in previous literature, thus confirming our findings. DUSP-16 (MKP-7) has previously been identified to regulate MAPK8 (JNK) in LPS-activated macrophages [59] and in activated endothelial cells [76]. Dusp16 (MKP-7) was also reported to have a critical role in the activation and functioning of T cells. Dusp16-deficient T cells had an exaggerated response to TCR activation and had enhanced proliferation properties [77]. DUSP16-deficient macrophages have also been reported to overproduce IL-12 in the context of TLR stimulation [78]. DUSP1 has been reported to play a key role in the feedback control and regulate MAPK8 during glucocorticoid-mediated repression of inflammatory gene expression [79]. DUSP1 is already known to regulate the expression of LPS-induced genes [14]. We also found DUSP 1 and DUSP16 to be important regulators of several MAP kinases from the interactome analysis.

Several novel DUSP-PK pairs were identified, and these require further characterization to confirm their role in immunity. Particularly interesting among these include pairs involving slingshot phosphatases which constitute a group of understudied phosphatases. SSH1-CDK13, SSH1-CDK19, SSH2-EPHB2 were identified with high significance. Slingshot phosphatases have been previously implicated in cancer progression [80,81] and have been known to mediate caspase-modulated actin polymerization towards bacterial clearance upon Legionella infection [82]. In our analysis, Slingshot phosphatase members SSH1, SSH2 were found to be downregulated exclusively in human DCs, while SSH3 was downregulated in both human DCs and MOs. Slingshot phosphatase members SSH1 and SSH2 were found to be downregulated exclusively in human DCs, while SSH3 was downregulated in both human DCs and MOs. DUSP10 and IGF1R expression were also found to be highly correlated. Insulin-like growth factors have been known to inhibit anti-tumoral responses of dendritic cells through the regulation of MAP kinases and have therefore been suggested to be targets for immunotherapy [83]. IGF1 has also been reported to influence activation of macrophages in response to high-fat diet or helminthic infection [84]. Additionally, we identified LRRK2 (leucine-rich repeat kinase 2) from the interactome analysis in activated DCs and MOs, to be potentially important in the context of immune signaling. LRRK2 was found to interact with DUSP1 and DUSP16. LRRK2 has been mainly associated with familial cases of Parkinson's disease [85] and has been known to play an important in innate immunity in the peripheral and central nervous system [86], especially in microglial inflammatory processes [87]. This present finding suggests the possibility of LRRK2 being additionally important in DUSP-mediated immune signaling.

4. Materials and Methods

4.1. Datasets

Studies pertaining to immune cells acquired using high-throughput techniques were searched using PubMed. The data matrices pertaining to each dataset were downloaded from the site of the publisher of these articles. Gene expression datasets were downloaded from Gene Expression Omnibus (GEO, https://www.ncbi.nlm.nih.gov/geo/) wherever applicable. The details of all the studies used in this study are provided in Table S11.

4.2. DUSP and Kinase Lists for Analysis

The list of DUSPs used for the analysis was sourced from Chen et al. [7]. The list of human and mouse kinases was sourced from UniProt (https://www.uniprot.org/docs/pkinfam) which used data from [4,88,89]. The master lists of dual specificity phosphatases and PKs used for data analysis in the current study are provided as Tables S12 and S13.

4.3. Classification of DUSP Family Members, Domain Analysis and Species Conservation Analysis of DUSP Sequences

The similarity tree between DUSPs was drawn using iTOL with alignment performed with Clustal Omega. Briefly, RefSeq accessions of the longest protein isoforms for all dual specificity phosphatases were retrieved from NCBI gene (https://www.ncbi.nlm.nih.gov/gene). The protein sequences were obtained with Batch Entrez (https://www.ncbi.nlm.nih.gov/sites/batchentrez) and aligned using Clustal Omega (https://www.ebi.ac.uk/Tools/msa/clustalo/) using default settings. The output alignment in PHYLIP (.ph) format and was visualized with Interactive Tree of Life (https://itol.embl.de/) with custom colors and tracks.

Domain analysis was carried out for the longest isoforms of all proteins belonging to DUSP subfamilies with multiple members using the SMART domain prediction tool (http://smart.embl-heidelberg.de/) [90], using the option "PFAM domains". Orthology data for all human genes were obtained from Homologene (Release 68, downloaded on October 4, 2018 from https://www.ncbi.nlm.nih.gov/homologene) for the analysis of sequence conservation across species. The counts for all genes in the Homologene database were obtained and the Taxonomy ID for each gene was mapped to the species type. The densities of ortholog counts for DUSP family members was plotted against the density of ortholog counts for all human genes in the background using R (v3.5.1) (https://cran.r-project.org/).

4.4. Landscape of DUSPs and Kinases in Immune Cells

Proteomic and transcriptomic data matrices were obtained from supplementary files of respective articles and accessions were converted into Entrez gene accession formats using bioDBnet:db2db (https://biodbnet-abcc.ncifcrf.gov/db/db2db.php) [91] and g:Profiler (https://biit.cs.ut.ee/gprofiler/gconvert.cgi) [92]. Gene expression data was also downloaded from Gene Expression Omnibus (GEO) wherever supplementary data was not available. Z-score-based normalization of data matrices from all studies was carried out using base R (v3.5.1). DUSP and kinase expression data were subsequently obtained from the normalized datasets. Cell type data and cell sorting information wherever available were retrieved for each of the expression datasets and appended with the expression data. Heatmaps were drawn in Morpheus (https://software.broadinstitute.org/morpheus/) with hierarchical clustering based on Euclidean distance metric, complete linkage method and clustering by rows and columns. We carried out correlation and interactome analysis to look at the interplay between DUSPs and PKs in naïve and activated immune cells. Correlation analysis between z-scores of kinase and DUSP profiles of immune cell proteomes was performed using "Spearman" method through R (v3.5.1). Heatmaps were drawn using Morpheus (https://software.broadinstitute.org/morpheus/).

4.5. Landscape of DUSPs and Kinases in Primary and Secondary Lymphoid Organs

We compiled tissue-based expression data from various datasets present in the Expression Atlas (https://www.ebi.ac.uk/gxa/home). Briefly, the pre-processed datasets from The FANTOM5 project [49], Genotype-Tissue Expression (GTEx) Project [50], The Human Protein Atlas [51,52], Illumina Body Map [53], NIH Roadmap Epigenomics Mapping Consortium [54] and the ENCODE project [55] were downloaded (Download date January 11,2019). The datasets were subjected to z-score-based normalization using R (v3.5.1). Data from primary and secondary lymphoid tissues were selected and DUSP and kinase expression data were compiled using in-house scripts. The data was plotted as heatmaps with Morpheus using the same parameters as in the previous section.

4.6. Baseline DUSP Interactome

We analyzed publicly available Protein-protein interaction (PPI) data to identify DUSP-kinase interactions. We chose the comPPI database (Compartmentalized Protein-Protein Interaction Database, v2.1.1, http://comppi.linkgroup.hu/home) to identify biologically significant high-confident interactions between proteins with similar subcellular localization patterns. ComPPI is an integrated database of protein subcellular localization and protein-protein interactions from multiple databases including BioGRID, CCSB, DIP, HPRD, IntAct and MatrixDB. [93]. The highly confident interactomes of each member of the dual specificity phosphatase family for Homo sapiens were fetched from comPPI, filtering for localization score and interaction score thresholds of 0.7 each. The accessions of interacting proteins were obtained through bioDBnet: db2db (https://biodbnet-abcc.ncifcrf.gov/db/db2db.php) and g: Profiler (https://biit.cs.ut.ee/gprofiler/gconvert.cgi) and the set of interactions were compiled and visualized in Cytoscape (version 3.7.0) to obtain an integrated DUSP interactome. The interactome was clustered into DUSP neighborhood networks using the AutoAnnotate (v1.2) package. The network statistics of the interactome were analyzed using the Network Analyzer tool in Cytoscape [94]. The network analysis identified several network parameters. For the sake of clear understanding, we define a few basic concepts of graph theory [95]. Typically, a network consists of nodes which are in this case genes/proteins and edges which describe the relationship between them. A directed graph is a triple ordered graph G = (V, E, f) where V represents vertices called node, E represents edges representing connections between nodes and f is a function that maps each element in E to an ordered pair of vertices in V. The ordered pair of vertices are known as directed edges and represented by E = (i, j). Network properties provide valuable insight into the organizational structure of a biological network and network centralities show how nodes can be ranked/prioritized according to their properties. One of the network centralities is betweenness centrality, which is a measure of how nodes that are intermediate between neighbors rank higher. This essentially means that without these nodes, there would be no other way that the two neighbors could communicate with each other. In the current study, the betweenness centralities measure were used to identify protein hubs central to the DUSP network, which could influence the flow of information triggered by DUSPs. Individual interactomes of each DUSP member were also analyzed to identify central proteins associated with DUSPs using the betweenness centrality measure.

4.7. Landscape of DUSPs and Kinases in Activated Immune Cells

Data matrices from proteomics and transcriptomics datasets containing gene/protein expression data were fetched from the supplementary tables or GEO for the following studies [29,32,33,61]. DUSP and kinase expression data were obtained from the normalized datasets. Fold-change ratios were calculated by dividing the ratio of intensity/RPKMs of activated/stimulated cells by the ratio of respective intensity/RPKM values of steady-state/unstimulated cells. The fold-change ratios were converted into log2(fold-change) by log transformation. Genes/proteins with log2(fold-change) of 1 (fold-ratio of 2) were considered to upregulated while those with −1 (fold-ratio of −1) were considered to be downregulated. Genes/proteins with ambiguous trends across studies were ignored (both up and down) and only those with overall trends of either up and down were considered for further analysis and genes. We carried out correlation and interactome analysis to investigate the interplay between DUSPs and PKs in activated immune cells. Correlation analysis between z-scores of kinase and DUSP expression profiles of activated immune cells was performed as described above.

4.8. Functional, Pathway and Network Analysis

The differentially expressed molecules in response to LPS were analyzed with Gene Ontology-based functional analysis through CLUEGO (ClueGO v2.5.2 + CluePedia v 1.5.2) [96] in Cytoscape (v 3.7.0) to identify processes affected by LPS. The parameters used included 'ClueGO: Functions' analysis mode, murine 'GO: Biological Process' (dated 18.01.2019), global network specificity, pV ≤ 0.05000 with GO Term grouping. Hypergeometric enrichment-based pathway analysis was performed using Reactome Pathways [97] in R/Bioconductor 3.5.1/3.7 [98] with clusterProfiler 3.8.1 [99] and reactome.db 1.64.0. Genesets with less than 10 genes were excluded from the analysis and p values were adjusted by Benjamini–Hochberg correction. Pathways reaching adjusted p-values ≤ 0.0075 were curated manually and plotted in R with ggplot package (v2 3.1.0) (https://cran.r-project.org/web/packages/ggplot2). Network analysis of activated DCs and MOs was performed using STRING in Cytoscape. The network properties were calculated using Network Analyzer.

5. Conclusions

Though the role of dual specificity phosphatases in innate and adaptive immunity is known, their interplay with kinases was not precisely understood. In the current study, we expanded the knowledge on the role of dual specificity phosphatase signaling in activated and steady-state cells through the analysis of high-resolution expression datasets. We confirmed the importance of several known DUSPs such as DUSP1 and DUSP10 in innate immunity. We also report potentially novel role of DUSPs such as the Slingshot phosphatases and PKs such as LRKK2 in immune signaling. These need to be further validated to confirm their roles. we also identified selective patterns of expression of a few DUSPs and PKs across hematopoietic cells which could be used as potential therapeutic targets. Furthermore, we also identified potential species-specific events of DUSP signaling which need to be further validated. Finally, we demonstrate the utility of meta-analysis of existing datasets to identify molecular mechanisms of various biological processes and fill existing gaps in understanding understudied proteins. The findings from this study will aid in the understanding of DUSP signaling in the context of innate immunity.

Supplementary Materials: Supplementary materials can be found at http://www.mdpi.com/1422-0067/20/9/2086/s1.

Author Contributions: Conceptualization, R.K.K.; methodology, R.K.K., Y.S.; formal analysis, Y.S., K.B.; data curation, S.M.P.; writing—original draft preparation, Y.S.; writing—review and editing, S.M.P., K.B., R.K.K., T.S.K.P. supervision, R.K.K. and T.S.K.P.; funding acquisition, R.K.K.

Funding: This research was funded by the Research Council of Norway (FRIMEDBIO "Young Research Talent" Grant 263168 to R.K.K.; and Centres of Excellence Funding Scheme Project 223255/F50 to CEMIR), Onsager fellowship from NTNU (to R.K.K.).

Int. J. Mol. Sci. **2019**, *20*, 2086

Acknowledgments: We thank the authors of the datasets used in this study for making their data publicly available.

Conflicts of Interest: The authors declare no conflict of interest

Abbreviations

DC	Dendritic cells
MO	Monocytes
LPS	Lipopolysaccharides
DUSP	Dual Specificity Phosphatase
PK	Protein Kinase
GO	Gene Ontology

References

1. Humphrey, S.J.; James, D.E.; Mann, M. Protein Phosphorylation: A Major Switch Mechanism for Metabolic Regulation. *Trends Endocrinol. Metab.* **2015**, *26*, 676–687. [CrossRef] [PubMed]
2. Cohen, P. The origins of protein phosphorylation. *Nat. Cell Biol.* **2002**, *4*, E127–E130. [CrossRef]
3. Ardito, F.; Giuliani, M.; Perrone, D.; Troiano, G.; Lo Muzio, L. The crucial role of protein phosphorylation in cell signaling and its use as targeted therapy (Review). *Int. J. Mol. Med.* **2017**, *40*, 271–280. [CrossRef] [PubMed]
4. Manning, G.; Whyte, D.B.; Martinez, R.; Hunter, T.; Sudarsanam, S. The protein kinase complement of the human genome. *Science* **2002**, *298*, 1912–1934. [CrossRef]
5. Griffith, M.; Griffith, O.L.; Coffman, A.C.; Weible, J.V.; McMichael, J.F.; Spies, N.C.; Koval, J.; Das, I.; Callaway, M.B.; Eldred, J.M.; et al. DGIdb: Mining the druggable genome. *Nat. Methods* **2013**, *10*, 1209–1210. [CrossRef]
6. Alonso, A.; Sasin, J.; Bottini, N.; Friedberg, I.; Friedberg, I.; Osterman, A.; Godzik, A.; Hunter, T.; Dixon, J.; Mustelin, T. Protein tyrosine phosphatases in the human genome. *Cell* **2004**, *117*, 699–711. [CrossRef] [PubMed]
7. Chen, M.J.; Dixon, J.E.; Manning, G. Genomics and evolution of protein phosphatases. *Sci. Signal.* **2017**, *10*, 474. [CrossRef]
8. Lang, R.; Hammer, M.; Mages, J. DUSP meet immunology: dual specificity MAPK phosphatases in control of the inflammatory response. *J. Immunol.* **2006**, *177*, 7497–7504. [CrossRef]
9. Liu, Y.; Shepherd, E.G.; Nelin, L.D. MAPK phosphatases—regulating the immune response. *Nat. Rev. Immunol.* **2007**, *7*, 202–212. [CrossRef] [PubMed]
10. Jeffrey, K.L.; Camps, M.; Rommel, C.; Mackay, C.R. Targeting dual-specificity phosphatases: manipulating MAP kinase signalling and immune responses. *Nat. Rev. Drug Discov.* **2007**, *6*, 391–403. [CrossRef] [PubMed]
11. Abraham, S.M.; Clark, A.R. Dual-specificity phosphatase 1: A critical regulator of innate immune responses. *Biochem. Soc. Trans.* **2006**, *34*, 1018–1023. [CrossRef] [PubMed]
12. Zhang, Y.; Reynolds, J.M.; Chang, S.H.; Martin-Orozco, N.; Chung, Y.; Nurieva, R.I.; Dong, C. MKP-1 is necessary for T cell activation and function. *J. Biol. Chem.* **2009**, *284*, 30815–30824. [CrossRef]
13. Korhonen, R.; Huotari, N.; Hommo, T.; Leppanen, T.; Moilanen, E. The expression of interleukin-12 is increased by MAP kinase phosphatase-1 through a mechanism related to interferon regulatory factor 1. *Mol. Immunol.* **2012**, *51*, 219–226. [CrossRef] [PubMed]
14. Hammer, M.; Mages, J.; Dietrich, H.; Servatius, A.; Howells, N.; Cato, A.C.; Lang, R. Dual specificity phosphatase 1 (DUSP1) regulates a subset of LPS-induced genes and protects mice from lethal endotoxin shock. *J. Exp. Med.* **2006**, *203*, 15–20. [CrossRef] [PubMed]
15. Qian, F.; Deng, J.; Cheng, N.; Welch, E.J.; Zhang, Y.; Malik, A.B.; Flavell, R.A.; Dong, C.; Ye, R.D. A non-redundant role for MKP5 in limiting ROS production and preventing LPS-induced vascular injury. *EMBO J.* **2009**, *28*, 2896–2907. [CrossRef] [PubMed]
16. Jeffrey, K.L.; Brummer, T.; Rolph, M.S.; Liu, S.M.; Callejas, N.A.; Grumont, R.J.; Gillieron, C.; Mackay, F.; Grey, S.; Camps, M.; Rommel, C.; et al. Positive regulation of immune cell function and inflammatory responses by phosphatase PAC-1. *Nat. Immunol.* **2006**, *7*, 274–283. [CrossRef] [PubMed]

17. Kovanen, P.E.; Bernard, J.; Al-Shami, A.; Liu, C.; Bollenbacher-Reilley, J.; Young, L.; Pise-Masison, C.; Spolski, R.; Leonard, W.J. T-cell development and function are modulated by dual specificity phosphatase DUSP5. *J. Biol. Chem.* **2008**, *283*, 17362–17369. [CrossRef]

18. Boulding, T.; Wu, F.; McCuaig, R.; Dunn, J.; Sutton, C.R.; Hardy, K.; Tu, W.; Bullman, A.; Yip, D.; Dahlstrom, J.E.; Rao, S. Differential Roles for DUSP Family Members in Epithelial-to-Mesenchymal Transition and Cancer Stem Cell Regulation in Breast Cancer. *PLoS ONE* **2016**, *11*, e0148065. [CrossRef] [PubMed]

19. Kaltenmeier, C.T.; Vollmer, L.L.; Vernetti, L.A.; Caprio, L.; Davis, K.; Korotchenko, V.N.; Day, B.W.; Tsang, M.; Hulkower, K.I.; Lotze, M.T.; Vogt, A. A Tumor Cell-Selective Inhibitor of Mitogen-Activated Protein Kinase Phosphatases Sensitizes Breast Cancer Cells to Lymphokine-Activated Killer Cell Activity. *J. Pharmacol. Exp. Ther.* **2017**, *361*, 39–50. [CrossRef] [PubMed]

20. Zhang, X.; Hyer, J.M.; Yu, H.; D'Silva, N.J.; Kirkwood, K.L. DUSP1 phosphatase regulates the proinflammatory milieu in head and neck squamous cell carcinoma. *Cancer Res.* **2014**, *74*, 7191–7197. [CrossRef] [PubMed]

21. Guo, K.; Li, J.; Tang, J.P.; Tan, C.P.; Hong, C.W.; Al-Aidaroos, A.Q.; Varghese, L.; Huang, C.; Zeng, Q. Targeting intracellular oncoproteins with antibody therapy or vaccination. *Sci. Transl. Med.* **2011**, *3*, 99ra85. [CrossRef]

22. Guo, K.; Tang, J.P.; Tan, C.P.; Wang, H.; Zeng, Q. Monoclonal antibodies target intracellular PRL phosphatases to inhibit cancer metastases in mice. *Cancer Biol. Ther.* **2008**, *7*, 750–757. [CrossRef] [PubMed]

23. Guo, K.; Li, J.; Wang, H.; Osato, M.; Tang, J.P.; Quah, S.Y.; Gan, B.Q.; Zeng, Q. PRL-3 initiates tumor angiogenesis by recruiting endothelial cells in vitro and in vivo. *Cancer Res.* **2006**, *66*, 9625–9635. [CrossRef]

24. Rios, P.; Nunes-Xavier, C.E.; Tabernero, L.; Kohn, M.; Pulido, R. Dual-specificity phosphatases as molecular targets for inhibition in human disease. *Antioxid. Redox Signal.* **2014**, *20*, 2251–2273. [CrossRef]

25. Subbannayya, Y.; Pinto, S.M.; Gowda, H.; Prasad, T.S. Proteogenomics for understanding oncology: recent advances and future prospects. *Expert. Rev. Proteomics* **2016**, *13*, 297–308. [CrossRef] [PubMed]

26. Lee, S.E.; Song, J.; Bosl, K.; Muller, A.C.; Vitko, D.; Bennett, K.L.; Superti-Furga, G.; Pandey, A.; Kandasamy, R.K.; Kim, M.S. Proteogenomic Analysis to Identify Missing Proteins from Haploid Cell Lines. *Proteomics* **2018**, *18*, e1700386. [CrossRef] [PubMed]

27. Pinto, S.M.; Verma, R.; Advani, J.; Chatterjee, O.; Patil, A.H.; Kapoor, S.; Subbannayya, Y.; Raja, R.; Gandotra, S.; Prasad, T.S.K. Integrated Multi-Omic Analysis of Mycobacterium tuberculosis H37Ra Redefines Virulence Attributes. *Front. Microbiol.* **2018**, *9*, 1314. [CrossRef]

28. Kandasamy, R.K.; Vladimer, G.I.; Snijder, B.; Muller, A.C.; Rebsamen, M.; Bigenzahn, J.W.; Moskovskich, A.; Sabler, M.; Stefanovic, A.; Scorzoni, S.; et al. A time-resolved molecular map of the macrophage response to VSV infection. *NPJ Syst. Biol. Appl.* **2016**, *2*, 16027. [CrossRef] [PubMed]

29. Rieckmann, J.C.; Geiger, R.; Hornburg, D.; Wolf, T.; Kveler, K.; Jarrossay, D.; Sallusto, F.; Shen-Orr, S.S.; Lanzavecchia, A.; Mann, M.; et al. Social network architecture of human immune cells unveiled by quantitative proteomics. *Nat. Immunol.* **2017**, *18*, 583–593. [CrossRef] [PubMed]

30. Chevrier, N.; Mertins, P.; Artyomov, M.N.; Shalek, A.K.; Iannacone, M.; Ciaccio, M.F.; Gat-Viks, I.; Tonti, E.; DeGrace, M.M.; Clauser, K.R.; et al. Systematic discovery of TLR signaling components delineates viral-sensing circuits. *Cell* **2011**, *147*, 853–867. [CrossRef]

31. Elpek, K.G.; Cremasco, V.; Shen, H.; Harvey, C.J.; Wucherpfennig, K.W.; Goldstein, D.R.; Monach, P.A.; Turley, S.J. The tumor microenvironment shapes lineage, transcriptional, and functional diversity of infiltrating myeloid cells. *Cancer Immunol. Res.* **2014**, *2*, 655–667. [CrossRef]

32. Gat-Viks, I.; Chevrier, N.; Wilentzik, R.; Eisenhaure, T.; Raychowdhury, R.; Steuerman, Y.; Shalek, A.K.; Hacohen, N.; Amit, I.; Regev, A. Deciphering molecular circuits from genetic variation underlying transcriptional responsiveness to stimuli. *Nat. Biotechnol.* **2013**, *31*, 342–349. [CrossRef]

33. Jovanovic, M.; Rooney, M.S.; Mertins, P.; Przybylski, D.; Chevrier, N.; Satija, R.; Rodriguez, E.H.; Fields, A.P.; Schwartz, S.; Raychowdhury, R.; et al. Immunogenetics. Dynamic profiling of the protein life cycle in response to pathogens. *Science* **2015**, *347*, 1259038. [CrossRef] [PubMed]

34. Kamal, A.H.M.; Fessler, M.B.; Chowdhury, S.M. Comparative and network-based proteomic analysis of low dose ethanol- and lipopolysaccharide-induced macrophages. *PLoS ONE* **2018**, *13*, e0193104. [CrossRef] [PubMed]

35. Mertins, P.; Przybylski, D.; Yosef, N.; Qiao, J.; Clauser, K.; Raychowdhury, R.; Eisenhaure, T.M.; Maritzen, T.; Haucke, V.; Satoh, T.; et al. An Integrative Framework Reveals Signaling-to-Transcription Events in Toll-like Receptor Signaling. *Cell Rep.* **2017**, *19*, 2853–2866. [CrossRef] [PubMed]

36. Mingueneau, M.; Kreslavsky, T.; Gray, D.; Heng, T.; Cruse, R.; Ericson, J.; Bendall, S.; Spitzer, M.H.; Nolan, G.P.; Kobayashi, K.; et al. The transcriptional landscape of alphabeta T cell differentiation. *Nat. Immunol.* **2013**, *14*, 619–632. [CrossRef] [PubMed]

37. Muller, M.M.; Lehmann, R.; Klassert, T.E.; Reifenstein, S.; Conrad, T.; Moore, C.; Kuhn, A.; Behnert, A.; Guthke, R.; Driesch, D.; et al. Global analysis of glycoproteins identifies markers of endotoxin tolerant monocytes and GPR84 as a modulator of TNFalpha expression. *Sci. Rep.* **2017**, *7*, 838. [CrossRef] [PubMed]

38. Shalek, A.K.; Satija, R.; Adiconis, X.; Gertner, R.S.; Gaublomme, J.T.; Raychowdhury, R.; Schwartz, S.; Yosef, N.; Malboeuf, C.; Lu, D.; et al. Single-cell transcriptomics reveals bimodality in expression and splicing in immune cells. *Nature* **2013**, *498*, 236–240. [CrossRef] [PubMed]

39. Coordinators, N.R. Database resources of the National Center for Biotechnology Information. *Nucleic Acids Res.* **2016**, *44*, D7–D19.

40. Arthur, J.S.; Ley, S.C. Mitogen-activated protein kinases in innate immunity. *Nat. Rev. Immunol.* **2013**, *13*, 679–692. [CrossRef]

41. Yuan, X.; Wu, H.; Bu, H.; Zhou, J.; Zhang, H. Targeting the immunity protein kinases for immuno-oncology. *Eur. J. Med. Chem.* **2019**, *163*, 413–427. [CrossRef]

42. Michel, T.; Poli, A.; Cuapio, A.; Briquemont, B.; Iserentant, G.; Ollert, M.; Zimmer, J. Human CD56bright NK Cells: An Update. *J. Immunol.* **2016**, *196*, 2923–2931. [CrossRef] [PubMed]

43. Tucker, K.L.; Sage, T.; Stevens, J.M.; Jordan, P.A.; Jones, S.; Barrett, N.E.; St-Arnaud, R.; Frampton, J.; Dedhar, S.; Gibbins, J.M. A dual role for integrin-linked kinase in platelets: regulating integrin function and alpha-granule secretion. *Blood* **2008**, *112*, 4523–4531. [CrossRef]

44. Bhattacharjee, A.; Pal, S.; Feldman, G.M.; Cathcart, M.K. Hck is a key regulator of gene expression in alternatively activated human monocytes. *J. Biol. Chem.* **2011**, *286*, 36709–36723. [CrossRef] [PubMed]

45. Boehm, T.; Hess, I.; Swann, J.B. Evolution of lymphoid tissues. *Trends Immunol.* **2012**, *33*, 315–321. [CrossRef] [PubMed]

46. Pabst, R. Plasticity and heterogeneity of lymphoid organs. What are the criteria to call a lymphoid organ primary, secondary or tertiary? *Immunol. Lett.* **2007**, *112*, 1–8. [CrossRef] [PubMed]

47. Girard-Madoux, M.J.H.; Gomez de Aguero, M.; Ganal-Vonarburg, S.C.; Mooser, C.; Belz, G.T.; Macpherson, A.J.; Vivier, E. The immunological functions of the Appendix: An example of redundancy? *Semin Immunol.* **2018**, *36*, 31–44. [CrossRef]

48. Kooij, I.A.; Sahami, S.; Meijer, S.L.; Buskens, C.J.; Te Velde, A.A. The immunology of the vermiform appendix: A review of the literature. *Clin. Exp. Immunol.* **2016**, *186*, 1–9. [CrossRef] [PubMed]

49. Kawaji, H.; Kasukawa, T.; Forrest, A.; Carninci, P.; Hayashizaki, Y. The FANTOM5 collection, a data series underpinning mammalian transcriptome atlases in diverse cell types. *Sci. Data* **2017**, *4*, 170113. [CrossRef] [PubMed]

50. Consortium, G.T. Human genomics. The Genotype-Tissue Expression (GTEx) pilot analysis: multitissue gene regulation in humans. *Science* **2015**, *348*, 648–660. [CrossRef] [PubMed]

51. Uhlen, M.; Zhang, C.; Lee, S.; Sjostedt, E.; Fagerberg, L.; Bidkhori, G.; Benfeitas, R.; Arif, M.; Liu, Z.; Edfors, F.; et al. A pathology atlas of the human cancer transcriptome. *Science* **2017**, *357*, 6352. [CrossRef] [PubMed]

52. Uhlen, M.; Fagerberg, L.; Hallstrom, B.M.; Lindskog, C.; Oksvold, P.; Mardinoglu, A.; Sivertsson, A.; Kampf, C.; Sjostedt, E.; Asplund, A.; et al. Proteomics. Tissue-based map of the human proteome. *Science* **2015**, *347*, 1260419. [CrossRef]

53. Barbosa-Morais, N.L.; Irimia, M.; Pan, Q.; Xiong, H.Y.; Gueroussov, S.; Lee, L.J.; Slobodeniuc, V.; Kutter, C.; Watt, S.; Colak, R.; et al. The evolutionary landscape of alternative splicing in vertebrate species. *Science* **2012**, *338*, 1587–1593. [CrossRef] [PubMed]

54. Roadmap Epigenomics Consortium; Kundaje, A.; Meuleman, W.; Ernst, J.; Bilenky, M.; Yen, A.; Heravi-Moussavi, A.; Kheradpour, P.; Zhang, Z.; Wang, J.; et al. Integrative analysis of 111 reference human epigenomes. *Nature* **2015**, *518*, 317–330. [CrossRef] [PubMed]

55. Lin, S.; Lin, Y.; Nery, J.R.; Urich, M.A.; Breschi, A.; Davis, C.A.; Dobin, A.; Zaleski, C.; Beer, M.A.; Chapman, W.C.; et al. Comparison of the transcriptional landscapes between human and mouse tissues. *Proc. Natl. Acad. Sci. USA* **2014**, *111*, 17224–17229. [CrossRef] [PubMed]

56. Wang, L.H.; Xiang, J.; Yan, M.; Zhang, Y.; Zhao, Y.; Yue, C.F.; Xu, J.; Zheng, F.M.; Chen, J.N.; Kang, Z.; et al. The mitotic kinase Aurora-A induces mammary cell migration and breast cancer metastasis by activating the Cofilin-F-actin pathway. *Cancer Res.* **2010**, *70*, 9118–9128. [CrossRef] [PubMed]

57. Lopes, L.J.S.; Tesser-Gamba, F.; Petrilli, A.S.; de Seixas Alves, M.T.; Garcia-Filho, R.J.; Toledo, S.R.C. MAPK pathways regulation by DUSP1 in the development of osteosarcoma: Potential markers and therapeutic targets. *Mol. Carcinog.* **2017**, *56*, 1630–1641. [CrossRef] [PubMed]

58. Theodosiou, A.; Smith, A.; Gillieron, C.; Arkinstall, S.; Ashworth, A. MKP5, a new member of the MAP kinase phosphatase family, which selectively dephosphorylates stress-activated kinases. *Oncogene* **1999**, *18*, 6981–6988. [CrossRef]

59. Matsuguchi, T.; Musikacharoen, T.; Johnson, T.R.; Kraft, A.S.; Yoshikai, Y. A novel mitogen-activated protein kinase phosphatase is an important negative regulator of lipopolysaccharide-mediated c-Jun N-terminal kinase activation in mouse macrophage cell lines. *Mol. Cell. Biol.* **2001**, *21*, 6999–7009. [CrossRef] [PubMed]

60. Bosl, K.; Giambelluca, M.; Haug, M.; Bugge, M.; Espevik, T.; Kandasamy, R.K.; Bergstrom, B. Coactivation of TLR2 and TLR8 in Primary Human Monocytes Triggers a Distinct Inflammatory Signaling Response. *Front. Physiol.* **2018**, *9*, 618. [CrossRef] [PubMed]

61. Amit, I.; Garber, M.; Chevrier, N.; Leite, A.P.; Donner, Y.; Eisenhaure, T.; Guttman, M.; Grenier, J.K.; Li, W.; Zuk, O.; et al. Unbiased reconstruction of a mammalian transcriptional network mediating pathogen responses. *Science* **2009**, *326*, 257–263. [CrossRef] [PubMed]

62. Vijayan, V.; Pradhan, P.; Braud, L.; Fuchs, H.R.; Gueler, F.; Motterlini, R.; Foresti, R.; Immenschuh, S. Human and murine macrophages exhibit differential metabolic responses to lipopolysaccharide—A divergent role for glycolysis. *Redox Biol.* **2019**, *22*, 101147. [CrossRef] [PubMed]

63. Sakuta, T.; Matsushita, K.; Yamaguchi, N.; Oyama, T.; Motani, R.; Koga, T.; Nagaoka, S.; Abeyama, K.; Maruyama, I.; Takada, H.; et al. Enhanced production of vascular endothelial growth factor by human monocytic cells stimulated with endotoxin through transcription factor SP-1. *J. Med. Microbiol.* **2001**, *50*, 233–237. [CrossRef]

64. Huang, C.Y.; Tan, T.H. DUSPs, to MAP kinases and beyond. *Cell Biosci.* **2012**, *2*, 24. [CrossRef] [PubMed]

65. Bhore, N.; Wang, B.J.; Chen, Y.W.; Liao, Y.F. Critical Roles of Dual-Specificity Phosphatases in Neuronal Proteostasis and Neurological Diseases. *Int. J. Mol. Sci.* **2017**, *18*, 1963. [CrossRef] [PubMed]

66. Alonso, A.; Pulido, R. The extended human PTPome: A growing tyrosine phosphatase family. *FEBS J.* **2016**, *283*, 1404–1429. [CrossRef] [PubMed]

67. Hatzihristidis, T.; Liu, S.; Pryszcz, L.; Hutchins, A.P.; Gabaldon, T.; Tremblay, M.L.; Miranda-Saavedra, D. PTP-central: A comprehensive resource of protein tyrosine phosphatases in eukaryotic genomes. *Methods* **2014**, *65*, 156–164. [CrossRef] [PubMed]

68. Rios, P.; Li, X.; Kohn, M. Molecular mechanisms of the PRL phosphatases. *FEBS J.* **2013**, *280*, 505–524. [CrossRef]

69. Huang, L.C.; Ross, K.E.; Baffi, T.R.; Drabkin, H.; Kochut, K.J.; Ruan, Z.; D'Eustachio, P.; McSkimming, D.; Arighi, C.; Chen, C.; et al. Integrative annotation and knowledge discovery of kinase post-translational modifications and cancer-associated mutations through federated protein ontologies and resources. *Sci. Rep.* **2018**, *8*, 6518. [CrossRef]

70. Pula, G.; Schuh, K.; Nakayama, K.; Nakayama, K.I.; Walter, U.; Poole, A.W. PKCdelta regulates collagen-induced platelet aggregation through inhibition of VASP-mediated filopodia formation. *Blood* **2006**, *108*, 4035–4044. [CrossRef]

71. Meinders, M.; Kulu, D.I.; van de Werken, H.J.; Hoogenboezem, M.; Janssen, H.; Brouwer, R.W.; van Ijcken, W.F.; Rijkers, E.J.; Demmers, J.A.; Kruger, I.; et al. Sp1/Sp3 transcription factors regulate hallmarks of megakaryocyte maturation and platelet formation and function. *Blood* **2015**, *125*, 1957–1967. [CrossRef]

72. Hitchcock, I.S.; Fox, N.E.; Prevost, N.; Sear, K.; Shattil, S.J.; Kaushansky, K. Roles of focal adhesion kinase (FAK) in megakaryopoiesis and platelet function: Studies using a megakaryocyte lineage specific FAK knockout. *Blood* **2008**, *111*, 596–604. [CrossRef]

73. Wu, J.; Mei, C.; Vlassara, H.; Striker, G.E.; Zheng, F. Oxidative stress-induced JNK activation contributes to proinflammatory phenotype of aging diabetic mesangial cells. *Am. J. Physiol. Renal. Physiol.* **2009**, *297*, F1622–F1631. [CrossRef] [PubMed]

74. Zhang, Y.; Blattman, J.N.; Kennedy, N.J.; Duong, J.; Nguyen, T.; Wang, Y.; Davis, R.J.; Greenberg, P.D.; Flavell, R.A.; Dong, C. Regulation of innate and adaptive immune responses by MAP kinase phosphatase 5. *Nature* **2004**, *430*, 793–797. [CrossRef]

75. Hammer, M.; Mages, J.; Dietrich, H.; Schmitz, F.; Striebel, F.; Murray, P.J.; Wagner, H.; Lang, R. Control of dual-specificity phosphatase-1 expression in activated macrophages by IL-10. *Eur. J. Immunol.* **2005**, *35*, 2991–3001. [CrossRef] [PubMed]

76. Nizamutdinova, I.T.; Kim, Y.M.; Lee, J.H.; Chang, K.C.; Kim, H.J. MKP-7, a negative regulator of JNK, regulates VCAM-1 expression through IRF-1. *Cell Signal.* **2012**, *24*, 866–872. [CrossRef] [PubMed]

77. Zhang, Y.; Nallaparaju, K.C.; Liu, X.; Jiao, H.; Reynolds, J.M.; Wang, Z.X.; Dong, C. MAPK phosphatase 7 regulates T cell differentiation via inhibiting ERK-mediated IL-2 expression. *J. Immunol.* **2015**, *194*, 3088–3095. [CrossRef] [PubMed]

78. Niedzielska, M.; Bodendorfer, B.; Munch, S.; Eichner, A.; Derigs, M.; da Costa, O.; Schweizer, A.; Neff, F.; Nitschke, L.; Sparwasser, T.; et al. Gene trap mice reveal an essential function of dual specificity phosphatase Dusp16/MKP-7 in perinatal survival and regulation of Toll-like receptor (TLR)-induced cytokine production. *J. Biol. Chem.* **2014**, *289*, 2112–2126. [CrossRef]

79. Peng, H.Z.; Yun, Z.; Wang, W.; Ma, B.A. Dual specificity phosphatase 1 has a protective role in osteoarthritis fibroblastlike synoviocytes via inhibition of the MAPK signaling pathway. *Mol. Med. Rep.* **2017**, *16*, 8441–8447. [CrossRef] [PubMed]

80. Aggelou, H.; Chadla, P.; Nikou, S.; Karteri, S.; Maroulis, I.; Kalofonos, H.P.; Papadaki, H.; Bravou, V. LIMK/cofilin pathway and Slingshot are implicated in human colorectal cancer progression and chemoresistance. *Virchows Arch.* **2018**, *472*, 727–737. [CrossRef] [PubMed]

81. Maimaiti, Y.; Maimaitiming, M.; Li, Y.; Aibibula, S.; Ainiwaer, A.; Aili, A.; Sun, Z.; Abudureyimu, K. SSH1 expression is associated with gastric cancer progression and predicts a poor prognosis. *BMC Gastroenterol.* **2018**, *18*, 12. [CrossRef]

82. Caution, K.; Gavrilin, M.A.; Tazi, M.; Kanneganti, A.; Layman, D.; Hoque, S.; Krause, K.; Amer, A.O. Caspase-11 and caspase-1 differentially modulate actin polymerization via RhoA and Slingshot proteins to promote bacterial clearance. *Sci. Rep.* **2015**, *5*, 18479. [CrossRef] [PubMed]

83. Huang, C.T.; Chang, M.C.; Chen, Y.L.; Chen, T.C.; Chen, C.A.; Cheng, W.F. Insulin-like growth factors inhibit dendritic cell-mediated anti-tumor immunity through regulating ERK1/2 phosphorylation and p38 dephosphorylation. *Cancer Lett.* **2015**, *359*, 117–126. [CrossRef] [PubMed]

84. Spadaro, O.; Camell, C.D.; Bosurgi, L.; Nguyen, K.Y.; Youm, Y.H.; Rothlin, C.V.; Dixit, V.D. IGF1 Shapes Macrophage Activation in Response to Immunometabolic Challenge. *Cell Rep.* **2017**, *19*, 225–234. [CrossRef] [PubMed]

85. Cook, D.A.; Kannarkat, G.T.; Cintron, A.F.; Butkovich, L.M.; Fraser, K.B.; Chang, J.; Grigoryan, N.; Factor, S.A.; West, A.B.; Boss, J.M.; et al. LRRK2 levels in immune cells are increased in Parkinson's disease. *NPJ Parkinsons Dis.* **2017**, *3*, 11. [CrossRef] [PubMed]

86. Lee, H.; James, W.S.; Cowley, S.A. LRRK2 in peripheral and central nervous system innate immunity: its link to Parkinson's disease. *Biochem. Soc. Trans.* **2017**, *45*, 131–139. [CrossRef] [PubMed]

87. Moehle, M.S.; Webber, P.J.; Tse, T.; Sukar, N.; Standaert, D.G.; DeSilva, T.M.; Cowell, R.M.; West, A.B. LRRK2 inhibition attenuates microglial inflammatory responses. *J. Neurosci.* **2012**, *32*, 1602–1611. [CrossRef] [PubMed]

88. Miranda-Saavedra, D.; Barton, G.J. Classification and functional annotation of eukaryotic protein kinases. *Proteins* **2007**, *68*, 893–914. [CrossRef]

89. Hunter, T. Signaling—2000 and beyond. *Cell* **2000**, *100*, 113–127. [CrossRef]

90. Letunic, I.; Bork, P. 20 years of the SMART protein domain annotation resource. *Nucleic Acids Res.* **2018**, *46*, D493–D496. [CrossRef] [PubMed]

91. Mudunuri, U.; Che, A.; Yi, M.; Stephens, R.M. bioDBnet: the biological database network. *Bioinformatics* **2009**, *25*, 555–556. [CrossRef] [PubMed]

92. Reimand, J.; Arak, T.; Adler, P.; Kolberg, L.; Reisberg, S.; Peterson, H.; Vilo, J. g:Profiler-a web server for functional interpretation of gene lists (2016 update). *Nucleic Acids Res.* **2016**, *44*, W83–W89. [CrossRef] [PubMed]

93. Veres, D.V.; Gyurko, D.M.; Thaler, B.; Szalay, K.Z.; Fazekas, D.; Korcsmaros, T.; Csermely, P. ComPPI: A cellular compartment-specific database for protein-protein interaction network analysis. *Nucleic Acids Res.* **2015**, *43*, D485–D493. [CrossRef] [PubMed]

94. Assenov, Y.; Ramirez, F.; Schelhorn, S.E.; Lengauer, T.; Albrecht, M. Computing topological parameters of biological networks. *Bioinformatics* **2008**, *24*, 282–284. [CrossRef] [PubMed]

95. Pavlopoulos, G.A.; Secrier, M.; Moschopoulos, C.N.; Soldatos, T.G.; Kossida, S.; Aerts, J.; Schneider, R.; Bagos, P.G. Using graph theory to analyze biological networks. *BioData Min.* **2011**, *4*, 10. [CrossRef]

96. Bindea, G.; Mlecnik, B.; Hackl, H.; Charoentong, P.; Tosolini, M.; Kirilovsky, A.; Fridman, W.H.; Pages, F.; Trajanoski, Z.; Galon, J. ClueGO: A Cytoscape plug-in to decipher functionally grouped gene ontology and pathway annotation networks. *Bioinformatics* **2009**, *25*, 1091–1093. [CrossRef]

97. Fabregat, A.; Jupe, S.; Matthews, L.; Sidiropoulos, K.; Gillespie, M.; Garapati, P.; Haw, R.; Jassal, B.; Korninger, F.; May, B.; et al. The Reactome Pathway Knowledgebase. *Nucleic Acids Res.* **2018**, *46*, D649–D655. [CrossRef] [PubMed]

98. Huber, W.; Carey, V.J.; Gentleman, R.; Anders, S.; Carlson, M.; Carvalho, B.S.; Bravo, H.C.; Davis, S.; Gatto, L.; Girke, T.; et al. Orchestrating high-throughput genomic analysis with Bioconductor. *Nat. Methods* **2015**, *12*, 115–121. [CrossRef] [PubMed]

99. Yu, G.; Wang, L.G.; Han, Y.; He, Q.Y. clusterProfiler: An R package for comparing biological themes among gene clusters. *OMICS* **2012**, *16*, 284–287. [CrossRef] [PubMed]

International Journal of
Molecular Sciences

MDPI

Review

Regulation of Dual-Specificity Phosphatase (DUSP) Ubiquitination and Protein Stability

Hsueh-Fen Chen, Huai-Chia Chuang * and Tse-Hua Tan *

Immunology Research Center, National Health Research Institutes, Zhunan 35053, Taiwan; 061013@nhri.edu.tw
* Correspondence: cinth@nhri.edu.tw (H.-C.C.); ttan@nhri.edu.tw (T.-H.T.); Tel.: +886-37-246166 (ext. 37612)
  (H.-C.C.); +886-37-246166 (ext. 37601) (T.-H.T.); Fax: +886-37-586642 (H.-C.C.); +886-37-586642 (T.-H.T.)

Received: 29 April 2019; Accepted: 29 May 2019; Published: 30 May 2019

Abstract: Mitogen-activated protein kinases (MAPKs) are key regulators of signal transduction and cell responses. Abnormalities in MAPKs are associated with multiple diseases. Dual-specificity phosphatases (DUSPs) dephosphorylate many key signaling molecules, including MAPKs, leading to the regulation of duration, magnitude, or spatiotemporal profiles of MAPK activities. Hence, DUSPs need to be properly controlled. Protein post-translational modifications, such as ubiquitination, phosphorylation, methylation, and acetylation, play important roles in the regulation of protein stability and activity. Ubiquitination is critical for controlling protein degradation, activation, and interaction. For DUSPs, ubiquitination induces degradation of eight DUSPs, namely, DUSP1, DUSP4, DUSP5, DUSP6, DUSP7, DUSP8, DUSP9, and DUSP16. In addition, protein stability of DUSP2 and DUSP10 is enhanced by phosphorylation. Methylation-induced ubiquitination of DUSP14 stimulates its phosphatase activity. In this review, we summarize the knowledge of the regulation of DUSP stability and ubiquitination through post-translational modifications.

Keywords: dual-specificity phosphatase; mitogen-activated protein kinase; ubiquitination; protein stability

1. The DUSP Family Phosphatases

Mitogen-activated protein kinases (MAPKs) are the important components of cell signaling pathways. MAPKs regulate physiological and pathological responses to various extracellular stimuli and environmental stresses [1–7]. The best-known members of the MAPK family are ERK, JNK, and p38 subgroups [5,6]. These kinases unusually require dual phosphorylation on both threonine and tyrosine residues within the conserved motif T-X-Y for kinase activity [8]. The MAPK signaling pathways are involved in the processes of gene transcription, mRNA translation, protein stability, protein localization, and enzyme activity, thus regulating various cellular functions including cell proliferation, cell differentiation, cell survival, and cell death [9,10]. MAPK signaling pathways are also involved in a number of diseases including inflammation and cancer [11,12]. Pathway outputs reflect the balance between the activation of upstream pathways and the inhibition of negative regulators. Inactivation of MAPKs are mediated by serine/threonine phosphatases, tyrosine phosphatase, and dual-specificity phosphatases (DUSPs) through dephosphorylation of threonine and/or tyrosine residues of the T-X-Y motif within the kinase activation loop [13]. The largest group of protein phosphatases that specifically regulates the MAPK activity in mammalian cells is the DUSP family phosphatases [13].

The DUSP family phosphatases dephosphorylate both threonine/serine and tyrosine residues of their substrates. All DUSPs have a common phosphatase domain, which contains conserved Asp, Cys, and Arg residues forming the catalytic site. A subfamily of DUSPs contains the MAP kinase-binding (MKB) motif or the kinase-interacting motif (KIM) that interacts with the common docking domain of MAPKs to mediate the enzyme–substrate interaction [14,15]. DUSPs containing the KIM domain are generally classified as typical DUSPs or MAP kinase phosphatases (MKPs), whereas DUSPs without the KIM domain are generally classified as atypical DUSPs (Table 1). However, there are a few exceptions. Three KIM-containing typical DUSPs, namely, DUSP2 (PAC1), DUSP5, and DUSP8, are not named as MKPs (Table 1). Two atypical DUSPs, DUSP14 (MKP6) and DUSP26 (MKP8), do not contain the KIM domain but still can dephosphorylate and inactivate MAPKs (Table 1). Typical DUSPs can be further grouped into three subgroups based on their predominant subcellular locations, that is, the nucleus, the cytoplasm, or both [15].

Table 1. Classification and domain structure of human dual-specificity phosphatases (DUSPs).

Classification	Gene Symbol	Alias	Domain Structure	MAPK Substrates
Typical DUSPs (also named MKPs)	DUSP1	MKP1, CL100, VH1, HVH1, PTPN10		JNK, p38 > ERK
	DUSP4	MKP2, VH2, HVH2, TYP		ERK, JNK > p38
	DUSP6	MKP3, PYST1		ERK
	DUSP7	PYST2, MKPX*		ERK
	DUSP9	MKP4		ERK > p38
	DUSP10	MKP5		JNK, p38
	DUSP16	MKP7		JNK (p38?)
Typical DUSPs (not named as MKPs)	DUSP2	PAC1		ERK, JNK, p38
	DUSP5	VH3, HVH3		ERK
	DUSP8	HB5, VH5, HVH-5, HVH8, (Mouse: M3/6)		JNK (p38?)
Atypical DUSPs	DUSP3	VHR		
	DUSP11	PIR1		
	DUSP12	YVH1		
	DUSP13	DUSP13A, DUSP13B, BEDP, MDSP, SKRP4, TMDP		
	DUSP15	VHY		
	DUSP18	DUSP20, LMW-DSP20		
	DUSP19	DUSP17, LMW-DSP3, SKRP1, TS-DSP1		
	DUSP21	LMW-DSP21		
	DUSP22	JKAP, JSP1, VHX, LMW-DSP2, MKPX*		
	DUSP23	DUSP25, VHZ, LDP-3, MOSP		
	DUSP24	STYXL1, MK-STYX		
	DUSP27			
	DUSP28	VHP, DUSP26#		
Atypical DUSPs (also named MKPs)	DUSP14	MKP6, MKP-L		JNK > ERK > p38
	DUSP26	MKP8, LDP-4, NATA1, SKRP3, NEAP, DUSP24#		p38 (ERK?)

Cdc25-homology	Kinase-interacting motif (KIM)	Phosphatase	Phosphatase (inactive)	PEST	Disintegrin	Unknown

*, MKPX is a duplicate name for both DUSP7 and DUSP22. #, DUSP24 and DUSP26 are renamed to DUSP26 and DUSP28, respectively. Domain structures are annotated from the Ensemble database.

DUSPs do not always require phosphatase activity to regulate the function of substrates. For example, DUSPs can control functions of MAPKs by sequestering them in the cytoplasm or nucleus [16–18]. Because both DUSPs and substrates of MAPKs interact with MAPKs via the common docking domain of MAPKs [19], DUSPs may also regulate MAPK signaling by competing with MAPK substrates for binding to MAPKs [15].

DUSPs are critical for the regulation of MAPK activity and are thus subject to complex regulation. Gene expression and phosphatase activity of DUSPs are regulated by gene transcription, protein modification, or protein stability. This review will focus on the regulation of DUSP protein stability and ubiquitination by post-translational modifications.

Int. J. Mol. Sci. **2019**, *20*, 2668

2. Negative Regulation of DUSPs by Lys48-Linked Ubiquitination and Proteasomal Degradation

Ubiquitination regulates many biological functions such as cell proliferation, cell apoptosis, and immune responses [20]. Ubiquitination is the modification of a protein by ubiquitin(s) on one or more lysine residues. Ubiquitination is mediated by an enzyme cascade involving three classes of enzymes: E1 (ubiquitin-activating enzyme), E2 (ubiquitin-conjugating enzyme), and E3 (ubiquitin ligase), resulting in covalent bonding of ubiquitin to lysine residues of protein substrates [21]. Ubiquitin contains seven lysine residues (Lys6, 11, 27, 29, 33, 48, and 63) that can act as ubiquitin acceptor forming ubiquitin chains with different topologies on protein substrates. The functions of Lys48-linked and Lys63-linked ubiquitinations are well characterized. Lys48-linked ubiquitination primarily controls proteasomal degradation; Lys63-linked ubiquitination controls several protein functions, including receptor endocytosis, protein trafficking, enzyme activity, and protein–protein interaction [22]. In addition, Lys48- and Lys63-linked ubiquitinations are associated with lysosomal degradation [23].

DUSP1 (MKP1) is a nuclear phosphatase [24]. DUSP1 binds to JNK and p38 with stronger affinity compared to its binding to ERK, leading to their dephosphorylation and inactivation [25]. Reciprocally, ERK induces DUSP1 proteasomal degradation by enhancing nuclear translocation and transcription activity of the transcription factor forkhead box M1 (FoxM1) [26], leading to the induction of the ubiquitin E3 ligase complex S-phase kinase-associated protein (Skp2)/cyclin-dependent kinase regulatory subunit 1 (Cks1) [27–30]. Besides ERK, other signaling molecules also control DUSP1 degradation. DUSP1 underwent an ubiquitin-mediated proteasomal degradation in response to glutamate-induced oxidative stress [31]. The involved E3-ubiquitin ligase was not identified; however, the process depends on the presence of PKCδ [31]. EGF plus lactoferrin induce a rapid proteasomal degradation of DUSP1, resulting in sustained ERK activation in human fibrosarcoma [32]. In rat cardiac myoblast H9c2 cells, the ubiquitin E3 ligase Atrogin-1 interacts with DUSP1 and promotes the ubiquitin-mediated proteasomal degradation of DUSP1, thereby leading to sustained activation of JNK signaling and subsequent cell apoptosis and ischemia/reperfusion injury [33]. Conversely, ubiquitin-specific peptidase 49 (USP49; a deubiquitinase) interacts with and deubiquitinates DUSP1, resulting in DUSP1 stabilization [34]. Angiotensin II-stimulated proteasome activity results in DUSP1 degradation and subsequent STAT1 activation in T cells, leading to induction of Th1 differentiation [35]; however, it is unclear how DUSP1 is regulated by angiotensin II. DUSP1 knockdown results in prolonged and enhanced STAT1 phosphorylation; it remains unclear whether DUSP1 can directly dephosphorylate STAT1.

DUSP4 (MKP2) preferentially inhibits ERK and JNK [36]. In senescent human fibroblasts, the phosphatase activity and protein levels of DUSP4 are increased due to impaired proteasomal activity [37]. 8-Bromo-cAMP (8-Br-cAMP) stimulation leads to reduction of the proteasomal degradation of DUSP4 in Leydig cells [38]; DUSP4 stabilization results in inhibition of ERK activity and subsequent reduction of the synthesis of P450scc steroidogenic enzyme, which is critical for steroid synthesis [38].

DUSP5 displays phosphatase activity toward ERK; DUSP5 overexpression results in both inactivation and nuclear translocation of ERK [16]. DUSP5 is a short-lived protein, which is ubiquitinated and subjected to proteasomal degradation [39]. DUSP5 degradation enhances the amplitude and duration of ERK signaling [40]. Reciprocally, ERK induces DUSP5 stability by decreasing DUSP5 ubiquitination [39]; the regulation is independent of ERK kinase activity but dependent on ERK–DUSP5 interaction [39].

DUSP6 (MKP3) preferentially inhibits ERK [41,42]. Reduction of DUSP6 by reactive oxygen species (ROS) is correlated with high ERK activity [43]. Similarly, in thyrocytes, B-Raf (V600E) mutation induces ROS generation, leading to proteasomal degradation of DUSP6 [44]. The B-Raf (V600E)-induced DUSP6 degradation results in ERK activation and cell senescence [44]. The anti-diabetic drug metformin accelerates the development of B-Raf (V600E)-mediated melanoma by inducing proteasomal degradation of DUSP6 through AMPK [45]. In breast cancer cells, PKCδ depletion results in ERK activation by inducing the level of the ubiquitin E3 ligase Nedd4, which induces DUSP6 degradation [46].

In contrast, thyroid-stimulating hormone (TSH) stabilizes DUSP6 by enhancing the expression of manganese superoxide dismutase (MnSOD), leading to prevention of senescence and induction of papillary thyroid carcinoma [44].

DUSP7, an ERK phosphatase, is ubiquitinated under hypoxic stress [47]. Hypoxia-inducible factors (HIFs) induce expression and cytoplasmic accumulation of the ubiquitin E3 ligase speckle-type POZ protein (SPOP) in clear cell renal cell carcinoma under hypoxic stress [47]. SPOP induces tumorigenesis by promoting ubiquitination and degradation of multiple regulators, including DUSP7 [47].

DUSP8 (M3/6) preferentially inactivates JNK and maybe p38 [48–50]. The protein synthesis inhibitor anisomycin [51] enhances the JNK pathway via activation of its upstream kinase SEK/MKK4 [52]. Anisomycin also stimulates JNK activity by inducing ubiquitination and degradation of DUSP8 [52]. In contrast, the proteasome inhibitor lactacystin prevents DUSP8 degradation, resulting in dephosphorylation and inactivation of JNK [52].

DUSP9 (MKP4), an ERK phosphatase, is associated with maintenance of the stemness of embryonic stem cells (ESCs) [53]. The long non-coding RNA (lncRNA) LincU directly binds and protects the DUSP9 protein from ubiquitin-mediated proteasomal degradation; the stabilized DUSP9 inhibits ERK activation, leading to preservation of naïve pluripotency of ESCs [53].

DUSP16 is also regulated by ubiquitin-mediated proteasomal degradation, and their ubiquitinations are regulated by phosphorylation (see below). In addition to Lys48-linked ubiquitination, one DUSP (DUSP14) is regulated by Lys63-linked ubiquitination (see Section 3.3).

3. Other Post-Translational Regulations of DUSP Ubiquitination and/or Stability

3.1. Phosphorylation

DUSP1 stability is differentially regulated by sustained or transient activation of ERK. Sustained activation of ERK phosphorylates DUSP1 on Ser296 and Ser323 residues [54]. The Ser296/323 phosphorylation of DUSP1 facilitates its interaction with the ubiquitin E3 ligase CUL1/SKP2/CKS1 complex, which targets DUSP1 for proteasomal degradation [27,54,55]. In contrast, ERK reduces DUSP1 degradation by phosphorylating two other residues, Ser359 and Ser364 [56,57]. Transient activation of ERK stimulates DUSP1 Ser359/364 phosphorylation, which enhances DUSP1 stability, and feedback attenuates ERK signaling [56]. One group reported that this ERK-induced DUSP1 stabilization may be independent of ubiquitination [57]. Furthermore, Krüpple-like transcription factor 5 (KLF5) promotes breast cancer cell survival partially through ERK-induced DUSP1 Ser359/364 phosphorylation, which is essential for DUSP1 protein stabilization [58].

DUSP1 stabilization results in inhibition of JNK and p38 activation, as well as subsequent inhibition of TNF-α and IL-6 production [59]. Glucocorticoids prevent animals from autoimmune diseases due to enhancement of DUSP1 expression and stability [60–62]. Insulin stimulation enhances DUSP1 phosphorylation, resulting in DUSP1 stabilization in vascular smooth muscle cells (VSMCs) [63]. This increase of DUSP1 leads to reduction of ERK activity and subsequent inhibition of cell migration [63]. The phosphorylated calcium/calmodulin kinase II (CaMKII) interacts with DUSP1 and prevents DUSP1 from proteasomal degradation [64]. Conversely, dephosphorylation of CaMKII leads to disruption of CaMKII–DUSP1 interaction, leading to proteasomal degradation of DUSP1 [64].

DUSP2 stability is induced by the atypical MAP kinase ERK4 [65]. Wild-type ERK4, but not catalytically inactive ERK4, binds to and stabilizes DUSP2 proteins [65]. This finding suggests that DUSP2 may be phosphorylated by ERK4, leading to DUSP2 stabilization.

DUSP4 is rapidly induced after ERK activation [66]. ERK interacts with and phosphorylates DUSP4 on Ser386 and Ser391 residues within the C-terminus [67], leading to prevention of DUSP4 from ubiquitin-mediated proteasomal degradation. The mechanism provides a negative feedback control of ERK activity [67,68]. Consistently, a short spliced isoform (encoding 303 amino acids) of human DUSP4 lacking the MAP kinase binding site is more susceptible to ubiquitination and proteasomal degradation than that of DUSP4 [69]. It is noted that one group reported no detectable effect of ERK on DUSP4 ubiquitination [57].

DUSP6 can also be a substrate of ERK. Upon serum stimulation, ERK phosphorylates DUSP6 on Ser159 and Ser197 residues in fibroblast cells [70]. The ERK-induced DUSP6 phosphorylation triggers ubiquitination and proteasomal degradation of DUSP6 [70]. EGF plus lactoferrin induce proteasomal degradation of DUSP6 [32]. P2X7 nucleotide or EGF also stimulates DUSP6 Ser197 phosphorylation by ERK, resulting in proteasomal degradation of DUSP6 in neurons and astrocytes [71]. Thus, ERK exerts a positive-feedback mechanism on its own kinase activity by promoting the degradation of DUSP6 [70,71]. In addition, insulin induces DUSP6 degradation through the ERK-mediated DUSP6 Ser159/197 phosphorylation in liver cells [72]. The reduction of DUSP6 by insulin signaling leads to downregulation of glucose-6-phosphatase, resulting in inhibition of glucose output of liver cells [72]. In addition to ERK-mediated degradation of DUSP6, mTOR signaling also induces the phosphorylation of DUSP6 on Ser159 residue and its subsequent proteasomal degradation [73]. Intracellular reactive oxygen species (ROS) accumulation such as hydrogen peroxide causes DUSP6 phosphorylation on Ser159 and Ser197 residues, leading to ubiquitination and degradation of DUSP6 in ovarian cancer cells [43].

DUSP6 is ubiquitinated and degraded by proteasome in the early phase of platelet-derived growth factor-B chains (PDGF-BB) stimulation; the process requires MEK-induced phosphorylation of DUSP6 on Ser174 residue [74]. In the later phase, DUSP6 is induced by ERK-mediated transcriptional expression, leading to inhibition of ERK activity [74]. Interestingly, both protein degradation and mRNA synthesis of DUSP6 are ERK-dependent, indicating both positive and negative regulation of DUSP6 by ERK [74]. Therefore, the regulation of DUSP6 by PDGF-BB stimulation exhibits a negative feedback control of PDGF-BB signaling [74].

DUSP10 is phosphorylated by mTORC2 on Ser224 and Ser230 residues upon insulin stimulation, leading to stabilization of DUSP10 and subsequent inactivation of p38 in glioblastoma cells [75].

DUSP16 preferentially inactivates JNK [17] and maybe p38 [76]. ERK phosphorylates Ser446 residue of DUSP16, resulting in enhancement of DUSP16 protein stability [77]. DUSP16 protein levels are rapidly decreased by ubiquitination and subsequent proteasomal degradation in quiescent cells [77]. ERK also phosphorylates DUSP16 on Ser446 residue [77,78]. This phosphorylation leads to stabilization of DUSP16 by preventing ubiquitination [77]. Induction of DUSP16 strongly suppresses JNK activation. Therefore, the activation of the ERK pathway can strongly inhibit JNK activation by stabilizing DUSP16 [77,78].

3.2. Oxidation

DUSP1 and DUSP4 proteins can be oxidized under oxidative stress. Oxidation of catalytic cysteine within the active site of DUSPs inactivates DUSP phosphatase activities and triggers their proteasomal degradation [55,79]. Superoxide induces DUSP1 proteasomal degradation, leading to JNK activation and subsequent cell death in lung cancer cells [80]. Metabolic disorder-induced oxidative stress causes DUSP1 S-glutathionylation and subsequent proteasomal degradation of DUSP1, resulting in monocyte migration and macrophage recruitment [81]. In addition, oxidation of DUSP1 induces its proteasomal degradation in monocytes upon asbestos stimulation, leading to induction of p38 activity and TNFα gene expression [82].

DUSP4 is redox sensitive. Under cadmium ion (Cd^{2+})-induced oxidative stress, DUSP4 is oxidized by glutathione disulfide (GSSG), leading to DUSP4 degradation and cell apoptosis [83]. In contrast, N-acetylcysteine (NAC) treatment upregulates the protein levels of DUSP4, protecting cells from cadmium ion (Cd^{2+})-induced apoptosis [83].

3.3. Methylation

DUSP14 (MKP6) is a MAP kinase phosphatase that inactivates JNK, ERK, and p38 in vitro [84]. DUSP14 is a negative regulator of T-cell receptor (TCR) signaling by directly inhibiting ERK in T cells [84]. DUSP14 also attenuates T-cell activation by directly dephosphorylating TAB1, leading to inhibition of TAB1–TAK1 complex and its downstream signaling molecules JNK and IKK [85]. Upon TCR signaling, DUSP14 interacts with the ubiquitin E3 ligase TRAF2, which promotes Lys63-linked ubiquitination on Lys103 residue of DUSP14 [86]. DUSP14 Lys63-linked ubiquitination is induced by methylation [87]. During TCR signaling, protein arginine methyltransferase 5 (PRMT5) interacts with DUSP14 and triggers its methylation on Arg17, Arg38, and Arg45 residues [87]. DUSP14 contains a TRAF2-binding motif, ^{27}IAQIT^{31}, which is adjacent to these methylation sites. DUSP14 methylation results in recruitment of the ubiquitin E3 ligase TRAF2, which in turn induces Lys63-linked ubiquitination on Lys103 residue of DUSP14 [86,87]. Methylation and subsequent ubiquitination stimulate the phosphatase activity of DUSP14. Taken together, methylation-induced ubiquitination of DUSP14 promotes the activation of DUSP14 phosphatase activity during TCR signaling, resulting in attenuation of T-cell activation [85–87] (Figure 1).

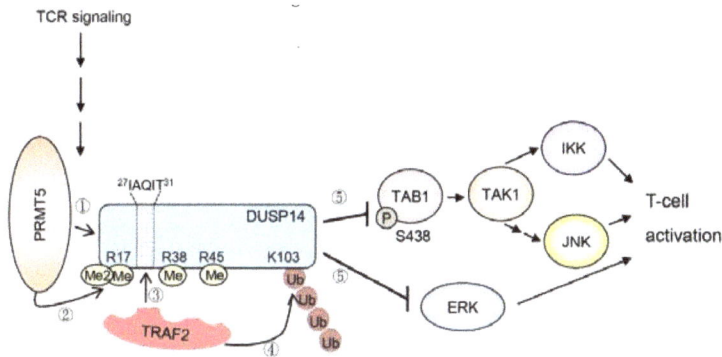

Figure 1. Upon T-cell receptor (TCR) signaling, the protein arginine methyltransferase PRMT5 interacts with DUSP14 and induces its methylation on Arg17, Arg38, and Arg45 residues. Arginine-methylated DUSP14 then interacts with the ubiquitin E3 ligase TRAF2, which binds to the motif containing IAQIT residues of DUSP14 and then promotes K63-linked ubiquitination on Lys103 residue of DUSP14. Methylation and subsequent ubiquitination enhance the phosphatase activity of DUSP14. Activated DUSP14 dephosphorylates TAB1, leading to sequential inactivation of TAK1 and downstream IKK/JNK activities. Activated DUSP14 also directly dephosphorylates ERK and attenuates the ERK signaling pathway. Arrows denote activation; T bars denote inhibition.

4. Dysregulation of DUSPs in Diseases

DUSPs are involved in immune cell homeostasis, inflammatory responses, metabolic regulation, and cancer development/progression [15,88]. For example, DUSP6 knockout mice show impaired T-cell glycolysis and increased T follicular helper cell (T$_{FH}$) differentiation [89]. DUSP6 knockout mice also show altered gut microbiome and transcriptome response against diet-induced obesity [90]. Moreover, DUSP6 downregulation is correlated with cancer progression of human pancreatic adenocarcinoma and lung cancer [91,92]. DUSP2 downregulation induces colon cancer stemness [93]. DUSP3 downregulation occurs in human non-small cell lung cancer patients [94]; consistently, DUSP3 deficiency results in enhanced cancer cell migration [95]. DUSP5 is also downregulated in human gastric and colorectal cancers [96,97]. In addition, DUSP22 knockout mice show enhanced T-cell-mediated immune responses and are more susceptible to experimental autoimmune encephalomyelitis (EAE) [98]; consistently, DUSP22 protein levels are decreased in T cells of human systemic lupus erythematosus (SLE) patients [99]. DUSP22 expression is also downregulated in human T-cell lymphoma [100,101]. Therefore, further studies of DUSP protein stability and/or ubiquitination may help understand the complex interplay between cell signaling pathways and disease pathogenesis.

5. Conclusions

The DUSP family phosphatases are key regulators of MAPK activity [13]. Because the half-lives of many DUSPs are only about 1 h, protein levels of DUSPs are tightly regulated by post-translational modifications [15]. The post-translational regulations of DUSP proteins are summarized in Figure 2. The studies for protein stability of DUSPs are summarized in Table 2.

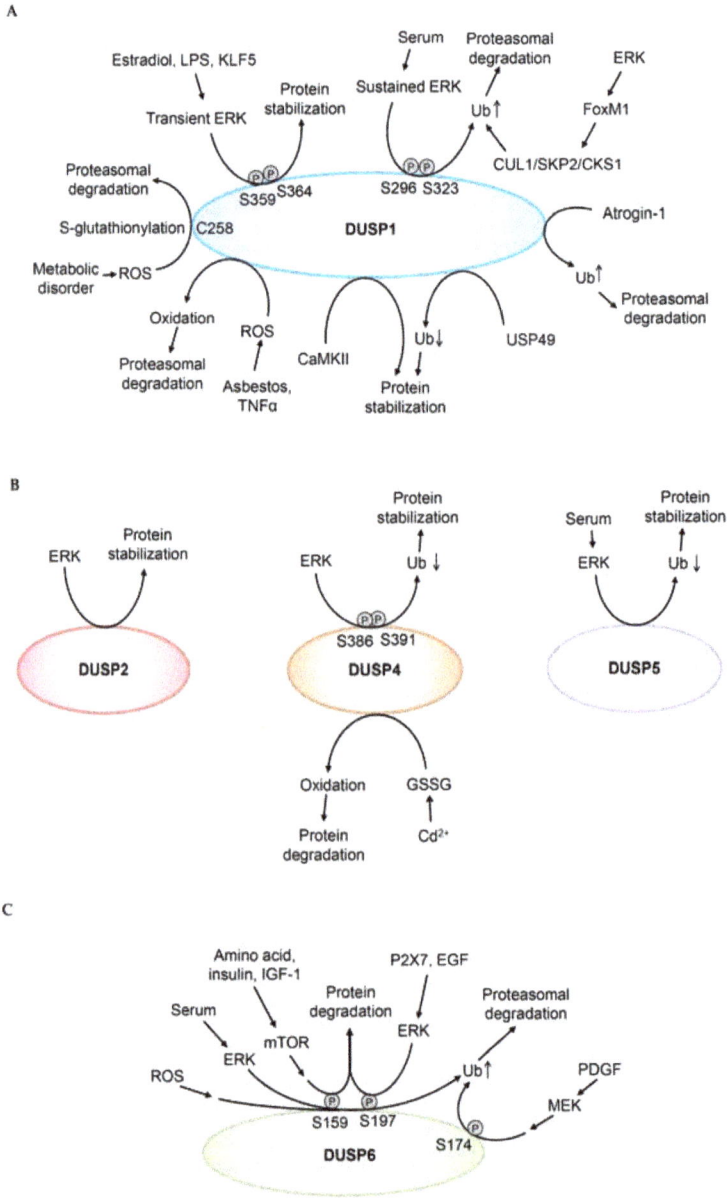

Figure 2. *Cont.*

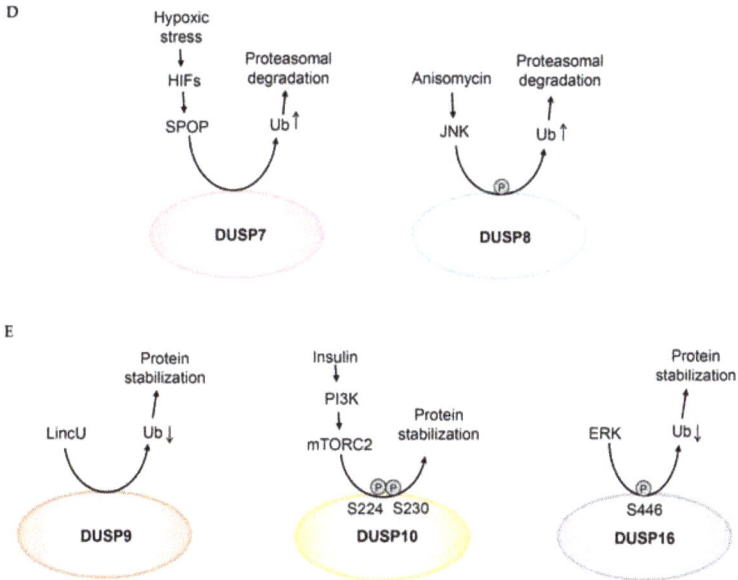

Figure 2. Post-translational modifications regulate DUSP protein stability. (**A**) Ubiquitination, oxidation, Cys258 S-glutathionylation, or Ser296/Ser323 phosphorylation of DUSP1 induces DUSP1 proteasomal degradation. Deubiquitination or phosphorylation of Ser359 and Ser364 residues enhances DUSP1 protein stability. (**B**) ERK induces DUSP2 protein stabilization; however, it is unclear whether ERK directly phosphorylates DUSP2. Phosphorylation of Ser386 and Ser391 residues of DUSP4 enhances its protein stability by inhibiting ubiquitin-mediated proteasomal degradation. Oxidation induces DUSP4 protein degradation. Reduced ubiquitination of DUSP5 enhances its protein stability. (**C**) Phosphorylation of Ser159, Ser174, or Ser197 residue induces proteasomal degradation of DUSP6. (**D**) Ubiquitination of DUSP7 or DUSP8 induces their proteasomal degradation. (**E**) Reduced ubiquitination of DUSP9 or DUSP16 enhances their protein stability. Phosphorylation of Ser224 and Ser230 residues enhances DUSP10 protein stability. Ub denotes ubiquitination of DUSPs. Oxidation indicates oxidation of DUSP1 or DUSP4.

Int. J. Mol. Sci. **2019**, *20*, 2668

Table 2. Regulation of DUSP protein stability or phosphatase activity.

	Stimuli	Stability	Modification	Modification Enzyme	Experimental Methods			Proteasome Inhibitor	Reference
					Protein Level	Half-Life	Ubiquitination		
DUSP1	Serum	↓	Phosphorylation↑ (human Ser296 †/Ser323 †); Ubiquitination↑	ERK; CUL1	✓	✓	✓	LLnL; MG132	[28,54]
	Estradiol	↑	Phosphorylation↑ (human Ser359 †/Ser364 †)	ERK	✓	✓	✓	LLnL; MG132; Lactacystin	[56]
	LPS	↑	Phosphorylation↑ (human Ser359 †/Ser364 †)	ERK	✓	✓	?	MG132; PS-341	[57]
	Pb²⁺	↓	Ubiquitination↑		✓	✓		LLnL; MG132	[27]
	Glutamate/PKCδ	↓	Ubiquitination↑		✓		✓	MG132; LLnL; Lactacystin	[31]
	Atrogin-1 upregulation	↓	Ubiquitination↑	Atrogin-1	✓	✓	✓	MG132	[33]
	USP49 upregulation	↑	Ubiquitination↓	USP49	✓		✓	MG132	[34]
	KLF5 upregulation	↑	Phosphorylation↑ (human Ser359 †/Ser364 †)	ERK	✓	✓		MG132	[58]
	LPS	↑	Phosphorylation↑ (human Ser359 †/Ser364 †)	ERK	✓	✓		MG132	[59]
	Insulin	↑	Phosphorylation↑		✓			MG132; Lactacystin	[n]
	Asbestos/ROS	↓	Oxidation↑		✓			MG132	[82]
	TNFα/ROS	↓	Oxidation↑		✓			MG132	[79]
	ROS	↓	S-gluatathionylation↑ (human Cys258 †)		✓			MG132	[81]
	Glucocorticoid	↑			✓			MG132; LLnL	[84]
	EGF plus Lactoferrin	↓			✓			MG132	[32]
	Angiotensin II/PKA	↓			✓			Bortezomib	[15]
	CaMKII inhibition	↓			✓	✓		MG132	[n]
	Luteolin/Superoxide	↓			✓	✓		MG132	[80]
DUSP2	ERK4	↑			✓	✓		MG132	[6]
DUSP4	LPS	↑	Phosphorylation↑ (human Ser386 †/Ser391 †)	ERK	✓	✓	?	MG132; PS-341	[57]
	ERK inhibitor	↓	Phosphorylation↓ (human Ser386 †/Ser391 †); Ubiquitination↑	ERK	✓	✓	✓	MG132	[6]
	Cd²⁺ / Oxidative stress	↓	Oxidation↑	GSSG	✓				[X]
	8-Bromo-cAMP	↑			✓	✓		MG132	[28]
	Senescence	↑			✓	✓		MG132	[37]
DUSP5	ERK2 binding	↑	Ubiquitination↓		✓	✓	✓	MG132	[39]

Table 2. *Cont.*

	Stimuli	Stability	Modification	Modification Enzyme	Experimental Methods				Reference
					Protein Level	Half-Life	Ubiquitination	Proteasome Inhibitor	
DUSP6	Serum	↓	Phosphorylation↑ (human Ser159 †/Ser197 †); Ubiquitination↑	ERK	✓	✓	✓	LLnL; Lactacystin	[70,72]
	ROS	↓	Phosphorylation↑ (human Ser159 †/Ser197 †); Ubiquitination↑		✓		✓	MG132	[43]
	PDGF	↓	Phosphorylation↑ (human Ser174 †); Ubiquitination↑		✓	✓	✓	MG132	[74]
	P2X7 nucleotide, EGF	↓	Phosphorylation↑ (human Ser197 †)	ERK	✓	✓		MG132	[71]
	Insulin	↓	Phosphorylation↑ (human Ser159 †/Ser197 †)	ERK	✓	✓			[72]
	Amino acid, insulin, IGF-1/mTOR	↓	Phosphorylation↑ (human Ser159 †)		✓	✓			[73]
	PKCδ downregulation	↑			✓			MG132	[46]
	TSH	↑			✓	✓		MG132	[44]
	Metformin/AMP-activated protein kinase	↓			✓			MG132	[45]
	EGF plus Lactoferrin	↓			✓			MG132	[32]
DUSP7	Hypoxic stress/HIFs	↓	Ubiquitination↑	SPOP	✓		✓	MG132	[47]
DUSP8	Anisomycin	↓	Phosphorylation↑; Ubiquitination↑	JNK	✓	✓	?	Lactacystin	[52]
DUSP9	LincU upregulation	↑	Ubiquitination↓		✓	✓	✓	MG132	[53]
DUSP10	Insulin	↑	Phosphorylation↑ (human Ser224 †/Ser230 †)	mTORC2	✓	✓			[75]
DUSP14	TCR signaling	(Activity↑)	Methylation↑ (human Arg17 †/Arg38 †/Arg45); Ubiquitination↑ (human Lys103 †)	PRMT5; TRAF2	✓		✓		[86,87]
DUSP16	ERK upregulation	↑	Phosphorylation↑ (human Ser446 †); Ubiquitination↓	ERK	✓	✓	✓	MG132; MG115	[77,78]

† denotes the amino acid residue is conserved in both human and mouse proteins. LPS denotes lipopolysaccharides. TSH denotes thyroid-stimulating hormone. GSSG denotes glutathione disulfide.

The protein stability of DUSP1, DUSP4, DUSP5, DUSP6, DUSP7, DUSP8, DUSP9, and DUSP16 are regulated by ubiquitin-mediated proteasomal degradation. Protein stability of DUSP2 and DUSP10 is increased by ERK and mTORC2, respectively [65,75]; however, it is unclear whether ubiquitination is involved in the degradation of these two phosphatases. To date, only three ubiquitin E3 ligases and one deubiquitinase (also named ubiquitin-specific peptidase (USP)) for degradation of DUSPs have been identified. The ubiquitin E3 ligases CUL1 and Atrogin-1 are responsible for DUSP1 ubiquitination; the ubiquitin E3 ligase SPOP is responsible for DUSP7 ubiquitination [33,47,54]. The deubiquitinase for DUSP1 is USP49 [34]. Additional ubiquitin E3 ligases and deubiquitinases for controlling proteasomal degradation of DUSPs await to be identified. The ubiquitination/proteasomal degradation of DUSPs are usually regulated by phosphorylation. Although MAPKs are dephosphorylated by DUSPs, MAPKs also reciprocally control protein stability of DUSPs and their downstream signaling pathways. One major kinase for DUSPs is ERK. Transient activation of ERK phosphorylates DUSP1 on Ser359 and Ser364 residues to stabilize DUSP1, providing a negative feedback that attenuates ERK activity [56]. In contrast, sustained ERK activity induces DUSP1 Ser296/323 phosphorylation and subsequent protein degradation, resulting in further enhancement of ERK signaling [27,54]. Consistently, both sustained ERK activation and decreased DUSP1 protein levels are observed in cancer cells [102,103]. Moreover, ERK phosphorylates DUSP6 on Ser159 and Ser197 residues, facilitating DUSP6 proteasomal degradation [70]. The MEK/ERK pathway also mediates DUSP6 Ser174 phosphorylation and subsequently induces DUSP6 degradation [74]. Unlike the inhibitory effect of ERK on DUSP6, ERK increases DUSP4 protein stability by phosphorylating DUSP4 on Ser386 and Ser391 residues [57]. Enhancement of DUSP4 protein levels leads to inhibition of ERK activity. Similarly, ERK protects DUSP16 from proteasomal degradation by phosphorylating DUSP16 on Ser446 residue [77]. DUSP16 preferentially dephosphorylates JNK and maybe p38 compared to ERK [17,76], suggesting that ERK-mediated DUSP16 induction negatively regulates the activation of JNK or maybe p38. Besides ERK, mTOR also regulates protein stability of DUSPs. mTORC2 phosphorylates DUSP10 on Ser224 and Ser230 residues and subsequently enhances DUSP10 protein stability, leading to reduction of p38 activity [75]. Moreover, mTOR signaling also induces DUSP6 Ser159 phosphorylation, resulting in proteasomal degradation of DUSP6 [73].

In addition to phosphorylation, oxidation of DUSPs also induces protein degradation. Asbestos or TNFα stimulation generates ROS that oxidizes DUSP1 and induces DUSP1 proteasomal degradation [79,82]. Cd^{2+}-induced oxidative stress triggers DUSP4 oxidation by glutathione disulfide (GSSG), resulting in protein degradation of DUSP4 [83].

Ubiquitination not only controls protein stability but also induces protein activity. The ubiquitin E3 ligase TRAF2-mediated Lys63-linked ubiquitination of DUSP14 enhances its phosphatase activity. It would be interesting to study whether Lys63-linked ubiquitination also regulates the activity of other DUSPs.

To our knowledge, eight DUSPs (DUSP1, DUSP4, DUSP5, DUSP6, DUSP7, DUSP8, DUSP9, and DUSP16) are known to be degraded by the proteasome, and it is important to understand whether other members of the DUSP family are also regulated by ubiquitin-mediated proteasomal degradation. Lastly, it would be useful to study whether other modifications, such as sumoylation, neddylation, acetylation, and glycosylation also control DUSP ubiquitination and protein stability.

Author Contributions: Conceptualization, T.-H.T.; writing—original draft preparation, H.-F.C. and H.-C.C.; writing—review and editing, H.-C.C. and T.-H.T.; supervision, H.-C.C. and T.-H.T.

Funding: This work was supported by grants from the National Health Research Institutes, Taiwan (IM-107-PP-01 and IM-107-SP-01, to T.-H.T.) and the Ministry of Science and Technology, Taiwan (107-2314-B-400-027 and 107-2321-B-400-013 to T.-H.T.; 107-2628-B-400-001 to H.-C.C.).

Acknowledgments: T.-H.T. is a recipient of the Taiwan Bio-Development Foundation (TBF) Chair in Biotechnology.

Int. J. Mol. Sci. **2019**, *20*, 2668

Abbreviations

Ub	ubiquitination
DUSP	dual-specificity phosphatase
MKP	MAP kinase phosphatase
MAPK	mitogen-activated protein kinase
KIM	kinase-interacting motif
Skp2	S-phase kinase-associated protein
Cks1	cyclin-dependent kinase regulatory subunit 1
FoxM1	forkhead box M1
STAT1	signal transducer and activator of transcription 1
LncRNA	long non-coding RNA
EGF	epidermal growth factor
PRMT5	protein arginine methyltransferase 5
T_{FH}	T follicular helper cell
EAE	experimental autoimmune encephalomyelitis
SLE	systemic lupus erythematosus
SPOP	Speckle-type POZ protein
PDGF-BB	platelet-derived growth factor-B chains

References

1. Widmann, C.; Gibson, S.; Jarpe, M.B.; Johnson, G.L. Mitogen-activated protein kinase: Conservation of a three-kinase module from yeast to human. *Physiol. Rev.* **1999**, *79*, 143–180. [CrossRef]
2. Raman, M.; Chen, W.; Cobb, M.H. Differential regulation and properties of MAPKs. *Oncogene* **2007**, *26*, 3100–3112. [CrossRef] [PubMed]
3. Chen, Y.R.; Meyer, C.F.; Tan, T.H. Persistent activation of c-Jun N-terminal kinase 1 (JNK1) in γ radiation-induced apoptosis. *J. Biol. Chem.* **1996**, *271*, 631–634. [CrossRef]
4. Chen, Y.R.; Wang, X.; Templeton, D.; Davis, R.J.; Tan, T.H. The role of c-Jun N-terminal kinase (JNK) in apoptosis induced by ultraviolet C and γ radiation: Duration of JNK activation may determine cell death and proliferation. *J. Biol. Chem.* **1996**, *271*, 31929–31936. [CrossRef]
5. MacCorkle, R.A.; Tan, T.H. Mitogen-activated protein kinases in cell-cycle control. *Cell Biochem. Biophys.* **2005**, *43*, 451–461. [CrossRef]
6. Cargnello, M.; Roux, P.P. Activation and function of the MAPKs and their substrates, the MAPK-activated protein kinases. *Microbiol. Mol. Biol. Rev.* **2011**, *75*, 50–83. [CrossRef] [PubMed]
7. Chen, Y.R.; Tan, T.H. The c-Jun N-terminal kinase pathway and apoptotic signaling. *Int. J. Oncol.* **2000**, *16*, 651–662. [CrossRef]
8. Marshall, C.J. MAP kinase kinase kinase, MAP kinase kinase and MAP kinase. *Curr. Opin. Genet. Dev.* **1994**, *4*, 82–89. [CrossRef]
9. Wada, T.; Penninger, J.M. Mitogen-activated protein kinases in apoptosis regulation. *Oncogene* **2004**, *23*, 2838–2849. [CrossRef]
10. Turjanski, A.G.; Vaque, J.P.; Gutkind, J.S. MAP kinases and the control of nuclear events. *Oncogene* **2007**, *26*, 3240–3253. [CrossRef]
11. Kyriakis, J.M.; Avruch, J. Mammalian mitogen-activated protein kinase signal transduction pathways activated by stress and inflammation. *Physiol. Rev.* **2001**, *81*, 807–869. [CrossRef]
12. Dhillon, A.S.; Hagan, S.; Rath, O.; Kolch, W. MAP kinase signalling pathways in cancer. *Oncogene* **2007**, *26*, 3279–3290. [CrossRef]
13. Caunt, C.J.; Keyse, S.M. Dual-specificity MAP kinase phosphatases (MKPs): Shaping the outcome of MAP kinase signalling. *FASEB J.* **2013**, *280*, 489–504. [CrossRef]
14. Farooq, A.; Zhou, M.M. Structure and regulation of MAPK phosphatases. *Cell. Signal.* **2004**, *16*, 769–779. [CrossRef]
15. Huang, C.Y.; Tan, T.H. DUSPs, to MAP kinases and beyond. *Cell Biosci.* **2012**, *2*, 24. [CrossRef]

16. Mandl, M.; Slack, D.N.; Keyse, S.M. Specific inactivation and nuclear anchoring of extracellular signal-regulated kinase 2 by the inducible dual-specificity protein phosphatase DUSP5. *Mol. Cell. Biol.* **2005**, *25*, 1830–1845. [CrossRef]

17. Masuda, K.; Shima, H.; Watanabe, M.; Kikuchi, K. MKP-7, a novel mitogen-activated protein kinase phosphatase, functions as a shuttle protein. *J. Biol. Chem.* **2001**, *276*, 39002–39011. [CrossRef]

18. Karlsson, M.; Mathers, J.; Dickinson, R.J.; Mandl, M.; Keyse, S.M. Both nuclear-cytoplasmic shuttling of the dual specificity phosphatase MKP-3 and its ability to anchor MAP kinase in the cytoplasm are mediated by a conserved nuclear export signal. *J. Biol. Chem.* **2004**, *279*, 41882–41891. [CrossRef] [PubMed]

19. Tanoue, T.; Adachi, M.; Moriguchi, T.; Nishida, E. A conserved docking motif in MAP kinases common to substrates, activators and regulators. *Nat. Cell Biol.* **2000**, *2*, 110–116. [CrossRef]

20. Welchman, R.L.; Gordon, C.; Mayer, R.J. Ubiquitin and ubiquitin-like proteins as multifunctional signals. *Nat. Rev. Mol. Cell Biol.* **2005**, *6*, 599–609. [CrossRef] [PubMed]

21. Dikic, I.; Wakatsuki, S.; Walters, K.J. Ubiquitin-binding domains -from structures to functions. *Nat. Rev. Mol. Cell Biol.* **2009**, *10*, 659–671. [CrossRef]

22. Malynn, B.A.; Ma, A. Ubiquitin makes its mark on immune regulation. *Immunity* **2010**, *33*, 843–852. [CrossRef]

23. Kwon, Y.T.; Ciechanover, A. The ubiquitin code in the ubiquitin-proteasome system and autophagy. *Trends Biochem. Sci.* **2017**, *42*, 873–886. [CrossRef] [PubMed]

24. Slack, D.N.; Seternes, O.M.; Gabrielsen, M.; Keyse, S.M. Distinct binding determinants for ERK2/p38α and JNK map kinases mediate catalytic activation and substrate selectivity of map kinase phosphatase-1. *J. Biol. Chem.* **2001**, *276*, 16491–16500. [CrossRef] [PubMed]

25. Patterson, K.I.; Brummer, T.; O'Brien, P.M.; Daly, R.J. Dual-specificity phosphatases: Critical regulators with diverse cellular targets. *Biochem. J.* **2009**, *418*, 475–489. [CrossRef]

26. Ma, R.Y.; Tong, T.H.; Cheung, A.M.; Tsang, A.C.; Leung, W.Y.; Yao, K.M. Raf/MEK/MAPK signaling stimulates the nuclear translocation and transactivating activity of FOXM1c. *J. Cell Sci.* **2005**, *118*, 795–806. [CrossRef] [PubMed]

27. Lin, Y.W.; Chuang, S.M.; Yang, J.L. ERK1/2 achieves sustained activation by stimulating MAPK phosphatase-1 degradation via the ubiquitin-proteasome pathway. *J. Biol. Chem.* **2003**, *278*, 21534–21541. [CrossRef] [PubMed]

28. Calvisi, D.F.; Pinna, F.; Meloni, F.; Ladu, S.; Pellegrino, R.; Sini, M.; Daino, L.; Simile, M.M.; De Miglio, M.R.; Virdis, P.; et al. Dual-specificity phosphatase 1 ubiquitination in extracellular signal-regulated kinase-mediated control of growth in human hepatocellular carcinoma. *Cancer Res.* **2008**, *68*, 4192–4200. [CrossRef] [PubMed]

29. Major, M.L.; Lepe, R.; Costa, R.H. Forkhead box M1B transcriptional activity requires binding of Cdk-cyclin complexes for phosphorylation-dependent recruitment of p300/CBP coactivators. *Mol. Cell. Biol.* **2004**, *24*, 2649–2661. [CrossRef]

30. Wang, I.C.; Chen, Y.J.; Hughes, D.; Petrovic, V.; Major, M.L.; Park, H.J.; Tan, Y.; Ackerson, T.; Costa, R.H. Forkhead box M1 regulates the transcriptional network of genes essential for mitotic progression and genes encoding the SCF (Skp2-Cks1) ubiquitin ligase. *Mol. Cell. Biol.* **2005**, *25*, 10875–10894. [CrossRef] [PubMed]

31. Choi, B.H.; Hur, E.M.; Lee, J.H.; Jun, D.J.; Kim, K.T. Protein kinase Cδ-mediated proteasomal degradation of MAP kinase phosphatase-1 contributes to glutamate-induced neuronal cell death. *J. Cell Sci.* **2006**, *119*, 1329–1340. [CrossRef] [PubMed]

32. Geetha, N.; Mihaly, J.; Stockenhuber, A.; Blasi, F.; Uhrin, P.; Binder, B.R.; Freissmuth, M.; Breuss, J.M. Signal integration and coincidence detection in the mitogen-activated protein kinase/extracellular signal-regulated kinase (ERK) cascade: Concomitant activation of receptor tyrosine kinases and of LRP-1 leads to sustained ERK phosphorylation via down-regulation of dual specificity phosphatases (DUSP1 and -6). *J. Biol. Chem.* **2011**, *286*, 25663–25674. [PubMed]

33. Xie, P.; Guo, S.; Fan, Y.; Zhang, H.; Gu, D.; Li, H. Atrogin-1/MAFbx enhances simulated ischemia/reperfusion-induced apoptosis in cardiomyocytes through degradation of MAPK phosphatase-1 and sustained JNK activation. *J. Biol. Chem.* **2009**, *284*, 5488–5496. [CrossRef] [PubMed]

34. Zhang, W.; Zhang, Y.; Zhang, H.; Zhao, Q.; Liu, Z.; Xu, Y. USP49 inhibits ischemia-reperfusion-induced cell viability suppression and apoptosis in human AC16 cardiomyocytes through DUSP1-JNK1/2 signaling. *J. Cell. Physiol.* **2019**, *234*, 6529–6538. [CrossRef]

35. Qin, X.Y.; Zhang, Y.L.; Chi, Y.F.; Yan, B.; Zeng, X.J.; Li, H.H.; Liu, Y. Angiotensin II regulates Th1 T cell differentiation through angiotensin II type 1 receptor-PKA-mediated activation of proteasome. *Cell. Physiol. Biochem.* **2018**, *45*, 1366–1376. [CrossRef]

36. Chen, P.; Hutter, D.; Yang, X.; Gorospe, M.; Davis, R.J.; Liu, Y. Discordance between the binding affinity of mitogen-activated protein kinase subfamily members for MAP kinase phosphatase-2 and their ability to activate the phosphatase catalytically. *J. Biol. Chem.* **2001**, *276*, 29440–29449. [CrossRef]

37. Torres, C.; Francis, M.K.; Lorenzini, A.; Tresini, M.; Cristofalo, V.J. Metabolic stabilization of MAP kinase phosphatase-2 in senescence of human fibroblasts. *Exp. Cell Res.* **2003**, *290*, 195–206. [CrossRef]

38. Gomez, N.V.; Gorostizaga, A.B.; Mori Sequeiros Garcia, M.M.; Brion, L.; Acquier, A.; Gonzalez-Calvar, S.I.; Mendez, C.F.; Podesta, E.J.; Paz, C. MAPK phosphatase-2 (MKP-2) is induced by hCG and plays a role in the regulation of CYP11A1 expression in MA-10 Leydig cells. *Endocrinology* **2013**, *154*, 1488–1500. [CrossRef]

39. Kucharska, A.; Rushworth, L.K.; Staples, C.; Morrice, N.A.; Keyse, S.M. Regulation of the inducible nuclear dual-specificity phosphatase DUSP5 by ERK MAPK. *Cell. Signal.* **2009**, *21*, 1794–1805. [CrossRef]

40. Kidger, A.M.; Rushworth, L.K.; Stellzig, J.; Davidson, J.; Bryant, C.J.; Bayley, C.; Caddye, E.; Rogers, T.; Keyse, S.M.; Caunt, C.J. Dual-specificity phosphatase 5 controls the localized inhibition, propagation, and transforming potential of ERK signaling. *Proc. Natl. Acad. Sci. USA* **2017**, *114*, E317–E326. [CrossRef]

41. Groom, L.A.; Sneddon, A.A.; Alessi, D.R.; Dowd, S.; Keyse, S.M. Differential regulation of the MAP, SAP and RK/p38 kinases by Pyst1, a novel cytosolic dual-specificity phosphatase. *EMBO J.* **1996**, *15*, 3621–3632. [CrossRef] [PubMed]

42. Muda, M.; Boschert, U.; Dickinson, R.; Martinou, J.C.; Martinou, I.; Camps, M.; Schlegel, W.; Arkinstall, S. MKP-3, a novel cytosolic protein-tyrosine phosphatase that exemplifies a new class of mitogen-activated protein kinase phosphatase. *J. Biol. Chem.* **1996**, *271*, 4319–4326. [CrossRef]

43. Chan, D.W.; Liu, V.W.; Tsao, G.S.; Yao, K.M.; Furukawa, T.; Chan, K.K.; Ngan, H.Y. Loss of MKP3 mediated by oxidative stress enhances tumorigenicity and chemoresistance of ovarian cancer cells. *Carcinogenesis* **2008**, *29*, 1742–1750. [CrossRef] [PubMed]

44. Kim, Y.H.; Choi, Y.W.; Han, J.H.; Lee, J.; Soh, E.Y.; Park, S.H.; Kim, J.H.; Park, T.J. TSH signaling overcomes B-RafV600E-induced senescence in papillary thyroid carcinogenesis through regulation of DUSP6. *Neoplasia* **2014**, *16*, 1107–1120. [CrossRef]

45. Martin, M.J.; Hayward, R.; Viros, A.; Marais, R. Metformin accelerates the growth of BRAF V600E-driven melanoma by upregulating VEGF-A. *Cancer Discov.* **2012**, *2*, 344–355. [CrossRef]

46. Lonne, G.K.; Masoumi, K.C.; Lennartsson, J.; Larsson, C. Protein kinase Cδ supports survival of MDA-MB-231 breast cancer cells by suppressing the ERK1/2 pathway. *J. Biol. Chem.* **2009**, *284*, 33456–33465. [CrossRef] [PubMed]

47. Li, G.; Ci, W.; Karmakar, S.; Chen, K.; Dhar, R.; Fan, Z.; Guo, Z.; Zhang, J.; Ke, Y.; Wang, L.; et al. SPOP promotes tumorigenesis by acting as a key regulatory hub in kidney cancer. *Cancer Cell* **2014**, *25*, 455–468. [CrossRef]

48. Cotsiki, M.; Oehrl, W.; Samiotaki, M.; Theodosiou, A.; Panayotou, G. Phosphorylation of the M3/6 dual-specificity phosphatase enhances the activation of JNK by arsenite. *Cell. Signal.* **2012**, *24*, 664–676. [CrossRef]

49. Muda, M.; Theodosiou, A.; Rodrigues, N.; Boschert, U.; Camps, M.; Gillieron, C.; Davies, K.; Ashworth, A.; Arkinstall, S. The dual specificity phosphatases M3/6 and MKP-3 are highly selective for inactivation of distinct mitogen-activated protein kinases. *J. Biol. Chem.* **1996**, *271*, 27205–27208. [CrossRef] [PubMed]

50. Chen, Y.R.; Shrivastava, A.; Tan, T.H. Down-regulation of the c-Jun N-terminal kinase (JNK) phosphatase M3/6 and activation of JNK by hydrogen peroxide and pyrrolidine dithiocarbamate. *Oncogene* **2001**, *20*, 367–374. [CrossRef] [PubMed]

51. Barbacid, M.; Vazquez, D. (^{3}H)anisomycin binding to eukaryotic ribosomes. *J. Mol. Biol.* **1974**, *84*, 603–623. [CrossRef]

52. Theodosiou, A.; Ashworth, A. Differential effects of stress stimuli on a JNK-inactivating phosphatase. *Oncogene* **2002**, *21*, 2387–2397. [CrossRef]

53. Jiapaer, Z.; Li, G.; Ye, D.; Bai, M.; Li, J.; Guo, X.; Du, Y.; Su, D.; Jia, W.; Chen, W.; et al. LincU preserves naive pluripotency by restricting ERK activity in embryonic stem cells. *Stem Cell Rep.* **2018**, *11*, 395–409. [CrossRef]

54. Lin, Y.W.; Yang, J.L. Cooperation of ERK and SCF^{Skp2} for MKP-1 destruction provides a positive feedback regulation of proliferating signaling. *J. Biol. Chem.* **2006**, *281*, 915–926. [CrossRef]

55. Moosavi, S.M.; Prabhala, P.; Ammit, A.J. Role and regulation of MKP-1 in airway inflammation. *Respir. Res.* **2017**, *18*, 154. [CrossRef]

56. Brondello, J.M.; Pouyssegur, J.; McKenzie, F.R. Reduced MAP kinase phosphatase-1 degradation after p42/p44MAPK-dependent phosphorylation. *Science* **1999**, *286*, 2514–2517. [CrossRef]

57. Crowell, S.; Wancket, L.M.; Shakibi, Y.; Xu, P.; Xue, J.; Samavati, L.; Nelin, L.D.; Liu, Y. Post-translational regulation of mitogen-activated protein kinase phosphatase (MKP)-1 and MKP-2 in macrophages following lipopolysaccharide stimulation: The role of the C termini of the phosphatases in determining their stability. *J. Biol. Chem.* **2014**, *289*, 28753–28764. [CrossRef]

58. Liu, R.; Zheng, H.Q.; Zhou, Z.; Dong, J.T.; Chen, C. KLF5 promotes breast cell survival partially through fibroblast growth factor-binding protein 1-pERK-mediated dual specificity MKP-1 protein phosphorylation and stabilization. *J. Biol. Chem.* **2009**, *284*, 16791–16798. [CrossRef]

59. Chen, P.; Li, J.; Barnes, J.; Kokkonen, G.C.; Lee, J.C.; Liu, Y. Restraint of proinflammatory cytokine biosynthesis by mitogen-activated protein kinase phosphatase-1 in lipopolysaccharide-stimulated macrophages. *J. Immunol.* **2002**, *169*, 6408–6416. [CrossRef]

60. Kassel, O.; Sancono, A.; Kratzschmar, J.; Kreft, B.; Stassen, M.; Cato, A.C. Glucocorticoids inhibit MAP kinase via increased expression and decreased degradation of MKP-1. *EMBO J.* **2001**, *20*, 7108–7116. [CrossRef]

61. Swantek, J.L.; Cobb, M.H.; Geppert, T.D. Jun N-terminal kinase/stress-activated protein kinase (JNK/SAPK) is required for lipopolysaccharide stimulation of tumor necrosis factor α (TNF-α) translation: Glucocorticoids inhibit TNF-α translation by blocking JNK/SAPK. *Mol. Cell. Biol.* **1997**, *17*, 6274–6282. [CrossRef]

62. Kontoyiannis, D.; Pasparakis, M.; Pizarro, T.T.; Cominelli, F.; Kollias, G. Impaired on/off regulation of TNF biosynthesis in mice lacking TNF AU-rich elements: Implications for joint and gut-associated immunopathologies. *Immunity* **1999**, *10*, 387–398. [CrossRef]

63. Jacob, A.; Smolenski, A.; Lohmann, S.M.; Begum, N. MKP-1 expression and stabilization and cGK Iα prevent diabetes-associated abnormalities in VSMC migration. *Am. J. Physiol. Cell Physiol.* **2004**, *287*, C1077–C1086. [CrossRef]

64. Ciccarelli, M.; Rusciano, M.R.; Sorriento, D.; Basilicata, M.F.; Santulli, G.; Campiglia, P.; Bertamino, A.; De Luca, N.; Trimarco, B.; Iaccarino, G.; et al. CaMKII protects MKP-1 from proteasome degradation in endothelial cells. *Cell. Signal.* **2014**, *26*, 2167–2174. [CrossRef]

65. Perander, M.; Al-Mahdi, R.; Jensen, T.C.; Nunn, J.A.; Kildalsen, H.; Johansen, B.; Gabrielsen, M.; Keyse, S.M.; Seternes, O.M. Regulation of atypical MAP kinases ERK3 and ERK4 by the phosphatase DUSP2. *Sci. Rep.* **2017**, *7*, 43471. [CrossRef]

66. Brondello, J.M.; Brunet, A.; Pouyssegur, J.; McKenzie, F.R. The dual specificity mitogen-activated protein kinase phosphatase-1 and -2 are induced by the p42/p44MAPK cascade. *J. Biol. Chem.* **1997**, *272*, 1368–1376. [CrossRef]

67. Peng, D.J.; Zhou, J.Y.; Wu, G.S. Post-translational regulation of mitogen-activated protein kinase phosphatase-2 (MKP-2) by ERK. *Cell Cycle* **2010**, *9*, 4650–4655.

68. Hijiya, N.; Tsukamoto, Y.; Nakada, C.; Tung Nguyen, L.; Kai, T.; Matsuura, K.; Shibata, K.; Inomata, M.; Uchida, T.; Tokunaga, A.; et al. Genomic loss of DUSP4 contributes to the progression of intraepithelial neoplasm of pancreas to invasive carcinoma. *Cancer Res.* **2016**, *76*, 2612–2625. [CrossRef]

69. Cadalbert, L.C.; Sloss, C.M.; Cunningham, M.R.; Al-Mutairi, M.; McIntire, A.; Shipley, J.; Plevin, R. Differential regulation of MAP kinase activation by a novel splice variant of human MAP kinase phosphatase-2. *Cell. Signal.* **2010**, *22*, 357–365. [CrossRef]

70. Marchetti, S.; Gimond, C.; Chambard, J.C.; Touboul, T.; Roux, D.; Pouyssegur, J.; Pages, G. Extracellular signal-regulated kinases phosphorylate mitogen-activated protein kinase phosphatase 3/DUSP6 at serines 159 and 197, two sites critical for its proteasomal degradation. *Mol. Cell. Biol.* **2005**, *25*, 854–864. [CrossRef]

71. Queipo, M.J.; Gil-Redondo, J.C.; Morente, V.; Ortega, F.; Miras-Portugal, M.T.; Delicado, E.G.; Perez-Sen, R. P2X7 nucleotide and EGF receptors exert dual modulation of the dual-specificity phosphatase 6 (MKP-3) in granule neurons and astrocytes, contributing to negative feedback on ERK signaling. *Front. Mol. Neurosci.* **2017**, *10*, 448. [CrossRef]

72. Feng, B.; Jiao, P.; Yang, Z.; Xu, H. MEK/ERK pathway mediates insulin-promoted degradation of MKP-3 protein in liver cells. *Mol. Cell. Endocrinol.* **2012**, *361*, 116–123. [CrossRef]

73. Bermudez, O.; Marchetti, S.; Pages, G.; Gimond, C. Post-translational regulation of the ERK phosphatase DUSP6/MKP3 by the mTOR pathway. *Oncogene* **2008**, *27*, 3685–3691. [CrossRef]

74. Jurek, A.; Amagasaki, K.; Gembarska, A.; Heldin, C.H.; Lennartsson, J. Negative and positive regulation of MAPK phosphatase 3 controls platelet-derived growth factor-induced Erk activation. *J. Biol. Chem.* **2009**, *284*, 4626–4634. [CrossRef]

75. Benavides-Serrato, A.; Anderson, L.; Holmes, B.; Cloninger, C.; Artinian, N.; Bashir, T.; Gera, J. mTORC2 modulates feedback regulation of p38 MAPK activity via DUSP10/MKP5 to confer differential responses to PP242 in glioblastoma. *Genes Cancer* **2014**, *5*, 393–406.

76. Tanoue, T.; Yamamoto, T.; Maeda, R.; Nishida, E. A novel MAPK phosphatase MKP-7 acts preferentially on JNK/SAPK and p38 α and β MAPKs. *J. Biol. Chem.* **2001**, *276*, 26629–26639. [CrossRef]

77. Katagiri, C.; Masuda, K.; Urano, T.; Yamashita, K.; Araki, Y.; Kikuchi, K.; Shima, H. Phosphorylation of Ser-446 determines stability of MKP-7. *J. Biol. Chem.* **2005**, *280*, 14716–14722. [CrossRef]

78. Masuda, K.; Shima, H.; Katagiri, C.; Kikuchi, K. Activation of ERK induces phosphorylation of MAPK phosphatase-7, a JNK specific phosphatase, at Ser-446. *J. Biol. Chem.* **2003**, *278*, 32448–32456. [CrossRef]

79. Kamata, H.; Honda, S.; Maeda, S.; Chang, L.; Hirata, H.; Karin, M. Reactive oxygen species promote TNFα-induced death and sustained JNK activation by inhibiting MAP kinase phosphatases. *Cell* **2005**, *120*, 649–661. [CrossRef]

80. Bai, L.; Xu, X.; Wang, Q.; Xu, S.; Ju, W.; Wang, X.; Chen, W.; He, W.; Tang, H.; Lin, Y. A superoxide-mediated mitogen-activated protein kinase phosphatase-1 degradation and c-Jun NH_2-terminal kinase activation pathway for luteolin-induced lung cancer cytotoxicity. *Mol. Pharmacol.* **2012**, *81*, 549–555. [CrossRef]

81. Kim, H.S.; Ullevig, S.L.; Zamora, D.; Lee, C.F.; Asmis, R. Redox regulation of MAPK phosphatase 1 controls monocyte migration and macrophage recruitment. *Proc. Natl. Acad. Sci. USA* **2012**, *109*, E2803–E2812. [CrossRef]

82. Tephly, L.A.; Carter, A.B. Differential expression and oxidation of MKP-1 modulates TNF-α gene expression. *Am. J. Respir. Cell Mol. Biol.* **2007**, *37*, 366–374. [CrossRef]

83. Barajas-Espinosa, A.; Basye, A.; Jesse, E.; Yan, H.; Quan, D.; Chen, C.A. Redox activation of DUSP4 by N-acetylcysteine protects endothelial cells from Cd^{2+}-induced apoptosis. *Free Radic. Biol. Med.* **2014**, *74*, 188–199. [CrossRef]

84. Marti, F.; Krause, A.; Post, N.H.; Lyddane, C.; Dupont, B.; Sadelain, M.; King, P.D. Negative-feedback regulation of CD28 costimulation by a novel mitogen-activated protein kinase phosphatase, MKP6. *J. Immunol.* **2001**, *166*, 197–206. [CrossRef]

85. Yang, C.Y.; Li, J.P.; Chiu, L.L.; Lan, J.L.; Chen, D.Y.; Chuang, H.C.; Huang, C.Y.; Tan, T.H. Dual-specificity phosphatase 14 (DUSP14/MKP6) negatively regulates TCR signaling by inhibiting TAB1 activation. *J. Immunol.* **2014**, *192*, 1547–1557. [CrossRef]

86. Yang, C.Y.; Chiu, L.L.; Tan, T.H. TRAF2-mediated Lys63-linked ubiquitination of DUSP14/MKP6 is essential for its phosphatase activity. *Cell. Signal.* **2016**, *28*, 145–151. [CrossRef]

87. Yang, C.Y.; Chiu, L.L.; Chang, C.C.; Chuang, H.C.; Tan, T.H. Induction of DUSP14 ubiquitination by PRMT5-mediated arginine methylation. *FASEB J.* **2018**, *32*, 6760–6770. [CrossRef]

88. Bermudez, O.; Pages, G.; Gimond, C. The dual-specificity MAP kinase phosphatases: Critical roles in development and cancer. *Am. J. Physiol. Cell Physiol.* **2010**, *299*, C189–C202. [CrossRef]

89. Hsu, W.C.; Chen, M.Y.; Hsu, S.C.; Huang, L.R.; Kao, C.Y.; Cheng, W.H.; Pan, C.H.; Wu, M.S.; Yu, G.Y.; Hung, M.S.; et al. DUSP6 mediates T cell receptor-engaged glycolysis and restrains T_{FH} cell differentiation. *Proc. Natl. Acad. Sci. USA* **2018**, *115*, E8027–E8036. [CrossRef]

90. Ruan, J.W.; Statt, S.; Huang, C.T.; Tsai, Y.T.; Kuo, C.C.; Chan, H.L.; Liao, Y.C.; Tan, T.H.; Kao, C.Y. Dual-specificity phosphatase 6 deficiency regulates gut microbiome and transcriptome response against diet-induced obesity in mice. *Nat. Microbiol.* **2016**, *2*, 16220. [CrossRef]

91. Okudela, K.; Yazawa, T.; Woo, T.; Sakaeda, M.; Ishii, J.; Mitsui, H.; Shimoyamada, H.; Sato, H.; Tajiri, M.; Ogawa, N.; et al. Down-regulation of DUSP6 expression in lung cancer: Its mechanism and potential role in carcinogenesis. *Am. J. Pathol.* **2009**, *175*, 867–881. [CrossRef]

92. Furukawa, T.; Sunamura, M.; Motoi, F.; Matsuno, S.; Horii, A. Potential tumor suppressive pathway involving DUSP6/MKP-3 in pancreatic cancer. *Am. J. Pathol.* **2003**, *162*, 1807–1815. [CrossRef]

93. Hou, P.C.; Li, Y.H.; Lin, S.C.; Lin, S.C.; Lee, J.C.; Lin, B.W.; Liou, J.P.; Chang, J.Y.; Kuo, C.C.; Liu, Y.M.; et al. Hypoxia-induced downregulation of DUSP-2 phosphatase drives colon cancer stemness. *Cancer Res.* **2017**, *77*, 4305–4316. [CrossRef]

94. Wang, J.Y.; Yeh, C.L.; Chou, H.C.; Yang, C.H.; Fu, Y.N.; Chen, Y.T.; Cheng, H.W.; Huang, C.Y.; Liu, H.P.; Huang, S.F.; et al. Vaccinia H1-related phosphatase is a phosphatase of ErbB receptors and is down-regulated in non-small cell lung cancer. *J. Biol. Chem.* **2011**, *286*, 10177–10184. [CrossRef]

95. Chen, Y.R.; Chou, H.C.; Yang, C.H.; Chen, H.Y.; Liu, Y.W.; Lin, T.Y.; Yeh, C.L.; Chao, W.T.; Tsou, H.H.; Chuang, H.C.; et al. Deficiency in VHR/DUSP3, a suppressor of focal adhesion kinase, reveals its role in regulating cell adhesion and migration. *Oncogene* **2017**, *36*, 6509–6517. [CrossRef]

96. Shin, S.H.; Park, S.Y.; Kang, G.H. Down-regulation of dual-specificity phosphatase 5 in gastric cancer by promoter CpG island hypermethylation and its potential role in carcinogenesis. *Am. J. Pathol.* **2013**, *182*, 1275–1285. [CrossRef]

97. Togel, L.; Nightingale, R.; Wu, R.; Chueh, A.C.; Al-Obaidi, S.; Luk, I.; Davalos-Salas, M.; Chionh, F.; Murone, C.; Buchanan, D.D.; et al. DUSP5 is methylated in CIMP-high colorectal cancer but is not a major regulator of intestinal cell proliferation and tumorigenesis. *Sci. Rep.* **2018**, *8*, 1767. [CrossRef]

98. Li, J.P.; Yang, C.Y.; Chuang, H.C.; Lan, J.L.; Chen, D.Y.; Chen, Y.M.; Wang, X.; Chen, A.J.; Belmont, J.W.; Tan, T.H. The phosphatase JKAP/DUSP22 inhibits T-cell receptor signalling and autoimmunity by inactivating Lck. *Nat. Commun.* **2014**, *5*, 3618. [CrossRef]

99. Chuang, H.C.; Chen, Y.M.; Hung, W.T.; Li, J.P.; Chen, D.Y.; Lan, J.L.; Tan, T.H. Downregulation of the phosphatase JKAP/DUSP22 in T cells as a potential new biomarker of systemic lupus erythematosus nephritis. *Oncotarget* **2016**, *7*, 57593–57605. [CrossRef]

100. Melard, P.; Idrissi, Y.; Andrique, L.; Poglio, S.; Prochazkova-Carlotti, M.; Berhouet, S.; Boucher, C.; Laharanne, E.; Chevret, E.; Pham-Ledard, A.; et al. Molecular alterations and tumor suppressive function of the DUSP22 (Dual Specificity Phosphatase 22) gene in peripheral T-cell lymphoma subtypes. *Oncotarget* **2016**, *7*, 68734–68748. [CrossRef]

101. Feldman, A.L.; Dogan, A.; Smith, D.I.; Law, M.E.; Ansell, S.M.; Johnson, S.H.; Porcher, J.C.; Ozsan, N.; Wieben, E.D.; Eckloff, B.W.; et al. Discovery of recurrent t(6;7)(p25.3;q32.3) translocations in ALK-negative anaplastic large cell lymphomas by massively parallel genomic sequencing. *Blood* **2011**, *117*, 915–919. [CrossRef] [PubMed]

102. Gioeli, D.; Mandell, J.W.; Petroni, G.R.; Frierson, H.F., Jr.; Weber, M.J. Activation of mitogen-activated protein kinase associated with prostate cancer progression. *Cancer Res.* **1999**, *59*, 279–284. [PubMed]

103. Hoshino, R.; Chatani, Y.; Yamori, T.; Tsuruo, T.; Oka, H.; Yoshida, O.; Shimada, Y.; Ari-i, S.; Wada, H.; Fujimoto, J.; et al. Constitutive activation of the 41-/43-kDa mitogen-activated protein kinase signaling pathway in human tumors. *Oncogene* **1999**, *18*, 813–822. [CrossRef] [PubMed]

International Journal of
Molecular Sciences

MDPI

Review

The Dual-Specificity Phosphatase 10 (DUSP10): Its Role in Cancer, Inflammation, and Immunity

Marta Jiménez-Martínez [1,2,3], Konstantinos Stamatakis [1,2,3] and Manuel Fresno [1,2,3,*]

1   Department of Cell Biology and Immunology, Centro de Biología Molecular 'Severo Ochoa' (CSIC-UAM), 28049 Madrid, Spain; mjimenez@cbm.csic.es (M.J.-M.); kstamatakis@cbm.csic.es (K.S.)
2   Department of Molecular Biology, Universidad Autónoma de Madrid, 28049 Madrid, Spain
3   Instituto de Investigación Sanitaria la Princesa (IIS-P), 28006 Madrid, Spain
*   Correspondence: mfresno@cbm.csic.es; Tel.: +34-911-964-565

Received: 13 February 2019; Accepted: 30 March 2019; Published: 1 April 2019

Abstract: Cancer is one of the most diagnosed diseases in developed countries. Inflammation is a common response to different stress situations including cancer and infection. In those processes, the family of mitogen-activated protein kinases (MAPKs) has an important role regulating cytokine secretion, proliferation, survival, and apoptosis, among others. MAPKs regulate a large number of extracellular signals upon a variety of physiological as well as pathological conditions. MAPKs activation is tightly regulated by phosphorylation/dephosphorylation events. In this regard, the dual-specificity phosphatase 10 (DUSP10) has been described as a MAPK phosphatase that negatively regulates p38 MAPK and c-Jun N-terminal kinase (JNK) in several cellular types and tissues. Several studies have proposed that extracellular signal-regulated kinase (ERK) can be also modulated by DUSP10. This suggests a complex role of DUSP10 on MAPKs regulation and, in consequence, its impact in a wide variety of responses involved in both cancer and inflammation. Here, we review DUSP10 function in cancerous and immune cells and studies in both mouse models and patients that establish a clear role of DUSP10 in different processes such as inflammation, immunity, and cancer.

Keywords: DUSP10; MAPK; inflammation; cancer

---

1. Introduction

The human genome contains a large number of genes that transcribe/translate to four families of protein tyrosine phosphatases (PTP), which are subclassified depending of their substrate, structure, regulation, and function. Of these genes, 61 encode for dual-specificity phosphatase subfamily and 10 of them are catalytically active MAP kinase phosphatases (MKPs), which are able to dephosphorylate dual specificity phospho-Tyrosine (pTyr) and phospho-Threonine (pThr) substrate, and one of them is catalytically inactive MKP [1]. DUSP10, also called MKP5, is a member of the MKPs subfamily involved in cell proliferation, differentiation, and migration [2]. Different studies have described a role for DUSP10 as a negative regulator of p38 and c-Jun N-terminal kinase (JNK) through their dephosphorylation [3]. However, new studies confer to DUSP10 the ability to regulate ERK1/2 activity [4]. Nonetheless, it cannot be discarded that the DUSP10 may regulate other proteins either through physical interaction or/and by its phosphatase activity. Up-to-date, enhanced expression of DUSP10 has been reported in several malignancies such as colon, prostate, and breast cancer and diseases such as multiple sclerosis, atherosclerosis, diabetes, celiac disease, and asthma. In addition, elevated DUSP10 expression is relevant to innate and adaptive responses by reducing an excessive inflammatory response. Here, we report an overview of DUSP10 function in different tumors and inflammatory diseases. These results show that DUSP10 is over-expressed in several major cancers, upregulated by numerous physiological stimuli and chemical compounds, and able to control the inflammatory response. The available data suggests that DUSP10 is a potent pro-tumorigenic and anti-inflammatory gene and therefore represents an attractive therapeutic target for the treatment of some cancers and inflammatory diseases.

2. Characteristics, Structure, and Function of DUSP10

Two transcripts of approximately 3.4 and 2.4 Kb have been identified within human genome *DUSP10* sequence. The long transcript is widely expressed in human tissues such as skeletal muscle and liver, and its expression is elevated by stress stimuli in cell culture [5]. In most mouse tissues, a 3.5 Kb transcript was highly expressed. However, the 2.7 Kb mRNA splice variant is specifically expressed in both mouse and rat testis [6] (Figure 1a). DUSP10 protein has two Cdc25 homology regions, a C-terminal catalytic domain and a particular 150 N-terminal amino acid sequence with unknown function, that differs from other family members [7] (Figure 1b).

(a)

Synonyms	MKP5
Location	1q41
Size	40.757 bp
Orientation	Minus strand
Transcrips	3,4 Kb (protein coding)
	2,4 Kb
Polypeptide Lenght	482 aa
Molecular Weight	52,64 kDa
Quaternary Structure	Monomere
Substrate Specificity	pTyr and pThr (and pSer)
Substrate Preference	p38 = JNK >> ERK
Subcellular Location	Cytoplasm and Nucleus

(b)

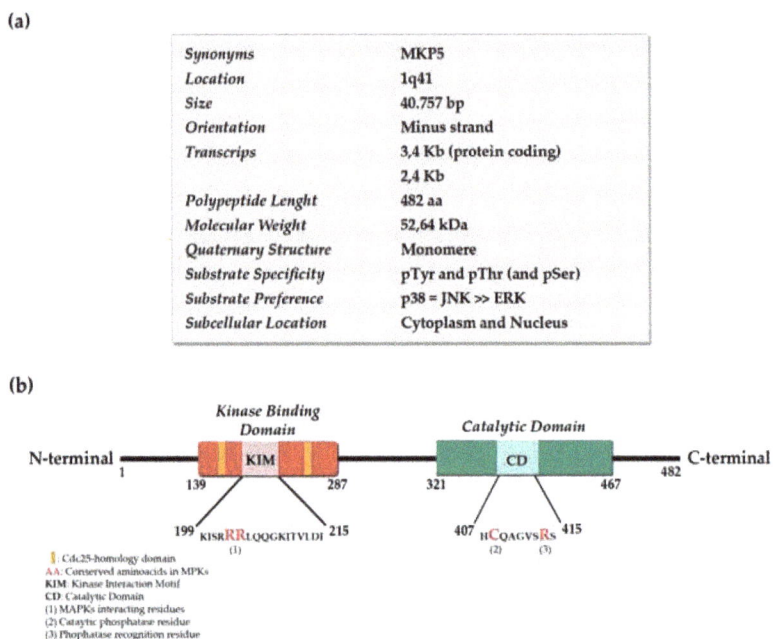

Figure 1. Characteristics and domain organization of human *DUSP10*. (**a**) Summary of the characteristics and information about the gene and protein; (**b**) domain organization of the protein and the most important and conserved polypeptide sequences for the phosphatase activity and interaction capacity.

The crystal structure characterization of human DUSP10 catalytic and kinase-binding domains was decisive to define its affinity for substrates. The MAP kinase-binding domain determines a main structural difference from other MPKs, which have little effect on p38 and JNK dephosphorylation. DUSP10 catalytic domain is particular because of an active conformation structure on its own [8,9]. The phosphorylation state of substrate is also important for DUSP10 affinity. For example, bi-phosphorylated p38α in Tyr-182 and Thr-180 is a more efficient substrate than mono-phosphorylated p38α, with pTyr being dephosphorylated slightly faster by DUSP10 than pThr [10]. Similarly, the crystal structure of p38α reveals a particular mode of interaction between the p38α docking site and the kinase-binding domain of DUSP10, which conserves the mechanism of molecular recognition in p38 [11,12]. Concomitantly, JNK1 interacts with DUSP10 using a different substrate-recognition mechanism than p38α through the catalytic domain, but not the kinase-binding domain of DUSP10 [13].

DUSP10 is a dual specificity phosphatase capable of dephosphorylating both pTyr and pThr residues of activated MAPKs with different affinities. It has been shown that isoforms of the p38 and JNK subfamilies are selectively and more effectively dephosphorylated than ERK subfamily members [7]. Additionally, it has been demonstrated that DUSP10 is able to remove phospho-residues and regulate ERK2 activity, with a *Km* more than 100-fold lower than a specific ERK phosphatase such as MKP3 [14]. Another study has shown that DUSP10 may act as a scaffold protein of ERK. DUSP10 negatively regulates ERK by retaining it in the cytoplasm, avoiding ERK enzymatic activity and downregulating the transcription of ERK-dependent genes [4]. However, current studies have also demonstrated that DUSP10 is able to dephosphorylate phospho-Serine (pSer) residue of non-MAPK substrates [15].

3. Expression and Regulation of DUSP10

DUSP10 mRNA expression has been detected in almost all tissues, but generally at low levels, except in liver and hematopoietic systems [5,16]. Our in silico UCSC Xena database analysis of available RNA-seq and microarray expression databases has confirmed that myeloid and T cells have the highest expression of DUSP10 [17]. Very few studies have addressed the transcriptional regulation of DUSP10 gene. The GTRD database analysis indicates (http://gtrd.biouml.org) (access on 6 January 2019) that DUSP10 mRNA transcription may be regulated by several transcription factors such as PPAR, EGR, STAT5b, RUNX1, VDR, NFAT5, and NF-κB, among others [18]. Several chromatin immunoprecipitation-mass sequencing (CHIP-seq) experiments reveal the binding of these factors on regulatory elements of DUSP10 gene (promoter region, as identified by H3K4me3 binding) [19]. In this regard, it has been proposed that DUSP10 gene has a particularity of being regulated by MAPKs, which promote DUSP10 expression, suggesting a negative feedback loop [20].

Described stimuli that trigger DUSP10 expression are summarized in Figure 2. Among them, epidermal growth factor (EGF) [21], vitamin D [22], and hyaluronic acid [23] can be found. Increases in DUSP10 have been also observed in hypoxia [24], and in response to some stresses such as anisomycin [4] or osmotic pressure [25]. Interestingly, several innate immunity receptor ligands such as pro-inflammatory LPS [16], verotoxin [26], and oxi-LDL [27] induce DUSP10 expression in macrophages. Surprisingly, a few anti-inflammatory compounds such as resveratrol, ginger, curcumin [28], andrographolide [26], and nepetoidin B [29] also increase DUSP10, suggesting that some of their anti-inflammatory activity can be mediated through DUSP10 suppression of MAPK pro-inflammatory activities. Although most studies have shown that LPS induces DUSP10 expression, a few (i.e., Reference [29]) show the contrary. Those differences may be ascribed to the different cell types used. It has also been shown that cytotoxic *Shigella* toxin (Stx1 or Verotoxin) induces DUSP10 expression as effectively as lipopolysaccharide (LPS) and up-regulates other genes such as cyclooxygenase-2 (*COX-2*) and early-growth response protein 1 (*EGR-1*) [26].

On the other hand, different groups have described how the DUSP10 gene is negatively regulated by mirRNAs, which are induced in different diseases and cancers. *DUSP10* is a direct target for miR-21, miR-30b, and miR-155, and has a specific binding site sequence in its 3′-untranslated region where this gene may be negatively regulated. A negative feedback loop in the regulation of *DUSP10* by miRNAs, may exist, since miR-21 and miR-155 transcription is dependent of AP-1 activity, a transcription factor downregulated by *DUSP10* expression [30]. In hepatocellular cancer and pancreatic cancer, miR-181 and miR92a, respectively, negatively regulate the DUSP10 expression, affecting the proliferation and migration of tumorigenic cells [31,32]. Enhanced mir-92a expression is also detected at the peak of the EAE and downregulates *DUSP10* transcription [33]. In hypoxic-ischemic brain damage, *DUSP10* is a target downregulated by miR-7a-2-3p in rats [34].

Only one pharmacological inhibitor (AS077234-4) has been described for DUSP10 to date. However, the information about this inhibitor and its exact mechanism of action is scarce [35].

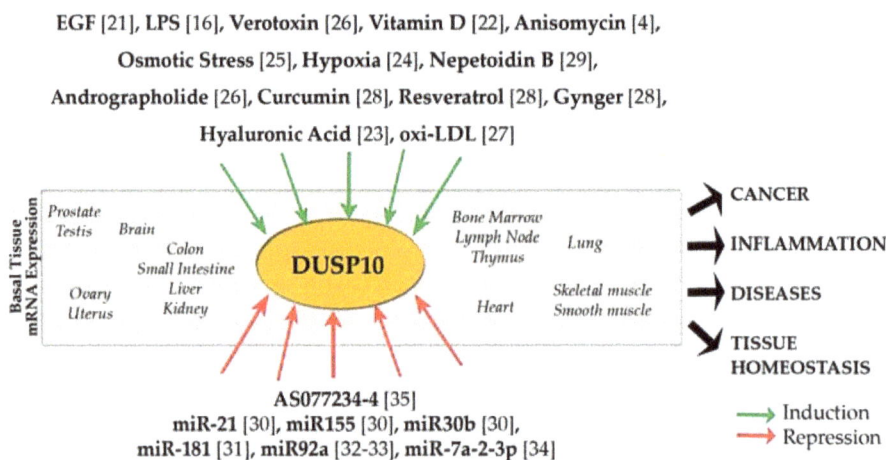

Figure 2. The effect of different signals on *DUSP10* mRNA expression affecting the development of cancer and diseases, controlling inflammatory responses, and altering tissue homeostasis. *DUSP10* mRNA is basally expressed in a wide variety of tissues and it is up- or downregulated by stimuli and inhibitors such as pharmacological agents or miRNAs.

4. DUSP10 in Inflammation and Immunity

Zhang et al. first identified DUSP10 as an inducible phosphatase during immune responses. In naïve mouse lymphocytes, *Dusp10* gene is constitutively expressed, but downregulated following T-cell receptor activation. In contrast, *Dusp10* is not constitutively expressed in macrophages and upregulated after LPS stimulation. They described that this phosphatase was not needed for development of the immune system, while is required in innate and adaptive immune responses, limiting p38 and JNK activity and regulating cytokine production [16]. Following this, several groups have focused on the activity of DUSP10 in immunity and inflammation in different disease settings. Thus, after infection, injury, disease induction, or treatment with several stimuli, DUSP10 can be upregulated or downregulated resulting in different cellular and molecular outcomes, leading to decreased or increased inflammation respectively (Figure 3).

A number of "natural" anti-inflammatory agents have been identified as able to upregulate DUSP10 expression, which in turn regulate inflammatory mediators through MAPKs. The treatment with Andrographolide (*Andrographis paniculata*) in endothelial cells induces DUSP10 expression, reducing hypoxia-induced HIF-1α expression and endothelin-1 secretion through decreased p38 activity [24]. Another compound is Nepetoidin B, which induces DUSP10 expression in macrophages, repressing LPS-induced nitric oxide (NO) production with a p38- and JNK-dependent effect, but no ERK [29]. Hyaluronic acid (HA) also induces DUSP10 expression, negatively regulating p38 and JNK activation by TNF-α in chondrocytes and promoting anti-inflammatory response in osteoarthritis [23].

As mentioned above, DUSP10 is able to dephosphorylate JNK or p38, but some reports have also described a role of DUSP10 through ERK and even dephosphorylation of some other non-MAPK targets. Initially, its effects on T-cells and macrophages were ascribed mostly to JNK deactivation [36]. However, DUSP10 negatively modulates p38 MAPK activity in macrophages that leads to downregulate inflammatory mediators such as TNF-α and IL-6 [37]. The *Shigella* toxin-induced TNF-α expression is reduced by inhibition of ERK in the human macrophage cell line THP-1. This shows the important role of MAPKs in inflammation and the necessity of regulating these pathways [26]. In response to infection of THP-1 cells with *Porphyromonas gingivalis*, a common pathogen associated with chronic adult periodontal disease, DUSP10 expression is decreased. Overexpression of DUSP10 in THP-1 cells reduced activation of ERK and JNK and, thus cytokine production, in response to *P. gingivalis* [20].

Manley et al. show that rhinoviral infection of airway epithelial cells downregulates DUSP10 expression leading to increased activation of p38 and JNK. Reduced DUSP10 levels potentiate the production of CXCL8 and CXCL1 but not IFN-β production in response to infection or IL-1β alone [38]. This suggests that decreases in DUSP10 lead to inflammation in airway diseases such as asthma, likely through regulating inflammasome activation [38]. ASC protein, a downstream common adaptor for different proteins within the inflammasome complex, induces chemokine expression through modulating ERK and JNK activation and, in addition, downregulates DUSP10 expression in THP-1 cells and mouse macrophages [20]. In contrast, a different virus infection (Influenza) induces DUSP10, reducing IRF3 nuclear accumulation that leads to downregulating type I IFN expression and T cell response in the lung epithelial cells. Interestingly, James et al. demonstrated that DUSP10 is able to specifically interact with dephosphorylate IRF3, independently of its interaction with MAPKs. They demonstrated that *Dusp10*-deficient macrophages increase the type I IFN expression and ISGs (IFN-stimulated genes) through IRF3 phosphorylation in response to TLR3 activation upon influenza virus inflection, which reduces its replication in *Dusp10* KO cells and increases innate immune responses [15]. In agreement with this, DUSP10 expression reduces the ISRE (IFN-stimulated response element) activity in the promoter region of RANTES [39]. Together, the above results imply specific roles for DUSP10 in viral infections likely depending on differential TLR signaling by each virus. In allergic airway inflammation, *Dusp10* is highly expressed in specific pathogenic Th2 cells. DUSP10 decreases p38 and GATA3 activity, thus negatively regulating cytokine production in response to IL-33, such as IL-13 and IL-5, and limits the inflammatory response [19].

A pro-inflammatory phenotype is observed in aging diabetic mesangial cells by an excessive oxidative stress, where decreased DUSP10 expression promotes a phosphorylated JNK activation and elevated MCP-1 production, which contributes to chronic inflammatory lesions and disease progression [40]. DUSP10 has been also proposed as a target for therapeutic treatment of atherosclerosis, since DUSP10 inhibition reduces NF-κB-induced TNF-α expression and increases TGF-β1 levels in this disease [27].

In prostatic cell types, DUSP10 overexpression is associated with a potent anti-inflammatory activity through modulation of p38-dependent signaling. DUSP10 overexpression, acting as ant-inflammatory protein, decreases pro-inflammatory responses mediated by cytokine-dependent NF-κB activation, COX-2 expression, and cytokine (IL-6 and IL-8) production in non-transformed prostatic epithelial cell lines [28]. Induced DUSP10 also decreases IL-6 expression through p38 inactivation after 1-alpha,25-dihydroxyvitamin D_3 (1,25D) or calcitriol treatment, promoting a decrease of prostatic inflammation [41].

The use of *Dusp10*-deficient mice allowed the elucidation of the role of this phosphatase on the regulation of immunity and inflammation, demonstrating that it plays an essential non-redundant role in regulating some aspects of immune function against infections and other diseases. The deficiency of Dusp10 expression in mice produces a higher vascular inflammatory response, increasing mainly p38 activation, erythrocyte extravasation, microcapillary thrombus formation, and neutrophil accumulation against LPS injection, which help to propose Dusp10 as a negative regulator of activation of neutrophil inflammatory response against injury by negatively regulating NADPH oxidase [42]. Deficient Dusp10 expression in a mouse model of sepsis-induced lung injury has also been involved in the positive production of pro-inflammatory cytokines such as IL-6, TNF-α, MIP-2, NO, and ROS levels in lung alveolar macrophages and promoting the resistance of mice to LPS-induced sepsis in acute lung injury [43]. DUSP10 is also implicated in metabolic regulation of adipose tissue inflammation. Dusp10-deficient mice develop adiposity, insulin resistance spontaneously, and glucose intolerance. Macrophages from Dusp10 deficient mice are activated, polarized to an M1 state, and infiltrate white adipose tissue, which shows enhanced p38 activity and reduced AKT activation [44].

Dusp10-deficient mice are resistant to experimental myelin-induced autoimmune encephalitis (EAE), an animal model of multiple sclerosis. DUSP10 was found to be a negative regulator of both Type 1 (IFN-γ) and Type 2 (TNF-α) cytokine expression in effector CD4 and CD8 T-cells, reducing AP-1 expression through regulating JNK activity. Thus, DUSP10 expression protects the host from injury due to excessive T-cell response to pathogens [16]. It has been shown that CD11c+ dendritic cells from Dusp10-deficient mice have enhanced antigen presentation activity to splenic CD4+ T-cells resulting increased IFN-γ production [45]. Moreover, this enhanced pro-inflammatory response observed in absence of Dusp10 may result beneficial or deleterious to the murine host, depending largely on the pathogenicity of the infecting parasite strain. Thus, DUSP10, is a strong regulator of the host immune response, regulating antigen presentation by both dendritic cells and T-cell responses.

Thus, the available data to date support that the function of this phosphatase is complex and cellular and disease context dependent (Figure 3).

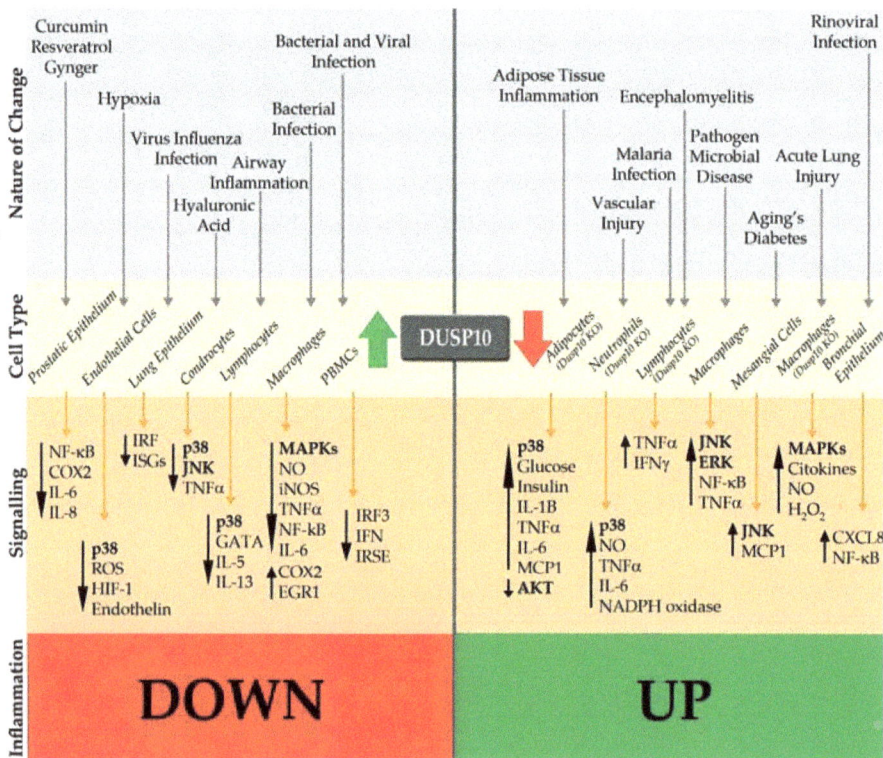

Figure 3. Summary of the role of DUSP10 in response to different infection stimuli and disease conditions. Depending on the nature of the insult/stimulus, DUSP10 expression can be up- (green arrow) or downregulated (red arrow) in each specific cell type. In consequence, it negatively or positively regulates different signal transduction cascades and effector molecules such as mitogen-activated protein kinases (MAPKs), cytokines, interleukins, etc., promoting down- or up-control of inflammation.

5. DUSP10 and Cancer

There have also been a few reports regarding the putative role of DUSP10 in cancer [46]. Most of those studies have found increased *DUSP10* mRNA in tumor tissue, thus suggesting a pro-tumorigenic role for this phosphatase. Nonetheless, there are some studies assigning a suppressive role of DUSP10 in cancer. The in silico UCSC Xena database (https://xena.ucsc.edu/) (access on 6 January 2019) analysis [17] indicates that basal *DUSP10* mRNA expression of normal tissues is low, except in liver and some blood cells (Figure 4a). However, *DUSP10* in most of the cancerous tissues has higher expression levels than that corresponding to normal tissue, although there is variability depending on the tumor type and even within a tumor type. This is particularly striking in some cancers such as acute myeloid leukemia, acute lymphoblastic leukemia, and, to a lesser extent, in hepatocellular carcinoma. Some types of metastatic melanoma and breast cancer carcinomas also show elevated *DUSP10* expression (Figure 4b).

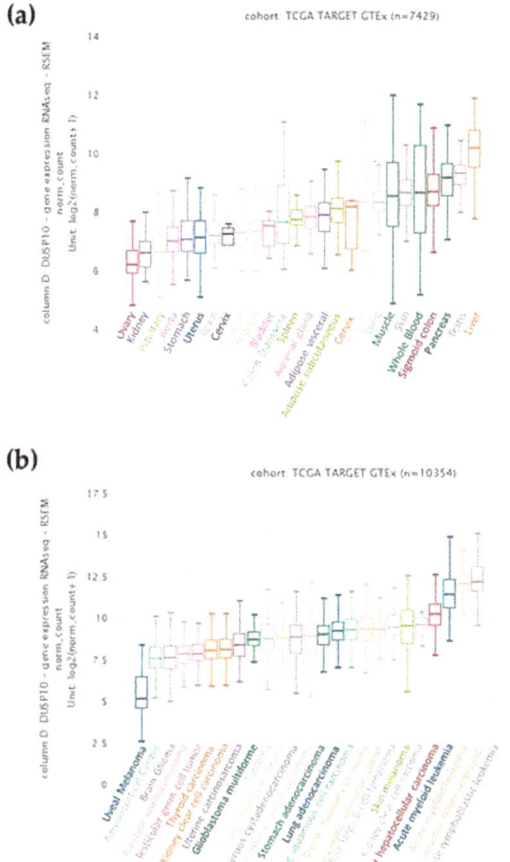

Figure 4. Analysis in silico of *DUSP10* mRNA levels in human samples. (**a**) Normal tissue and (**b**) tumors using the UCSC Xena Browser-GTEx cohort and the TCGA-ARGET cohort, respectively.

It is well known how p38 and JNK MAPKs are implicated in a variety of cell processes such as migration, proliferation, differentiation, and survival. The regulation of this signaling is very important for tumorigenicity. In consequence, it is crucial to regulate this pathway to control cancer progression and to develop therapeutic strategies [47]. A study with mice lacking *Dusp10* demonstrated the role of this phosphatase in muscular dystrophy, a degenerative skeletal muscle disease. In this case, Shi et al. described that Dusp10 is a negative regulator of muscle stem cell function in mice, decreasing cell proliferation and myogenesis by selective p38 and JNK dephosphorylation [48].

On the other hand, several research studies in cancer during the last years have reported *DUSP10* upregulation in colon cancerous tissue and to a lesser extent in other cancers such as lung, breast, prostate, and glioblastoma [4]. However, further basic and clinical investigations in specific diseases

Int. J. Mol. Sci. **2019**, *20*, 1626

would be required to determine the role of DUSP10 in cancer. Below, we will summarize main studies about the presence of DUSP10 in cancer reported to date.

5.1. Hepatocellular and Pancreatic Cancer

In human hepatocellular cancer (HC), low expression of p73 (TP53INP1) diminishes its binding as transcription factor to DUSP10 promoter, downregulating the phosphatase expression, and promoting metastasis via increased ERK activation [49]. This is despite higher *Dusp10* mRNA levels observed in these tumors (see Figure 4b). Along similar lines, miR-181 family represses DUSP10 transcription by a direct and sequence-specific manner the miRNA binds to the 3' UTR of *DUSP10* mRNA. This leads to increased p38 phosphorylation, which promotes HC cell migration [50]. Thus, in HC, low expression of DUSP10 seems to be associated to migration and metastasis. In human pancreatic cancer (PC), JNK activation through DUSP10 downregulation, which is directly targeted by miR-92a, promotes the proliferation of cancerous cells [51].

5.2. Gastrointestinal Cancer

DUSP10 mRNA is frequently over-expressed in colon carcinoma and different polymorphisms (SNPs) have been identified near or within the sequence of *DUSP10* gene, which are correlated with colorectal cancer (CRC) risk. In a Chinese population, two SNPs (rs908858 and rs11118838) lead to the haplotypes 'ACTCAACTA' or 'GCCCACCCA', which correlate with an increased or a decreased CRC risk, respectively [31]. These polymorphisms and SNP rs12724393 are associated with less CRC risk in females of Chinese population; while only the rs908858 is linked with a decreased risk in males [32]. Another SNP (rs6687758) located downstream of DUSP10 is associated with tooth agenesis, which in the family history of cancer positively correlated with a specific haplotype for this polymorphism [52]. Moreover, the last described SNP is a common genetic variant near DUSP10 that increases CRC risk in alcohol consumers in Korean population [53]. These studies support the idea that enhanced DUSP10 is related with CRC risk, and specific polymorphisms that modify its levels could alter the activity of direct DUSP10-targets or other important pathways for the tumour.

In mouse models of intestinal inflammation-associated with dextran-sulphate sodium (DSS) induced colitis, it has been demonstrated that c-GMP-dependent protein kinase (PKG2) induces *Dusp10* expression in colon and damages luminal epithelium and the intestinal immune system, by reducing phospho-JNK and suppressing apoptosis [54]. In apparent contradiction, Png et al. reported that *Dusp10* KO mice were more resistant to DSS-induced colitis by increasing ERK1/2 activation and KLF5 expression [55]. This reduction of colitis eventually led to a decrease in the colon polyp number after azoxymethane (AOM) carcinogen and DSS treatment. In other reports, p53 binding to JNK prevents its dephosphorylation by up-regulating DUSP10 and promoting apoptosis in CRC cell lines upon genotoxic stress [56].

5.3. Breast Cancer

In the context of breast cancer (BC), MKPs subfamily is differentially expressed depending on the specific subtype of malignancy. However, DUSP10 is an upregulated gene in the major BC studies. For example, DUSP10 has been identified as an induced gene in HER2-positive breast tumors [57]. In agreement with this, microarray phosphatome profiling of BC patients has shown that DUSP10 is over-expressed in a specific BC genetic profile, with both estrogen receptor (ER)-negative and epidermal grown factor receptor 2 overexpressing BC [58]. Another study revealed that ER-positive and p53 wildtype BC patients have a worse overall survival associated to AGR2 over-expression, which directly induces DUSP10 expression [59]. In addition, DUSP10 can be induced by oxidative stresses such as H_2O_2 and binds to over-oxidized peroxiredoxin; this complex preserves the DUSP10 activity, leading to p38 inactivation and reducing cancer-associated senescence in BC [60].

5.4. Prostate Cancer

In prostate cancer (PC), studies to date have been focused on the relationship between DUSP10 expression with 1,25D, the hormonally active metabolite of vitamin D, used to treat this disease. Thus, it has been identified by cDNA microarray analysis that *DUSP10* expression is increased in primary cultures of human prostatic epithelial normal and cancer cells, but not in normal prostatic stromal cells, after 1,25D treatment [22]. The upregulation of DUSP10 by 1,25D occurs in primary prostatic adenocarcinoma, while DUSP10 expression is absent in PC cell lines derived metastases, leading to pro-carcinogenic inflammation. This *DUSP10* up-regulation is due to a specific vitamin D receptor (VDR) response element within the DUSP10 promoter [41]. That link between them is also observed in human peripheral blood mononuclear cells (PBMCs) from patients, where DUSP10 is induced in response to 1,25D through VDR [61,62]. Besides inducing DUSP10, 1,25D treatment also inhibits prostaglandin synthesis through decreasing COX-2 and prostaglandin E_2 and $F_{2\alpha}$ receptor (EP and FP) expression and increasing 15-prostaglandin dehydrogenase. Furthermore, the combination of 1,25D and non-steroidal anti-inflammatory drugs (NSAIDs) synergistically reduces PC cell growth, and the authors propose that this pathway is a possible therapeutic approach to PC [28]. In consequence, DUSP10 has an anti-inflammatory effect in PC, although its elevated expression promotes carcinogenesis. However, equilibrium between an anti-inflammatory microenvironment and DUSP10 expression levels could be important to regulate PC.

5.5. Other Cancers and Diseases

DUSP10 is expressed in the majority of meningiomas of all grades, regulating cell proliferation and tumor progression. This tumorigenic phenotype is due to DUSP10 dephosphorylates p38, in which inactivation is determinant in the pathogenesis of meningiomas [63]. The use of specific DUSP10 inhibitor (AS077234-4) decreases the DUSP10 transcription and, in consequence, an induction of oligodendrocyte differentiation of cortical precursors is observed. Authors suggest DUSP10 inhibition as a possible target for treating dysfunctional myelin deposition-associated diseases such as multiple sclerosis [35]. Highly expressed DUSP10 is also found in a common bone marrow disease such as acute myeloid leukemia (AML) [64] and its expression correlates with disease subtypes [65]. In fact, our analysis of available databases indicates that AML cells have the highest DUSP10 mRNA expression of all cancer types. The importance of this for AML pathogenesis and treatment remains to be determined and deserve further investigations.

6. Conclusions

The literature discussed above reveals important roles for DUSP10 in immunity, inflammation, and cancer. The relationship between cancer and inflammation has been proposed years ago, representing an essential communication of cytokines and immune cells with neoplastic cells that occurs in tumor microenvironment [66]. Several studies have demonstrated how DUSP10 is an important phosphatase expressed in a wide variety of tissues and upon different conditions. We should underline that a variety of stress stimuli such as sepsis, injury, chronic inflammation, and viral infection modulates DUSP10 expression. Depending on the nature of this change, DUSP10 is inhibited or increased and regulates the innate and immune response leading to a pro- or anti-inflammatory activity. A large number of experimental data report that DUSP10 has an anti-inflammatory role. Particularly, DUSP10 is induced in many types of cancers, and its enhanced expression confers in general a pro-tumorigenic capacity to neoplastic cells for proliferation, migration, and differentiation. However, in some cases, DUSP10 inflammation modulating activity may also suppress tumor development in an indirect way in some inflammation-driven cancers pointing our complex role of DUSP10 in cancer. Therefore, investigation into the role of DUSP10 in inflammation in cancer would be of great interest. In addition, the action of DUSP10 on cancer, inflammation, and immunity may take place through inactivation of its 'classical' targets p38 and JNK, but also through ERK or by directly targeting other proteins like IRF3 in a MAPK-independent manner. Summarizing, DUSP10 expression has an essential role in controlling tumor development and inflammatory response through MAPKs or other novel and different targets. Moreover, DUSP10 is a good anti-cancer and anti-inflammatory therapeutic target, but more studies are needed to elucidate its role in each specific cellular and physiological context.

Author Contributions: M.F. conceived the idea. M.J.-M. wrote the manuscript and collected the data. K.S. and M.F. contributed to critical editing the manuscript. All authors approved the final version before submitting.

Funding: This research was funded by grants from from "Ministerio de Ciencia e Innovación" (SAF2013-42850-R and SAF2016-75988-R) "Comunidad de Madrid" (S2017/BMD-3671. INFLAMUNE-CM), "Fondo de Investigaciones Sanitarias" (BIOIMID) to M.F. K.S. was recipient of a Spanish Association Against Cancer oncology investigator grant (AECC AIO)".

Conflicts of Interest: The authors declare no conflict of interest.

References

1. Alonso, A.; Pulido, R. The extended human PTPome: A growing tyrosine phosphatase family. *FEBS J.* **2016**, *283*, 2197–2201. [CrossRef]
2. Mishra, A.; Oules, B.; Pisco, A.O.; Ly, T.; Liakath-Ali, K.; Walko, G.; Viswanathan, P.; Tihy, M.; Nijjher, J.; Dunn, S.J.; et al. A protein phosphatase network controls the temporal and spatial dynamics of differentiation commitment in human epidermis. *Elife* **2017**, *6*, e27356. [CrossRef] [PubMed]
3. Keyse, S.M. Dual-specificity MAP kinase phosphatases (MKPs) and cancer. *Cancer Metastasis Rev.* **2008**, *27*, 253–261. [CrossRef]
4. Nomura, M.; Shiiba, K.; Katagiri, C.; Kasugai, I.; Masuda, K.; Sato, I.; Sato, M.; Kakugawa, Y.; Nomura, E.; Hayashi, K.; et al. Novel function of MKP-5/DUSP10, a phosphatase of stress-activated kinases, on ERK-dependent gene expression, and upregulation of its gene expression in colon carcinomas. *Oncol. Rep.* **2012**, *28*, 931–936. [CrossRef]
5. Tanoue, T.; Moriguchi, T.; Nishida, E. Molecular cloning and characterization of a novel dual specificity phosphatase, MKP-5. *J. Biol. Chem.* **1999**, *274*, 19949–19956. [CrossRef]
6. Masuda, K.; Shima, H.; Kikuchi, K.; Watanabe, Y.; Matsuda, Y. Expression and comparative chromosomal mapping of MKP-5 genes DUSP10/Dusp10. *Cytogenet. Genome Res.* **2000**, *90*, 71–74. [CrossRef]
7. Theodosiou, A.; Smith, A.; Gillieron, C.; Arkinstall, S.; Ashworth, A. MKP5, a new member of the MAP kinase phosphatase family, which selectively dephosphorylates stress-activated kinases. *Oncogene* **1999**, *18*, 6981–6988. [CrossRef] [PubMed]
8. Tao, X.; Tong, L. Crystal structure of the MAP kinase binding domain and the catalytic domain of human MKP5. *Protein Sci.* **2007**, *16*, 880–886. [CrossRef]
9. Jeong, D.G.; Yoon, T.S.; Kim, J.H.; Shim, M.Y.; Jung, S.K.; Son, J.H.; Ryu, S.E.; Kim, S.J. Crystal structure of the catalytic domain of human MAP kinase phosphatase 5: Structural insight into constitutively active phosphatase. *J. Mol. Biol.* **2006**, *360*, 946–955. [CrossRef]
10. Zhang, Y.Y.; Mei, Z.Q.; Wu, J.W.; Wang, Z.X. Enzymatic activity and substrate specificity of mitogen-activated protein kinase p38alpha in different phosphorylation states. *J. Biol. Chem.* **2008**, *283*, 26591–26601. [CrossRef]
11. Goldsmith, E.J. Three-dimensional docking in the MAPK p38alpha. *Sci. Signal.* **2011**, *4*, pe47. [CrossRef] [PubMed]
12. Zhang, Y.Y.; Wu, J.W.; Wang, Z.X. A distinct interaction mode revealed by the crystal structure of the kinase p38alpha with the MAPK binding domain of the phosphatase MKP5. *Sci. Signal.* **2011**, *4*, ra88. [CrossRef] [PubMed]
13. Liu, X.; Zhang, C.S.; Lu, C.; Lin, S.C.; Wu, J.W.; Wang, Z.X. A conserved motif in JNK/p38-specific MAPK phosphatases as a determinant for JNK1 recognition and inactivation. *Nat. Commun.* **2016**, *7*, 10879. [CrossRef] [PubMed]
14. Zhou, B.; Wang, Z.X.; Zhao, Y.; Brautigan, D.L.; Zhang, Z.Y. The specificity of extracellular signal-regulated kinase 2 dephosphorylation by protein phosphatases. *J. Biol. Chem.* **2002**, *277*, 31818–31825. [CrossRef] [PubMed]

15. James, S.J.; Jiao, H.; Teh, H.Y.; Takahashi, H.; Png, C.W.; Phoon, M.C.; Suzuki, Y.; Sawasaki, T.; Xiao, H.; Chow, V.T.; et al. MAPK Phosphatase 5 Expression Induced by Influenza and Other RNA Virus Infection Negatively Regulates IRF3 Activation and Type I Interferon Response. *Cell Rep.* **2015**, *10*, 1722–1734. [CrossRef]

16. Zhang, Y.; Blattman, J.N.; Kennedy, N.J.; Duong, J.; Nguyen, T.; Wang, Y.; Davis, R.J.; Greenberg, P.D.; Flavell, R.A.; Dong, C. Regulation of innate and adaptive immune responses by MAP kinase phosphatase 5. *Nature* **2004**, *430*, 793–797. [CrossRef] [PubMed]

17. Goldman, M.; Craft, B.; Kamath, A.; Brooks, A.N.; Zhu, J.; Haussler, D. The UCSC Xena Platform for cancer genomics data visualization and interpretation. *bioRxiv* **2018**, 326470. [CrossRef]

18. Yevshin, I.; Sharipov, R.; Kolmykov, S.; Kondrakhin, Y.; Kolpakov, F. GTRD: A database on gene transcription regulation-2019 update. *Nucleic Acids Res.* **2019**, *47*, D100–D105. [CrossRef]

19. Yamamoto, T.; Endo, Y.; Onodera, A.; Hirahara, K.; Asou, H.K.; Nakajima, T.; Kanno, T.; Ouchi, Y.; Uematsu, S.; Nishimasu, H.; et al. DUSP10 constrains innate IL-33-mediated cytokine production in ST2(hi) memory-type pathogenic Th2 cells. *Nat. Commun.* **2018**, *9*, 4231. [CrossRef]

20. Taxman, D.J.; Holley-Guthrie, E.A.; Huang, M.T.; Moore, C.B.; Bergstralh, D.T.; Allen, I.C.; Lei, Y.; Gris, D.; Ting, J.P. The NLR adaptor ASC/PYCARD regulates DUSP10, mitogen-activated protein kinase (MAPK), and chemokine induction independent of the inflammasome. *J. Biol. Chem.* **2011**, *286*, 19605–19616. [CrossRef]

21. Finch, A.R.; Caunt, C.J.; Perrett, R.M.; Tsaneva-Atanasova, K.; McArdle, C.A. Dual specificity phosphatases 10 and 16 are positive regulators of EGF-stimulated ERK activity: Indirect regulation of ERK signals by JNK/p38 selective MAPK phosphatases. *Cell. Signal.* **2012**, *24*, 1002–1011. [CrossRef] [PubMed]

22. Peehl, D.M.; Shinghal, R.; Nonn, L.; Seto, E.; Krishnan, A.V.; Brooks, J.D.; Feldman, D. Molecular activity of 1,25-dihydroxyvitamin D3 in primary cultures of human prostatic epithelial cells revealed by cDNA microarray analysis. *J. Steroid Biochem. Mol. Biol.* **2004**, *92*, 131–141. [CrossRef] [PubMed]

23. Furuta, J.; Ariyoshi, W.; Okinaga, T.; Takeuchi, J.; Mitsugi, S.; Tominaga, K.; Nishihara, T. High molecular weight hyaluronic acid regulates MMP13 expression in chondrocytes via DUSP10/MKP5. *J. Orthop. Res.* **2017**, *35*, 331–339. [CrossRef] [PubMed]

24. Lin, H.C.; Su, S.L.; Lin, W.C.; Lin, A.H.; Yang, Y.C.; Lii, C.K.; Chen, H.W. Andrographolide inhibits hypoxia-induced hypoxia-inducible factor 1alpha and endothelin 1 expression through the heme oxygenase 1/CO/cGMP/MKP-5 pathways in EA.hy926 cells. *Environ. Toxicol.* **2018**, *33*, 269–279. [CrossRef] [PubMed]

25. Liovic, M.; Lee, B.; Tomic-Canic, M.; D'Alessandro, M.; Bolshakov, V.N.; Lane, E.B. Dual-specificity phosphatases in the hypo-osmotic stress response of keratin-defective epithelial cell lines. *Exp. Cell Res.* **2008**, *314*, 2066–2075. [CrossRef] [PubMed]

26. Leyva-Illades, D.; Cherla, R.P.; Galindo, C.L.; Chopra, A.K.; Tesh, V.L. Global transcriptional response of macrophage-like THP-1 cells to Shiga toxin type 1. *Infect. Immun.* **2010**, *78*, 2454–2465. [CrossRef]

27. Luo, L.J.; Liu, F.; Wang, X.Y.; Dai, T.Y.; Dai, Y.L.; Dong, C.; Ge, B.X. An essential function for MKP5 in the formation of oxidized low density lipid-induced foam cells. *Cell. Signal.* **2012**, *24*, 1889–1898. [CrossRef] [PubMed]

28. Nonn, L.; Duong, D.; Peehl, D.M. Chemopreventive anti-inflammatory activities of curcumin and other phytochemicals mediated by MAP kinase phosphatase-5 in prostate cells. *Carcinogenesis* **2007**, *28*, 1188–1196. [CrossRef]

29. Wu, X.; Gao, H.; Sun, W.; Yu, J.; Hu, H.; Xu, Q.; Chen, X. Nepetoidin B, a Natural Product, Inhibits LPS-stimulated Nitric Oxide Production via Modulation of iNOS Mediated by NF-kappaB/MKP-5 Pathways. *Phytother. Res.* **2017**, *31*, 1072–1077. [CrossRef]

30. Vlad, G.; Suciu-Foca, N. Induction of antigen-specific human T suppressor cells by membrane and soluble ILT3. *Exp. Mol. Pathol.* **2012**, *93*, 294–301. [CrossRef]

31. Zhang, T.; Li, X.; Du, Q.; Gong, S.; Wu, M.; Mao, Z.; Gao, Z.; Long, Y.; Jin, T.; Geng, T.; et al. DUSP10 gene polymorphism and risk of colorectal cancer in the Han Chinese population. *Eur. J. Cancer Prev.* **2014**, *23*, 173–176. [CrossRef] [PubMed]

32. Duan, X.; Gao, Y.; Yang, H.; Feng, T.; Jin, T.; Long, Y.; Chen, C. Polymorphisms in the DUSP10 gene are associated with sex-specific colorectal cancer risk in a Han population. *Int. J. Clin. Exp. Pathol.* **2015**, *8*, 2018–2025. [PubMed]

33. Rezaei, N.; Talebi, F.; Ghorbani, S.; Rezaei, A.; Esmaeili, A.; Noorbakhsh, F.; Hakemi, M.G. MicroRNA-92a Drives Th1 Responses in the Experimental Autoimmune Encephalomyelitis. *Inflammation* **2018**. [CrossRef] [PubMed]

34. Zhang, Z.B.; Tan, Y.X.; Zhao, Q.; Xiong, L.L.; Liu, J.; Xu, F.F.; Xu, Y.; Bobrovskaya, L.; Zhou, X.F.; Wang, T.H. miRNA-7a-2-3p Inhibits Neuronal Apoptosis in Oxygen-Glucose Deprivation (OGD) Model. *Front. Neurosci.* **2019**, *13*, 16. [CrossRef] [PubMed]

35. Gobert, R.P.; Joubert, L.; Curchod, M.L.; Salvat, C.; Foucault, I.; Jorand-Lebrun, C.; Lamarine, M.; Peixoto, H.; Vignaud, C.; Fremaux, C.; et al. Convergent functional genomics of oligodendrocyte differentiation identifies multiple autoinhibitory signaling circuits. *Mol. Cell. Biol.* **2009**, *29*, 1538–1553. [CrossRef] [PubMed]

36. Lang, R.; Hammer, M.; Mages, J. DUSP meet immunology: Dual specificity MAPK phosphatases in control of the inflammatory response. *J. Immunol.* **2006**, *177*, 7497–7504. [CrossRef] [PubMed]

37. Hommo, T.; Pesu, M.; Moilanen, E.; Korhonen, R. Regulation of Inflammatory Cytokine Production by MKP-5 in Macrophages. *Basic Clin. Pharmacol. Toxicol.* **2015**, *117*, 96–104. [CrossRef] [PubMed]

38. Manley, G.C.A.; Stokes, C.A.; Marsh, E.K.; Sabroe, I.; Parker, L.C. DUSP10 Negatively Regulates the Inflammatory Response to Rhinovirus through IL-1beta Signalling. *J. Virol.* **2019**, *93*, e01659-18. [CrossRef] [PubMed]

39. McCoy, C.E.; Carpenter, S.; Palsson-McDermott, E.M.; Gearing, L.J.; O'Neill, L.A. Glucocorticoids inhibit IRF3 phosphorylation in response to Toll-like receptor-3 and -4 by targeting TBK1 activation. *J. Biol. Chem.* **2008**, *283*, 14277–14285. [CrossRef] [PubMed]

40. Wu, J.; Mei, C.; Vlassara, H.; Striker, G.E.; Zheng, F. Oxidative stress-induced JNK activation contributes to proinflammatory phenotype of aging diabetic mesangial cells. *Am. J. Physiol. Renal Physiol.* **2009**, *297*, F1622–F1631. [CrossRef] [PubMed]

41. Nonn, L.; Peng, L.; Feldman, D.; Peehl, D.M. Inhibition of p38 by vitamin D reduces interleukin-6 production in normal prostate cells via mitogen-activated protein kinase phosphatase 5: Implications for prostate cancer prevention by vitamin D. *Cancer Res.* **2006**, *66*, 4516–4524. [CrossRef] [PubMed]

42. Qian, F.; Deng, J.; Cheng, N.; Welch, E.J.; Zhang, Y.; Malik, A.B.; Flavell, R.A.; Dong, C.; Ye, R.D. A non-redundant role for MKP5 in limiting ROS production and preventing LPS-induced vascular injury. *EMBO J.* **2009**, *28*, 2896–2907. [CrossRef] [PubMed]

43. Qian, F.; Deng, J.; Gantner, B.N.; Flavell, R.A.; Dong, C.; Christman, J.W.; Ye, R.D. Map kinase phosphatase 5 protects against sepsis-induced acute lung injury. *Am. J. Physiol. Lung Cell. Mol. Physiol.* **2012**, *302*, L866–L874. [CrossRef] [PubMed]

44. Zhang, Y.; Nguyen, T.; Tang, P.; Kennedy, N.J.; Jiao, H.; Zhang, M.; Reynolds, J.M.; Jaeschke, A.; Martin-Orozco, N.; Chung, Y.; et al. Regulation of Adipose Tissue Inflammation and Insulin Resistance by MAPK Phosphatase 5. *J. Biol. Chem.* **2015**, *290*, 14875–14883. [CrossRef] [PubMed]

45. Cheng, Q.; Zhang, Q.; Xu, X.; Yin, L.; Sun, L.; Lin, X.; Dong, C.; Pan, W. MAPK phosphotase 5 deficiency contributes to protection against blood-stage Plasmodium yoelii 17XL infection in mice. *J. Immunol.* **2014**, *192*, 3686–3696. [CrossRef] [PubMed]

46. Rios, P.; Nunes-Xavier, C.E.; Tabernero, L.; Kohn, M.; Pulido, R. Dual-specificity phosphatases as molecular targets for inhibition in human disease. *Antioxid. Redox Signal.* **2014**, *20*, 2251–2273. [CrossRef] [PubMed]

47. Wagner, E.F.; Nebreda, A.R. Signal integration by JNK and p38 MAPK pathways in cancer development. *Nat. Rev. Cancer* **2009**, *9*, 537–549. [CrossRef]

48. Shi, H.; Verma, M.; Zhang, L.; Dong, C.; Flavell, R.A.; Bennett, A.M. Improved regenerative myogenesis and muscular dystrophy in mice lacking Mkp5. *J. Clin. Investig.* **2013**, *123*, 2064–2077. [CrossRef]

49. Ng, K.Y.; Chan, L.H.; Chai, S.; Tong, M.; Guan, X.Y.; Lee, N.P.; Yuan, Y.; Xie, D.; Lee, T.K.; Dusetti, N.J.; et al. TP53INP1 Downregulation Activates a p73-Dependent DUSP10/ERK Signaling Pathway to Promote Metastasis of Hepatocellular Carcinoma. *Cancer Res.* **2017**, *77*, 4602–4612. [CrossRef]

50. Song, M.K.; Park, Y.K.; Ryu, J.C. Polycyclic aromatic hydrocarbon (PAH)-mediated upregulation of hepatic microRNA-181 family promotes cancer cell migration by targeting MAPK phosphatase-5, regulating the activation of p38 MAPK. *Toxicol. Appl. Pharmacol.* **2013**, *273*, 130–139. [CrossRef]

51. He, G.; Zhang, L.; Li, Q.; Yang, L. miR-92a/DUSP10/JNK signalling axis promotes human pancreatic cancer cells proliferation. *Biomed. Pharmacother.* **2014**, *68*, 25–30. [CrossRef]

52. Williams, M.A.; Biguetti, C.; Romero-Bustillos, M.; Maheshwari, K.; Dinckan, N.; Cavalla, F.; Liu, X.; Silva, R.; Akyalcin, S.; Uyguner, Z.O.; et al. Colorectal Cancer-Associated Genes Are Associated with Tooth Agenesis and May Have a Role in Tooth Development. *Sci. Rep.* **2018**, *8*, 2979. [CrossRef] [PubMed]

53. Song, N.; Shin, A.; Oh, J.H.; Kim, J. Effects of interactions between common genetic variants and alcohol consumption on colorectal cancer risk. *Oncotarget* **2018**, *9*, 6391–6401. [CrossRef] [PubMed]

54. Wang, R.; Kwon, I.K.; Singh, N.; Islam, B.; Liu, K.; Sridhar, S.; Hofmann, F.; Browning, D.D. Type 2 cGMP-dependent protein kinase regulates homeostasis by blocking c-Jun N-terminal kinase in the colon epithelium. *Cell Death Differ.* **2014**, *21*, 427–437. [CrossRef] [PubMed]

55. Png, C.W.; Weerasooriya, M.; Guo, J.; James, S.J.; Poh, H.M.; Osato, M.; Flavell, R.A.; Dong, C.; Yang, H.; Zhang, Y. DUSP10 regulates intestinal epithelial cell growth and colorectal tumorigenesis. *Oncogene* **2016**, *35*, 206–217. [CrossRef] [PubMed]

56. Gowda, P.S.; Zhou, F.; Chadwell, L.V.; McEwen, D.G. p53 binding prevents phosphatase-mediated inactivation of diphosphorylated c-Jun N-terminal kinase. *J. Biol. Chem.* **2012**, *287*, 17554–17567. [CrossRef] [PubMed]

57. Lucci, M.A.; Orlandi, R.; Triulzi, T.; Tagliabue, E.; Balsari, A.; Villa-Moruzzi, E. Expression profile of tyrosine phosphatases in HER2 breast cancer cells and tumors. *Anal. Cell. Oncol.* **2010**, *32*, 361–372. [CrossRef]

58. Manzano, R.G.; Martinez-Navarro, E.M.; Forteza, J.; Brugarolas, A. Microarray phosphatome profiling of breast cancer patients unveils a complex phosphatase regulatory role of the MAPK and PI3K pathways in estrogen receptor-negative breast cancers. *Int. J. Oncol.* **2014**, *45*, 2250–2266. [CrossRef]

59. Hrstka, R.; Bouchalova, P.; Michalova, E.; Matoulkova, E.; Muller, P.; Coates, P.J.; Vojtesek, B. AGR2 oncoprotein inhibits p38 MAPK and p53 activation through a DUSP10-mediated regulatory pathway. *Mol. Oncol.* **2016**, *10*, 652–662. [CrossRef]

60. Turner-Ivey, B.; Manevich, Y.; Schulte, J.; Kistner-Griffin, E.; Jezierska-Drutel, A.; Liu, Y.; Neumann, C.A. Role for Prdx1 as a specific sensor in redox-regulated senescence in breast cancer. *Oncogene* **2013**, *32*, 5302–5314. [CrossRef]

61. Vukic, M.; Neme, A.; Seuter, S.; Saksa, N.; de Mello, V.D.; Nurmi, T.; Uusitupa, M.; Tuomainen, T.P.; Virtanen, J.K.; Carlberg, C. Relevance of vitamin D receptor target genes for monitoring the vitamin D responsiveness of primary human cells. *PLoS ONE* **2015**, *10*, e0124339. [CrossRef] [PubMed]

62. Saksa, N.; Neme, A.; Ryynanen, J.; Uusitupa, M.; de Mello, V.D.; Voutilainen, S.; Nurmi, T.; Virtanen, J.K.; Tuomainen, T.P.; Carlberg, C. Dissecting high from low responders in a vitamin D3 intervention study. *J. Steroid Biochem. Mol. Biol.* **2015**, *148*, 275–282. [CrossRef] [PubMed]

63. Johnson, M.D.; Reeder, J.E.; O'Connell, M. p38MAPK activation and DUSP10 expression in meningiomas. *J. Clin. Neurosci.* **2016**, *30*, 110–114. [CrossRef] [PubMed]

64. Arora, D.; Kothe, S.; van den Eijnden, M.; Hooft van Huijsduijnen, R.; Heidel, F.; Fischer, T.; Scholl, S.; Tolle, B.; Bohmer, S.A.; Lennartsson, J.; et al. Expression of protein-tyrosine phosphatases in Acute Myeloid Leukemia cells: FLT3 ITD sustains high levels of DUSP6 expression. *Cell Commun. Signal.* **2012**, *10*, 19. [CrossRef] [PubMed]

65. Kabir, N.N.; Ronnstrand, L.; Kazi, J.U. Deregulation of protein phosphatase expression in acute myeloid leukemia. *Med. Oncol.* **2013**, *30*, 517. [CrossRef] [PubMed]

66. Hanahan, D.; Weinberg, R.A. Hallmarks of cancer: The next generation. *Cell* **2011**, *144*, 646–674. [CrossRef]

International Journal of
Molecular Sciences

MDPI

Article

Dysregulation of Lipid Metabolism in Mkp-1 Deficient Mice during Gram-Negative Sepsis

Jinhui Li [1], Xiantao Wang [2], William E. Ackerman IV [3], Abel J. Batty [1], Sean G. Kirk [1], William M. White [1], Xianxi Wang [1], Dimitrios Anastasakis [2], Lobelia Samavati [4], Irina Buhimschi [1,5], Leif D. Nelin [1,5], Markus Hafner [2] and Yusen Liu [1,5,*]

1   Center for Perinatal Research, The Research Institute at Nationwide Children's Hospital, Columbus, OH 43215, USA; jinhui.liu@nationwidechildrens.org (J.L.); Abel.Batty@nationwidechildrens.org (A.J.B.); Sean.Kirk@nationwidechildrens.org (S.G.K.); white.2729@buckeyemail.osu.edu (W.M.W.); xianxi.wang@nationwidechildrens.org (X.W.); Irina.Buhimschi@nationwidechildrens.org (I.B.); leif.nelin@nationwidechildrens.org (L.D.N.)
2   Laboratory of Muscle Stem Cells and Gene Regulation, National Institute of Arthritis and Musculoskeletal and Skin Disease, National Institutes of Health, Bethesda, MD 20892, USA; xiantao.wang@nih.gov (X.W.); dimitrios.anastasakis@nih.gov (D.A.); markus.hafner@nih.gov (M.H.)
3   Department of Obstetrics and Gynecology, The Ohio State University College of Medicine, Columbus, OH 43210, USA; William.Ackerman@nationwidechildrens.org
4   Center for Molecular Medicine and Genetics, Wayne State University School of Medicine, Detroit, MI 48201, USA; lsamavat@med.wayne.edu
5   Department of Pediatrics, The Ohio State University College of Medicine, Columbus, Ohio 43205, USA
*   Correspondence: yusen.liu@nationwidechildrens.org; Tel.: +1-614-355-6728

Received: 9 November 2018; Accepted: 30 November 2018; Published: 6 December 2018

Abstract: Mitogen-activated protein kinase phosphatase (Mkp)-1 exerts its anti-inflammatory activities during Gram-negative sepsis by deactivating p38 and c-Jun N-terminal kinase (JNK). We have previously shown that Mkp-$1^{+/+}$ mice, but not Mkp-$1^{-/-}$ mice, exhibit hypertriglyceridemia during severe sepsis. However, the regulation of hepatic lipid stores and the underlying mechanism of lipid dysregulation during sepsis remains an enigma. To understand the molecular mechanism underlying the sepsis-associated metabolic changes and the role of Mkp-1 in the process, we infected Mkp-$1^{+/+}$ and Mkp-$1^{-/-}$ mice with *Escherichia coli* i.v., and assessed the effects of Mkp-1 deficiency on tissue lipid contents. We also examined the global gene expression profile in the livers via RNA-seq. We found that in the absence of *E. coli* infection, Mkp-1 deficiency decreased liver triglyceride levels. Upon *E. coli* infection, Mkp-$1^{+/+}$ mice, but not Mkp-$1^{-/-}$ mice, developed hepatocyte ballooning and increased lipid deposition in the livers. *E. coli* infection caused profound changes in the gene expression profile of a large number of proteins that regulate lipid metabolism in wildtype mice, while these changes were substantially disrupted in Mkp-$1^{-/-}$ mice. Interestingly, in Mkp-$1^{+/+}$ mice *E. coli* infection resulted in downregulation of genes that facilitate fatty acid synthesis but upregulation of Cd36 and Dgat2, whose protein products mediate fatty acid uptake and triglyceride synthesis, respectively. Taken together, our studies indicate that sepsis leads to a substantial change in triglyceride metabolic gene expression programs and Mkp-1 plays an important role in this process.

Keywords: *E. coli* infection; sepsis; liver steatosis; hypertriglyceridemia; Mkp-1

---

1. Introduction

Severe sepsis and septic shock are a major cause of death in the United States, accounting for 215,000 deaths and 750,000 hospitalizations annually [1]. The mortality rate of septic shock still approaches 50% despite improvements in critical care medicine [2]. Metabolic dysregulation has been reported in septic patients [3–6] as well as in experimental animals such as sheep [7], dogs [8], and rodents following sepsis induction [9,10]. Septic patients or animals with Gram-negative bacteria infection often develop hyperlipidemia [11–13], which is often referred to as the lipemia of sepsis [14]. Since triglyceride-rich lipoproteins can bind and sequester endotoxin, hyperlipidemia of sepsis is considered a component of the innate, non-adaptive host immune response to infection [14]. In addition to the ability to sequester lipopolysaccharide (LPS), certain fatty acid species interfere with the inflammatory signaling pathway mediated by nuclear factor (NF)-κB [15]. However, derivatives of fatty acids such as arachidonic acid and prostaglandins are important mediators of the systemic inflammatory response, raising the possibility that hyperlipidemia may also be contributing factor to the pathophysiology of sepsis [16]. This is supported by a small study that found a higher incidence of hypertriglyceridemia among non-survivors than among survivors [6]. As septic patients and animals have decreased lipoprotein lipase activity in peripheral tissues such as heart, muscle, and adipose tissue as the result of elevated tumor necrosis factor (TNF)-α levels in the circulation, hypertriglyceridemia during sepsis has been attributed to defective triglyceride clearance in peripheral tissues [17–20]. However, increased very low density lipoprotein (VLDL)-mediated triglyceride efflux from the liver may also contribute to sepsis-mediated hypertriglyceridemia [21]. In fact, hepatocytes isolated from mice challenged with endotoxin exhibit increased secretion of VLDL, likely as a result of increased expression of apolipoprotein (Apo) b48, Apob100, and microsomal triglyceride transfer proteins [22], the proteins critical to VLDL assembly. One study has shown that TNF-α challenge in mice elevates the hepatic expression of sterol regulatory element-binding transcription factor (Srebf) 1 (also referred to as SREBP-1), a master regulator of lipogenesis and fatty acid synthase (Fasn), which controls the synthesis of saturated fatty acids [23]. However, another study has reported a decreased expression of lipogenesis genes in endotoxin-challenged mice [24]. The mechanisms of hepatic lipid regulation during sepsis remain poorly understood.

Mkp-1 exerts its anti-inflammatory effects by dephosphorylating p38 and JNK during gram-negative bacteria infection [25–30]. Previously, we have found that hypertriglyceridemia occurs in *E. coli*-infected *Mkp-1*$^{+/+}$ mice, but not in *Mkp-1*$^{-/-}$ mice [10], suggesting that Mkp-1 plays an essential role in the sepsis-induced hypertriglyceridemia. To understand the molecular mechanisms by which Mkp-1 regulates lipid metabolism during sepsis, we assessed the liver lipid contents and the global gene expression profiles in wildtype and Mkp-1 deficient mice before and after sepsis induction. We found that *E. coli* infection enhanced liver triglyceride synthesis in an Mkp-1-dependent manner with an attenuation of the transcriptional program responsible for fatty acid synthesis and fatty acid oxidation. Our studies also suggest that sepsis likely exacerbates liver lipid accumulation through both increased liver fatty acid uptake and decreased fatty acid β-oxidation.

2. Results

2.1. Changes in p38 Activity and Dysregulation of Lipid Metabolism Caused by Mkp-1 Deficiency and E. coli Infection

Since p38 is a preferred substrate of Mkp-1 [31] and plays a critical role in the regulation of the inflammatory response [25,30,32,33], we assessed p38 activity in the livers of *Mkp-1*$^{+/+}$ and *Mkp-1*$^{-/-}$ mice both before and after *E. coli* infection (Figure 1A). In the absence of infection liver p38 activity was higher in the *Mkp-1*$^{-/-}$ mice than in the *Mkp-1*$^{+/+}$ mice. In *E. coli*-infected wildtype mice, liver p38 activity became virtually undetectable 24 h post infection. In contrast, liver p38 activity remained high in *E. coli*-infected *Mkp-1*$^{-/-}$ mice. Quantification of p38 activity in the four groups further highlights the essential role of Mkp-1 both in the regulation of basal p38 activity in uninfected mice and particularly in the deactivation of p38 following *E. coli* infection (Figure 1A, right panel).

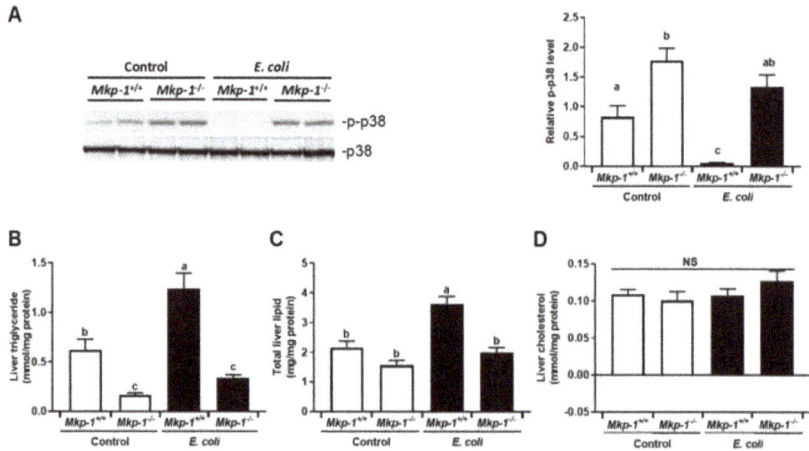

Figure 1. Liver p38 activity and lipid contents before and after *E. coli* infection. *Mkp-1+/+* and *Mkp-1−/−* mice were either infected i.v. with *E. coli* at a dose of 2.5×10^7 colony forming units (CFU)/g of body weight or injected with PBS. Mice were euthanized after 24 h, and livers were harvested. (**A**) Phospho-p38 levels in the livers of control and *E. coli*-infected mice. Parts of the livers were homogenized to extract soluble proteins. Protein samples (40 μg) from distinct animals were resolved to detect phospho-p38 by Western blotting, using a polyclonal antibody against phospho-p38. The membranes were stripped and reblotted with p38 antibody to verify comparable loading. Images shown in the left panel are representative results. Bar graph in the right panel represent quantification of phosphor-p38 (p-p38) levels in each groups ($n = 4$). Groups marked with distinct letters above the bars indicate significant differences ($p < 0.05$, two-way Analysis of variance (ANOVA)); (**B**) Liver triglyceride; (**C**) Total liver lipid; (**D**) Liver cholesterol. Values in B-D represent means ± SE from 10 different animals, and data were analyzed by two-way ANOVA with Tukey post-test to evaluate main and interactive effects. NS: not significant.

We have previously shown that *Mkp-1+/+* mice but not *Mkp-1−/−* mice developed hyperglyceridemia in response to *E. coli* infection [10]. To understand how Mkp-1 deficiency and sepsis affect liver lipid contents, we measured triglyceride, total lipid, and cholesterol levels in the livers. Consistent with a previous report [34], uninfected *Mkp-1−/−* mice had lower liver triglyceride than uninfected *Mkp-1+/+* mice (Figure 1B). Surprisingly, *E. coli* infection significantly increased total liver triglyceride levels in *Mkp-1+/+* mice, but not in *Mkp-1−/−* mice (Figure 1C). *E. coli* infection also increased total liver lipid content in the *Mkp-1+/+* mice, but not in *Mkp-1−/−* mice (Figure 1C), while liver cholesterol content did not differ between groups (Figure 1D). The differences in liver lipid contents were corroborated by histology analyses (Figure 2A). Diffused hepatocellular swelling and clearing, an indication of hepatic glycogen deposition, were seen in both un-infected *Mkp-1+/+* and *Mkp-1−/−* mice, although the hepatocellular swelling and clearing were more prevalent in *Mkp-1+/+* mice than in *Mkp-1−/−* mice (Figure 2A, Top row). Upon *E. coli* infection, dramatic differences were seen in the histology between the *Mkp-1+/+* and *Mkp-1−/−* livers (Figure 2A, bottom row). Histological examination revealed moderate to marked, centrilobular to midzonal, hepatocyte vacuolation (often referred to as hepatocyte ballooning) in the livers of *E. coli*-infected *Mkp-1+/+* mice, indicating hepatic lipidosis. Additionally, multifocal-random moderate to marked necrosuppurative hepatitis with rare intralesional rod-shaped bacteria were seen throughout the *Mkp-1+/+* liver sections. Furthermore, in *Mkp-1+/+* liver sections numerous neutrophils and necrotic cellular debris were randomly distributed throughout hepatic parenchyma and occasionally formed abscesses. In contrast, *E. coli*-infected *Mkp-1−/−* livers exhibited marked midzonal hepatocellular swelling and clearing, suggesting considerable glycogen levels in these mice. Additionally, multifocal-random and marked necrosuppurative hepatitis was seen in the

$Mkp\text{-}1^{-/-}$ liver sections. Hepatic abscesses in the $Mkp\text{-}1^{-/-}$ livers were predominantly comprised of numerous bacterial rods intermixed with cellular necrotic debris and neutrophils. Quantitation of hepatocyte ballooning using the Brunt liver steatosis scores system confirmed more lipid droplets in liver sections of uninfected $Mkp\text{-}1^{+/+}$ mice that in those of uninfected $Mkp\text{-}1^{-/-}$ mice (Figure 2B). *E. coli* infection further enhanced hepatocyte lipidosis in $Mkp\text{-}1^{+/+}$ mice but not in $Mkp\text{-}1^{-/-}$ mice. As expected, *E. coli* infection resulted in a prominent increase in inflammatory infiltrate score, which is further exacerbated in $Mkp\text{-}1^{-/-}$ mice (Figure 2C).

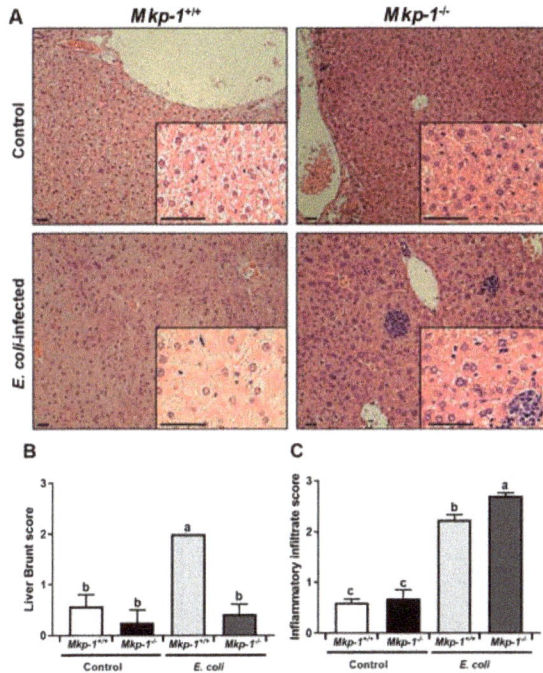

Figure 2. Liver histology of $Mkp\text{-}1^{+/+}$ and $Mkp\text{-}1^{-/-}$ mice before and after *E. coli* infection. Control and *E. coli*-infected mice (2.5 × 10⁷ CFU/g of body weight, i.v.) were euthanized 24 h post infection. Small portions of the livers were excised and fixed in formalin, paraffinized, and sectioned for histological assessment. Liver lipid level was scored according to Brunt steatosis scoring system. (**A**) Histology of livers from control and *E. coli*-infected mice. Images are representative H&E-stained liver sections. Outset magnification ×100; inset ×400; (**B**) Liver lipid score evaluated by the Brunt steatosis scoring system; (**C**) Hepatic inflammatory infiltrate score. Hepatic inflammatory infiltrate score was evaluated by counting neutrophils in 20 randomly selected optical fields. Two-way ANOVA was conducted to detect group differences. Groups without a common superscript were significantly different ($p < 0.05$).

2.2. The Impact of Mkp-1 Deficiency and E. coli Infection on Global Liver Gene Expression Profile

To explore the mechanisms of lipid dysregulation caused by *Mkp-1* deficiency and sepsis, RNA-seq was conducted using hepatic RNA extracts. Without *E. coli* infection, *Mkp-1* deficiency increased the mRNA expression of 241 genes and decreased the mRNA expression of 116 genes (Figure 3A). In $Mkp\text{-}1^{+/+}$ mice, *E. coli* infection enhanced the expression of 2519 genes and lowered the expression of 2850 genes (Figure 3B). In contrast, *E. coli* infection caused the upregulation of 3666 genes and downregulation of 3585 genes in $Mkp\text{-}1^{-/-}$ mice (Figure 3C). Impressively, compared to *E. coli*-infected wildtype mice, 2750 genes were upregulated and 2671 genes were downregulated in *E. coli*-infected $Mkp\text{-}1^{-/-}$ mice, suggesting profound exacerbation of the host transcriptional responses in the livers of $Mkp\text{-}1^{-/-}$ mice (Figure 3D).

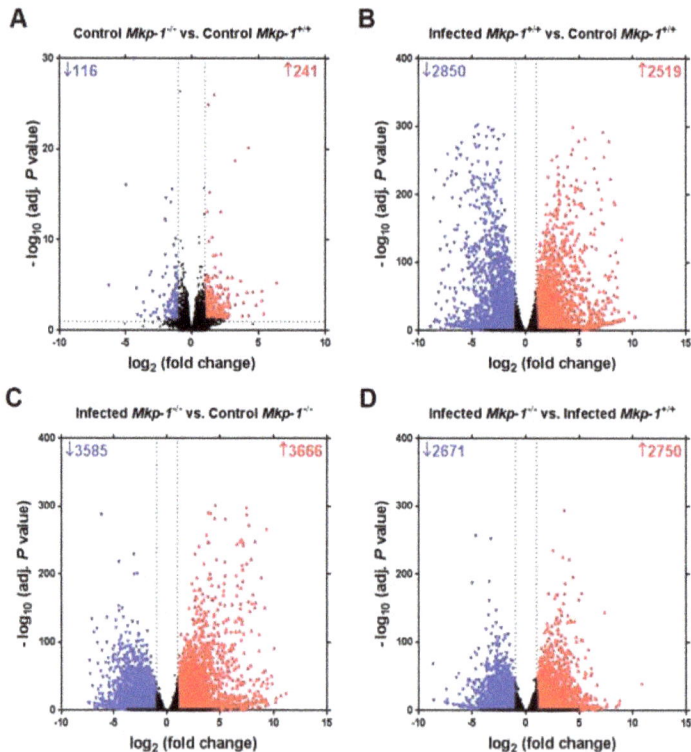

Figure 3. Differentially expressed genes in $Mkp\text{-}1^{+/+}$ and $Mkp\text{-}1^{-/-}$ mice before and following *E. coli* infection. $Mkp\text{-}1^{+/+}$ and $Mkp\text{-}1^{-/-}$ mice were either infected i.v. with *E. coli* at a dose of 2.5×10^7 CFU/g of body weight or injected with PBS (controls). Mice were euthanized after 24 h, and total RNA was isolated from the livers of four mice using Trizol for RNA-seq analyses. Volcano plots show the extent of differentially expressed gene (adjusted p value < 0.05, absolute value of $\log_2$ fold change 2) in each of four comparisons: control $Mkp\text{-}1^{+/+}$ vs. control $Mkp\text{-}1^{-/-}$ (**A**), infected $Mkp\text{-}1^{+/+}$ vs. control $Mkp\text{-}1^{+/+}$ (**B**), infected $Mkp\text{-}1^{-/-}$ vs. control $Mkp\text{-}1^{-/-}$ (**C**), and infected $Mkp\text{-}1^{-/-}$ vs. infected $Mkp\text{-}1^{+/+}$ (**D**).

We also categorized the differentially expressed genes in the liver, by Panther pathway analysis. In the absence of infection, genes of several pathways were differentially expressed in the livers of $Mkp\text{-}1^{-/-}$ mice relative to those of wildtype mice, including the integrin signaling pathway, the p53 pathway, and the p38 pathway (Figure 4A). Consistent with the idea that *E. coli* infection elicits a landscape change in inflammatory gene expression, upon *E. coli* infection genes in the Toll-like receptor (TLR) pathway and inflammatory cytokine and chemokine pathways were substantially altered in the wildtype mice (Figure 4B). In the *Mkp-1* knockout mice the inflammatory response, indicated by alteration in the chemokine and cytokine signaling pathways, TLR pathways, as well as interleukin signaling pathways, were dramatically enhanced after *E. coli* infection (Figure 4C). Moreover, blood coagulation, cholesterol biosynthesis, and inflammatory cytokine and chemokine pathways were more robustly altered by *Mkp-1* deficiency in the *E. coli*-infected mice (Figure 4D). Interestingly, the cholesterol biosynthesis as well as blood coagulation genes were significantly altered upon *Mkp-1* deletion in the *E. coli*-infected mice.

We also analyzed the RNA-seq data using DESeq2 algorithm to identify the pathways differentially affected in $Mkp\text{-}1^{+/+}$ and $Mkp\text{-}1^{-/-}$ mice after *E. coli* infection (Figure 5A). Among the pathways that were preferentially affected by *E. coli* infection in the two strain of mice were various metabolic processes, including retinol metabolism, steroid synthesis, and carbon metabolism (Figure 5B), suggesting an important function of Mkp-1 in broad metabolic functions.

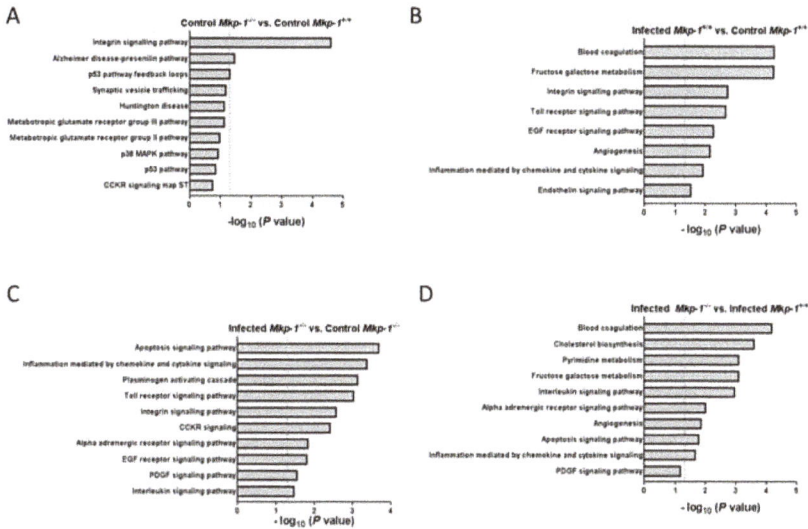

Figure 4. Pathway over-representation analysis for differentially expressed transcripts in control and *E. coli*-infected *Mkp-1+/+* and *Mkp-1−/−* mice. RNA-seq data were analyzed with the Panther pathway classification system. Bar plots depict overrepresented Panther pathways affected by Mkp-1 knockout and/or *E. coli* infection in the following comparisons: control *Mkp-1−/−* vs. control *Mkp-1+/+* (**A**), *E. coli*-infected *Mkp-1+/+* vs. control *Mkp-1+/+* (**B**), *E. coli*-infected *Mkp-1−/−* vs. control *Mkp-1−/−* (**C**), and *E. coli*-infected *Mkp-1−/−* vs. *E. coli*-infected *Mkp-1+/+* (**D**). Dotted lines indicate $p = 0.05$ on the $-\log_{10}$ scale.

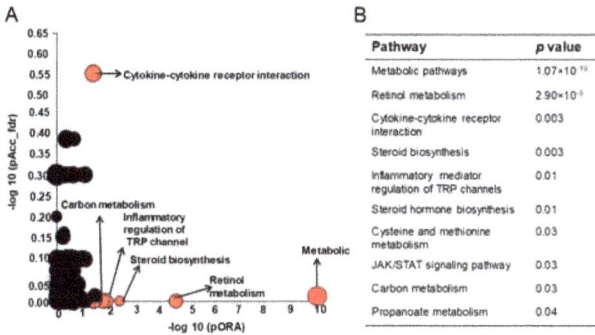

Figure 5. Differential affected pathways in the livers of *Mkp-1+/+* and *Mkp-1−/−* mice by *E. coli* infection. The RNA-seq data were analyzed using the iPathwayGuide analytic platform to identify the pathways differentially modulated in *Mkp-1+/+* and *Mkp-1−/−* mice by *E. coli* infection. (**A**) Pathways differentially modulated in *Mkp-1+/+* and *Mkp-1−/−* mice. Pathway analysis was done on the differentially expressed genes (>2-fold change either direction) using iPathwayGuide analysis tool. Each pathway is represented by a single dot in the graph for over-representation on the horizontal axis (pORA) and the perturbation on the vertical axis (pAcc), with the size of the dot proportional to the number of genes differentially modulated in that pathway. Differentially modulated pathways (FDR < 5%) were shown in red and pathways that did not reach statistical significance were shown in black. Note: In both axis, *p*-values are shown on the $-\log_{10}$ scale; (**B**) The top 10 pathways differentially modulated in *Mkp-1+/+* and *Mkp-1−/−* mice by *E. coli* infection according to the pORA values.

2.3. E. coli Infection Caused a Major Shift in Gene Expression of the Fatty Acid and Glucose Metabolic Programs and Mkp-1 Deficiency Disrupts This Shift

Because of the striking differences in the triglyceride contents between *Mkp-1*$^{+/+}$ and *Mkp-1*$^{-/-}$ mice, particularly after *E. coli* infection, we focused on the proteins involved in triglyceride and fatty acid metabolisms. We found that fatty acid metabolism and hepatic glycolysis/gluconeogenesis pathways were profoundly altered by both *Mkp-1* deficiency and *E. coli* infection (Figure 6). The heat map presented in Figure 6 depicts the expression levels of carbohydrate metabolism genes significantly altered by *E. coli* infection and/or *Mkp-1* knockout. A number of genes involved in fatty acid transport, such as fatty acid-binding protein 1 (Fabp1) and four apolipoproteins (Apoa1, 2, 5 and Apoc3), were downregulated by *E. coli* infection in *Mkp-1*$^{+/+}$ mice. Additionally, *E. coli* infection also attenuated the expression of genes involved in fatty acid synthesis and utilization, such as stearoyl-CoA desaturase 1 (Scd1), long chain fatty acid-CoA ligase 1 (Acsl1), acyl-CoA thioesterase (Acot) 1 and 4, peroxisome proliferator-activated receptor (PPAR) γ coactivator 1-α and β (Ppargc1a and b), Forkhead box protein O1 (Foxo1), acetyl-CoA acyltransferase 1 (Acaa1b), and CCAAT/enhancer-binding protein (Cebp) α (Cebpa) in *Mkp-1*$^{+/+}$ mice. Furthermore, a number of perilipin (Plin) genes, including Plin 2, 3, and 4 were upregulated in *Mkp-1*$^{+/+}$ mice. Interestingly, many genes involved in glycolysis were upregulated by *E. coli* infection in both *Mkp-1*$^{+/+}$ and *Mkp-1*$^{-/-}$ mice, including pyruvate dehydrogenase kinase (Pdk) 3/4, and 6-phosphofructo-2-kinase/fructose-2,6-biphosphatase 3 (Pfkfb3). Most importantly, *E. coli* infection-induced changes of the majority of these genes were markedly altered by *Mkp-1* deficiency. The profound changes in the transcriptome of the lipid and carbohydrate metabolic pathways suggest that a major shift in carbon/energy metabolism occurred in wildtype mice upon *E. coli* infection and that the Mkp-1 protein is required for this metabolic shift.

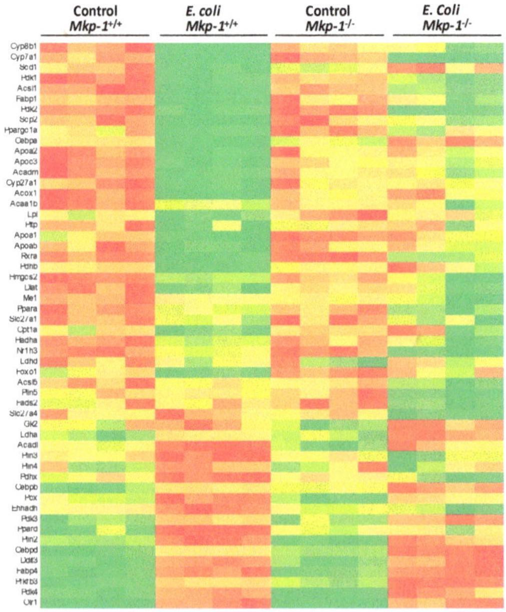

Figure 6. Expression of fatty acid metabolism genes significantly altered by *E. coli* infection and/or *Mkp-1* knockout. Heat map showing relative changes in expression of select transcripts in individual mice as assessed using RNA-seq. The color gradient ranges from red (highest levels of expression) to green (lowest expression levels), with yellow representing intermediate levels. Note: Each lane represents a different animal, and only significantly affected genes were shown in the heat map ($p < 0.05$, Student's *t*-test, $n = 4$).

2.4. E. coli Infection Lowers the Expression of Mammalian Target of Rapamycin (mTOR) and Lipogenic Genes

The hepatic mTOR/Akt signaling promotes hepatic lipid biosynthesis by activating SREBP-1 and increasing PPARγ expression [35]. RNA-seq data demonstrated that infected *Mkp-1*$^{-/-}$ mice had downregulated mTOR (designated as Mtor for mouse) expression relative to infected *Mkp-1*$^{+/+}$ mice, although the expression levels in uninfected mice were similar in *Mkp-1*$^{+/+}$ and *Mkp-1*$^{-/-}$ mice (Figure 7A). The expression of Pparg (the murine ortholog of PPARγ) was lower in uninfected *Mkp-1*$^{-/-}$ mice than in uninfected *Mkp-1*$^{+/+}$ mice, although *E. coli* infection resulted in a decrease in Pparg expression in both groups. The expression of three other lipogenic regulators, Ppargc1a (murine ortholog of PPARγ coactivator (PGC)-1α)), Ppargc1b (murine ortholog of PGC-1β), and Srebf1 (murine ortholog of SREBP-1), were similar between *Mkp-1*$^{+/+}$ and *Mkp-1*$^{-/-}$ mice. *E. coli* infection resulted in a decrease in their expression in both *Mkp-1*$^{+/+}$ and *Mkp-1*$^{-/-}$ groups. The differences observed by RNA-seq for both Mtor and Pparg expression were confirmed by quantitative reverse transcription PCR (qRT-PCR) (Figure 7B). Additionally, RNA-seq also identifies some differences in the expression of five lipogenic genes, Fasn, Scd1, acetyl-CoA carboxylase α (designated as Acaca for the murine ortholog), acetyl-CoA carboxylase β (designated as Acacb for the murine ortholoog), and diglyceride acyltransferase 2 (Dgat2) (Figure 7C). In *Mkp-1*$^{+/+}$ mice, *E. coli* infection downregulated the expression of Fasn and Scd1 (Figure 7C), but modestly increased the mRNA expression of Dgat2, an enzyme responsible for synthesis of triglyceride from fatty acid and glycerol [36,37]. In contrast, Scd1 expression was enhanced in *Mkp-1*$^{-/-}$ mice following *E. coli* infection. Acetyl-CoA carboxylase, particularly Acaca, catalyzes the carboxylation of acetyl-CoA to malonyl-CoA, the rate-limiting step in fatty acid synthesis. Although the levels of liver Acaca mRNA in *Mkp-1*$^{+/+}$ and *Mkp-1*$^{-/-}$ mice were similar, *E. coli* infection led to a decrease in liver Acaca mRNA expression in both groups (Figure 7C). Unlike Acaca which primarily regulates fatty acid synthesis, Acacb plays an important role in fatty acid oxidation. The expression levels of Acacb mRNA did not substantially differ except between un-infected *Mkp-1*$^{+/+}$ and *E. coli*-infected *Mkp-1*$^{-/-}$ mice (Figure 7C).

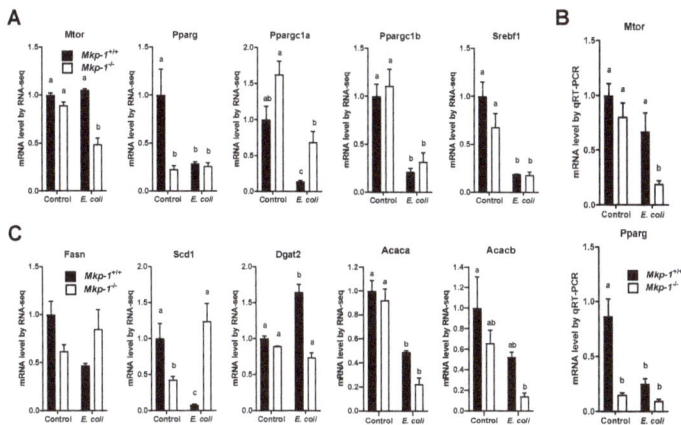

Figure 7. Hepatic mRNA expression of lipogenesis genes before and after *E. coli* infection. *Mkp-1*$^{+/+}$ and *Mkp-1*$^{-/-}$ mice were either infected i.v. with *E. coli* at a dose of 2.5×10^7 CFU/g of body weight or injected with PBS. Mice were euthanized after 24 h, and total RNA was isolated from the livers using Trizol. mRNA expression for different genes was assessed based on the RNA-seq data set, or quantified via qRT-PCR. Expression in un-infected *Mkp-1*$^{+/+}$ mice was set as 1. Values represent means ± S.E. from 4 animals for RNA-seq and 4–7 animals for qRT-PCR in each group. (**A**) Expression levels of lipogenic regulator genes based on RNA-seq; (**B**) Expression levels of Mtor and Pparg based on qRT-PCR; (**C**) Expression levels of lipogenic genes based on RNA-seq. Values in the graphs were compared by two-way ANOVA. Groups marked with distinct letters above the bars indicate significant differences ($p < 0.05$).

To assess the levels of the lipogenic proteins in the livers, Western blot analyses were performed using liver homogenates (Figure 8). Fasn protein levels in *E. coli*-infected *Mkp-1*$^{+/+}$ mice appeared lower although the difference was not statistically significant (Figure 8). Scd protein levels mirrored the differences in Scd1 mRNA levels. In the absence of *E. coli* infection, Scd1 protein levels were substantially lower in *Mkp-1*$^{-/-}$ mice than in *Mkp-1*$^{+/+}$ mice. While liver Scd1 protein levels dramatically plummeted in *Mkp-1*$^{+/+}$ mice following *E. coli* infection, liver Scd1 protein levels were significantly increased in *Mkp-1*$^{-/-}$ mice. Dgat2 protein levels were significantly increased in *Mkp-1*$^{+/+}$ mice following *E. coli* infection, but did not significantly change in *Mkp-1*$^{-/-}$ mice.

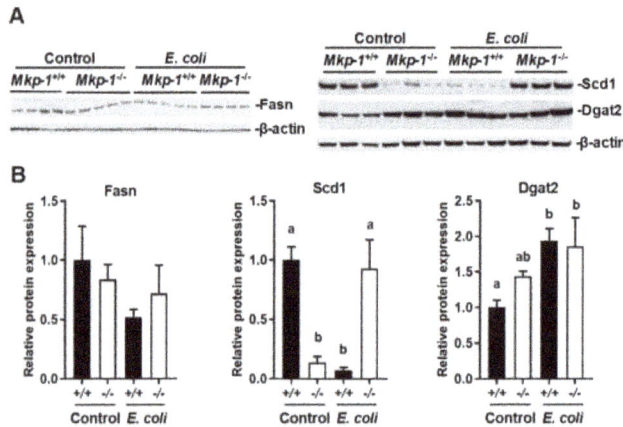

Figure 8. Levels of liver lipogenic proteins in control and *E. coli*-infected mice. Control and *E. coli*-infected mice (2.5×10^7 CFU/g of body weight, i.v.) were euthanized 24 h post infection. (**A**) Representative results of Western blot analyses. Liver protein extracts from control or *E. coli*-infected *Mkp-1*$^{+/+}$ and *Mkp-1*$^{-/-}$ mice were subjected to Western blotting with the indicated antibodies. The housekeeping protein β-actin was used as a control for normalization of protein loading; (**B**) Quantification of protein expression. The images were scanned by densitometry and expression levels normalized to β-actin. The expression level in control *Mkp-1*$^{+/+}$ mice was set as 1. The results were analyzed by two-way ANOVA. Groups marked with distinct letters above the bars indicate significant differences ($p < 0.05$).

2.5. E. coli Infection Increased the Expression of Genes Involved in Liver Fatty Acid Uptake, and Lowered the Expression of Genes Involved in Mitochondrial and Peroxisomal Fatty Acid Oxidation

Fatty acid uptake is an important mechanism for lipid accumulation in the liver. In response to the decline of blood glucose level, fat tissues release fatty acids through lipolysis. Cluster of differentiation 36 (Cd36) plays an important role in fatty acid uptake from the blood stream for lipogenesis in the liver [38]. To understand the mechanism underlying the differential effects of *E. coli* infection on liver and blood triglyceride contents of *Mkp-1*$^{+/+}$ and *Mkp-1*$^{-/-}$ mice, we measured Cd36 levels by qRT-PCR (Figure 9). Liver Cd36 mRNA levels were similar between control *Mkp-1*$^{+/+}$ and *Mkp-1*$^{-/-}$ mice, and *E. coli* infection caused a significant increase in Cd36 mRNA levels in *Mkp-1*$^{+/+}$ mice but had little effect on Cd36 mRNA levels in *Mkp-1*$^{-/-}$ mice.

We also assessed the expression of genes involved in fatty acid β-oxidation through qRT-PCR (Figure 10A). Carnitine palmitoyltransferase I and II (Cpt1a and Cpt2) are essential enzymes for β-oxidation of long chain fatty acid in mitochondria [39,40], while acyl-CoA oxidase 1 (Acox1) mediates fatty acid β-oxidation in peroxisome [39,41]. Cpt1a mRNA expression was significantly decreased in both *Mkp-1*$^{+/+}$ and *Mkp-1*$^{-/-}$ mice upon *E. coli* infection, although expression levels in control conditions were similar in the two genotypes of mice. Liver Cpt2 mRNA expression levels were lower in control *Mkp-1*$^{-/-}$ mice than in control *Mkp-1*$^{+/+}$ mice. *E. coli* infection decreased Cpt2 mRNA levels in *Mkp-1*$^{-/-}$ mice but had little effect in *Mkp-1*$^{+/+}$ mice. Acox1 mRNA levels were similar in

Mkp-1$^{+/+}$ and *Mkp-1$^{-/-}$* mice, and were decreased in both genotypes upon *E. coli* infection. Cpt1a protein levels, detected by Western blotting, were similar in un-infected *Mkp-1$^{+/+}$* and *Mkp-1$^{-/-}$* mice, and *E. coli* infection caused a significant decrease in both *Mkp-1$^{+/+}$* and *Mkp-1$^{-/-}$* mice (Figure 10B). Taken together, these results suggest that fatty acid oxidation in the liver was decreased with *E. coli* infection while fatty acid uptake was increased in *Mkp-1$^{+/+}$* but not in *Mkp-1$^{-/-}$* mice.

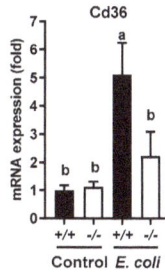

Figure 9. Liver mRNA expression levels of Cd36 in control and *E. coli*-infected mice. Control and *E. coli*-infected mice (2.5×10^7 CFU/g of body weight, i.v.) were euthanized 24 h post infection. Total RNA was isolated from the livers using Trizol. Cd36 mRNA levels were assessed via qRT-PCR. Expression in un-infected *Mkp-1$^{+/+}$* mice was set as 1. Values represent means ± S.E. from 4–7 animals in each group. The results were analyzed by two-way ANOVA. Groups marked with distinct letters above the bars indicate significant differences ($p < 0.05$).

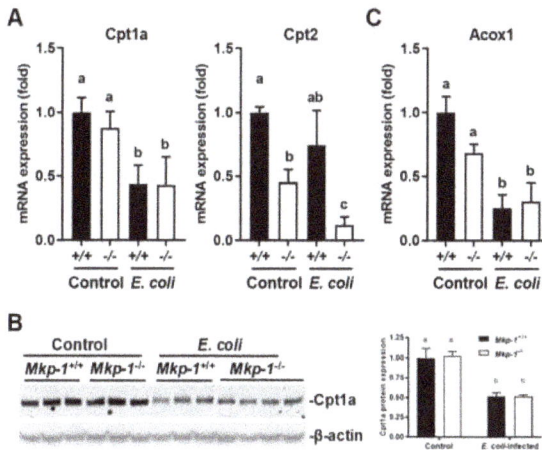

Figure 10. Hepatic expression of fatty acid β-oxidation proteins or genes before and after *E. coli* infection. *Mkp-1$^{+/+}$* and *Mkp-1$^{-/-}$* mice were either infected i.v. with *E. coli* at a dose of 2.5×10^7 CFU/g of body weight or injected with PBS. Livers were harvested 24 h post infection to isolate total RNA or protein. The mRNA and protein levels were assessed by qRT-PCR or Western blot analyses. (**A**) mRNA levels of fatty acid oxidation proteins assessed by qRT-PCR; (**B**) Representative Western blot of Cpt1a protein. Forty μg of protein for each sample was used for Western blot analysis. Each lane represents an individual animal. The membrane was stripped and reblotted with β-actin antibody to verify comparable loading. The densities of the bands were determined by densitometry, normalized to β-actin levels, and the relative expression level of Cpt1a protein in each group was depicted as means ± SE in the graph on the right ($n = 3$–4 mice per group); (**C**) Acox1 mRNA levels assessed by qRT-PCR. Expression in un-infected *Mkp-1$^{+/+}$* mice was set as 1. Values represent means ± S.E. of 3–5 different animals in each group. Values were compared by two-way ANOVA. Groups marked with distinct letters above the bars indicate significant differences ($p < 0.05$).

*2.6. Effects of E. coli Infection and Mkp-1 Deficiency on the Expression of Phosphoenolpyruvate
Carboxykinase 1 (Pck1) Protein*

The cytosolic Pck1 protein, often referred to as phosphoenolpyruvate carboxykinase-cytosolic
isoform (PEPCK-c), is an important regulator in gluconeogenesis [42–44]. Hepatic Pck1 facilitates
gluconeogenesis by synthesizing phosphoenolpyruvate from oxaloacetate [45]. Previously, it has
been shown that liver Pck1 expression is enhanced in *Mkp-1*$^{-/-}$ mice in control conditions [46].
Our RNA-seq analysis indicates that liver Pck1 mRNA levels were significantly increased in both
Mkp-1$^{+/+}$ and *Mkp-1*$^{-/-}$ mice following *E. coli* infection. Although the basal Pck1 mRNA levels
in *Mkp-1*$^{+/+}$ and *Mkp-1*$^{-/-}$ mice were similar (Figure 11A), liver Pck1 protein levels in control
Mkp-1$^{-/-}$ mice were significantly higher than those in control *Mkp-1*$^{+/+}$ mice (Figure 11B). *E. coli*
infection further enhanced the expression of Pck1 protein in both *Mkp-1*$^{+/+}$ and *Mkp-1*$^{-/-}$ mice
(Figure 11B). The enhanced expression of Pck1 suggests that both *Mkp-1* knockout and *E. coli* infection
stimulate gluconeogenesis.

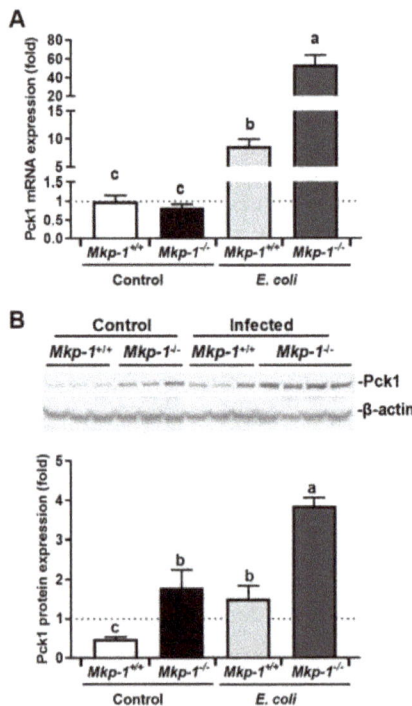

Figure 11. Hepatic PCK1 expression in *Mkp-1*$^{+/+}$ and *Mkp-1*$^{-/-}$ mice before and after *E. coli* infection.
Mkp-1$^{+/+}$ and *Mkp-1*$^{-/-}$ mice were either infected i.v. with *E. coli* at a dose of 2.5 $\times$ 10^7 CFU/g of
body weight or injected with PBS. Mice were euthanized after 24 h to extract total RNA and protein.
(**A**) Pck1 mRNA expression levels were quantitated via qRT-PCR. Expression in un-infected *Mkp-1*$^{+/+}$
mice was set as 1. Values represent means $\pm$ S.E. from 4–5 different animals in each group; (**B**) Liver
Pck1 protein levels. Liver protein extracts from control or *E. coli*-infected *Mkp-1*$^{+/+}$ and *Mkp-1*$^{-/-}$
mice were subjected to Western blotting with a mouse monoclonal antibody against Pck1. Each lane
represents an individual animal. The membranes were stripped and then used to blotting with β-actin
antibody. Representative results from Western blot analyses are shown. The densities of the bands were
determined by densitometry, and normalized to β-actin levels, and the relative expression levels of
Pck1 protein in each group were depicted as means $\pm$ SE in the graph below. Values were compared by
two-way ANOVA. Groups marked with distinct letters above the bars indicate significant differences
($p < 0.05$).

3. Discussion

We have previously shown that $Mkp\text{-}1^{-/-}$ mice exhibited significant increases in both inflammatory cytokine production and mortality after *E. coli* infection relative to $Mkp\text{-}1^{+/+}$ mice [10]. Our earlier studies have also shown that *E. coli* infection triggers a dramatic increase in blood triglyceride levels in $Mkp\text{-}1^{+/+}$ mice but not in $Mkp\text{-}1^{-/-}$ mice [10], suggesting profound alterations in lipid metabolism in the $Mkp\text{-}1^{-/-}$ mice following *E. coli* infection. To understand the role of Mkp-1 in the regulation of metabolism during sepsis, we analyzed the lipid contents and global gene expression profiles in the livers of $Mkp\text{-}1^{+/+}$ and $Mkp\text{-}1^{-/-}$ mice either in control conditions or after *E. coli* infection. Here we report that *E. coli* infection increased liver lipid content in $Mkp\text{-}1^{+/+}$ mice (Figure 1), indicating that hyperlipidemia after *E. coli* infection was not the result of depletion of hepatic lipid stores. The increase in liver lipid content in $Mkp\text{-}1^{+/+}$ mice was supported by hepatocyte ballooning after *E. coli* infection (Figure 2). RNA-seq analysis of the liver tissues detected a profound difference in gene expression profiles between wildtype and $Mkp\text{-}1^{-/-}$ mice in the liver following *E. coli* infection (Figure 3). Remarkably, over 5000 genes (>20% of all murine genes) exhibited a >2-fold difference in expression levels between the two groups of mice after *E. coli* infection (Figure 3D), highlighting the critical role of Mkp-1 in the liver response to sepsis. Interestingly, *E. coli* infection caused profound changes in the expression of many genes involved in lipid metabolism, including fatty acid uptake, utilization, and synthesis in $Mkp\text{-}1^{+/+}$ mice (Figure 6). However, in $Mkp\text{-}1^{-/-}$ mice the *E. coli* infection-induced changes in lipid metabolism-related genes were profoundly disrupted. In other words, unlike in the wildtype livers, in the absence of Mkp-1 the livers were unable to adjust their lipid metabolic program (Figure 6).

3.1. The Critical Role of Mkp-1 as a p38 Regulator in the Liver

Consistent with the notion that p38 is the preferred substrate of Mkp-1, we found that base line p38 activity was higher in uninfected $Mkp\text{-}1^{-/-}$ livers than in $Mkp\text{-}1^{+/+}$ livers (Figure 1A). This indicates that Mkp-1 is critical in the control of p38 activity in the liver under normal conditions. Interestingly, in pre-clinical studies and clinical trials a common side effect of the p38 inhibitors is hepatotoxicity [47], suggesting that base line p38 activity is required for protection of the liver. Elevated base line p38 activity could be directly or indirectly implicated in the changes seen in the metabolic pathways and transcriptome. Remarkably, we found that liver p38 activity is dramatically decreased 24 h after *E. coli* infection in $Mkp\text{-}1^{+/+}$ mice (Figure 1A), while it remained high in $Mkp\text{-}1^{-/-}$ mice. Considering that *Mkp-1* is a gene highly inducible by extracellular stimuli [29,32,33,48,49], a systemic *E. coli* infection likely enhanced the expression of *Mkp-1* in the liver of the $Mkp\text{-}1^{+/+}$ leading to the de-phosphorylation of p38. In the absence of the *Mkp-1* gene, it is not surprising that p38 remained high in the livers of the *E. coli*-infected $Mkp\text{-}1^{-/-}$ mice. Since p38 plays an important role in inflammatory responses [50], persistently high p38 activity in the livers of *E. coli*-infected $Mkp\text{-}1^{-/-}$ mice provides a plausible explanation for the differential expression of genes involved in inflammation (Figures 4 and 5).

3.2. Mkp-1 and Hyperlipidemia of Sepsis

In a prior study we reported that *E. coli* infection caused a 13-fold increase in blood triglyceride levels in $Mkp\text{-}1^{+/+}$ mice, but not the $Mkp\text{-}1^{-/-}$ mice [10]. This is consistent with the work of others reporting that bacterial endotoxin causes 'lipemia of sepsis' by increasing VLDL-mediated triglyceride release from hepatocytes [14], and by limiting lipoprotein lipase-mediated VLDL clearance in peripheral tissues [17,18,51]. Lipoproteins, such as triglyceride-rich VLDL, not only provide fuel to the host to fight against bacterial infection, but also sequester and neutralize endotoxin [14,52]. Therefore, hypertriglyceridemia during sepsis may be an adaptive host response to bacterial infections. Our studies indicate that *E. coli* infection induced increases in not only blood triglyceride [10] but also liver triglyceride content (Figure 1) in a Mkp-1-dependent manner, raised a strong possibility that the liver actually accelerates glyceride synthesis during sepsis. While it is unclear how the wildtype mice manage to increase blood triglyceride levels, while also increasing liver triglyceride contents, we can speculate. In our experimental setting we observed that *Mkp-1+/+* mice usually stop feeding 2–3 h after *E. coli* infection, which is likely the result of an acute phase response and cytokine storm. Infected mice usually abstain from feeding in the first few days then then resume feeding as they recover. The septic mice were sacrificed at 24 h for the measurement of blood and liver lipid contents. Thus, the acute increase in triglyceride levels in blood and liver of *E. coli*-infected *Mkp-1+/+* mice were unlikely due to increased consumption of food [53]. Instead, the increased blood triglyceride content is probably due to increased catabolism of adipose and muscle tissues, as TNF-α produced following *E. coli* infection has been shown to suppress lipogenesis and enhance lipolysis in these tissues [54,55]. We speculate that increased liver triglyceride following infection in the wildtype mice is more likely the result of enhanced hepatic lipogenesis from fatty acids and glycerol taken up from the blood rather than from fatty acids synthesized de novo from acetyl-CoA in the liver for the following reasons. First, a number of lipogenic genes involved in triglyceride synthesis from acetyl-CoA, including Pparg (murine ortholog of PPARγ), Ppargc1a/b (murine ortholog of PGC1α/β), Srebf1, Fasn, Scd1, and Acaca that catalyzes the rate-limiting step in fatty acid synthesis, were substantially downregulated (Figures 6 and 7). Second, the expression of Dgat2, the liver enzyme responsible for triglyceride synthesis from fatty acids and glyceride [56], was increased in *Mkp-1+/+* mice after *E. coli* infection (Figures 7 and 8). Finally, mRNA expression of Cd36, the liver protein mediating fatty acid uptake [38], is markedly upregulated in *Mkp-1+/+* mice after *E. coli* infection. Fatty acids taken up from the blood can be converted in the liver to triglyceride by Dgat2 [56]. We think that the liver likely synthesizes triglyceride for consumption by other organs, because both the mitochondrial fatty acid β-oxidation-related Cpt1a protein [39,40]) and peroxisomal fatty acid β-oxidation protein Acox1 [39,41,57] are downregulated in *Mkp-1+/+* mice after *E. coli* infection (Figure 10). It is worth noting that a recent study has shown that p38 inhibition enhanced parenteral nutrition-induced hepatic steatosis and attenuated the expression of Cpt1a, Acox1, and Ppargc1a in a rat model [58]. The dramatic decrease in p38 activity in *E. coli*-infected *Mkp-1+/+* livers is consistent with the decreased expression of the Cpt1a, Acox1, and Ppargc1a as well as the increased triglyceride content in *E. coli*-infected *Mkp-1+/+* mice.

Several factors can explain why *Mkp-1−/−* mice failed to develop hyperlipidemia after *E. coli* infection like the *Mkp-1+/+* mice did. First, it has been shown that *Mkp-1−/−* mice on chow diet have lower adipose mass and higher metabolic activity with enhanced glucogenic activity under normal conditions [34]. Because *Mkp-1−/−* mice had lower fat mass, *E. coli* infection might not be able to trigger the release of fatty acids and glycerol from adipose tissue into the circulation. It is also possible that exacerbated production of large quantities of pro- and anti-inflammatory cytokines in *Mkp-1−/−* mice disrupted the normal triglyceride mobilization program in the adipose tissues. Alternatively, in the absence of Mkp-1, the livers might not be able to esterify the fatty acids released by adipose tissues into VLDL triglycerides in the liver. However, considering the normal Dgat2 protein levels in the livers of *E. coli*-infected *Mkp-1−/−* mice (Figure 8), we think this is unlikely.

Previously, it has been shown that liver-specific *Mkp-1* deletion enhances gluconeogenesis [34,46,59,60]. Furthermore, it has been found that *Mkp-1−/−* livers exhibit decreased Pparg and Srebf1 expression and increased Ppargc1a and Pck1 expression [46,60]. Our analyses, for the most part, corroborated their findings (Figure 6), although in our hands the difference in the liver expression of Ppargc1a and Srebf1 expression between *Mkp-1+/+* and *Mkp-1−/−* mice was not significant. Consistent with the elevated Pck1 expression in liver-specific *Mkp-1* knockout mice [46], we also found that liver Pck1 protein expression was significantly higher in the un-infected *Mkp-1−/−* mice than in un-infected *Mkp-1+/+* mice (Figure 11). The dramatically enhanced Pck1 expression suggests that in the absence of Mkp-1 animals adapt a more active 'wasting' program to meet their metabolic needs, partially explaining how *Mkp-1−/−* mice utilize less glycogen following *E. coli* infection [10].

While our analyses shed insight into the mechanisms by which Mkp-1 facilitates hyperlipidemia during sepsis through a transcriptome lens, there are some limitations that need to be considered.

RNA-seq performed here only gives a snap-shot of the transcriptome 24 h post-infection and prior to infection, the transitional transcriptome might be equally important in understanding the role of Mkp-1 in lipid metabolism during sepsis. Additionally, there were some disparities between RNA-seq data set, the qRT-PCR data, and/or Western blotting results. Despite these limitations, our analyses provide novel information on the function of Mkp-1 in the orchestration of lipid metabolic changes to facilitate immune defense. Our results clearly indicate that Mkp-1 plays an important role in the regulation of both the inflammatory response and metabolic programming during host defense against bacterial infection.

4. Materials and Methods

4.1. Experimental Animals and E. coli Infection

The present study was approved by the Institutional Animal Care and Use Committee of the Research Institute at Nationwide Children's Hospital (01505AR, 9 January 2017). *Mkp-1*$^{-/-}$ mice and the *Mkp-1*$^{+/+}$ controls on a C57/129 mixed background [61] were kindly provided by Bristol-Myers Squibb Pharmaceutical Research Institute. In the absence of challenge, *Mkp-1*$^{-/-}$ mice exhibit no sign of abnormality or growth retardation. All mice were housed with a 12 h alternating light-dark cycle at room temperature, and had free access to standard chow diet and water throughout the study. *Mkp-1*$^{-/-}$ and *Mkp-1*$^{+/+}$ mice were infected with or without *E. coli* as previously described [10]. Briefly, *E. coli* (O55:B5, ATCC 12014), purchased from American Type Culture Collection (Manassas, VA, USA), was grown in Luria broth at 37 °C for 18 h. The bacteria were pelleted by centrifugation, washed three times with phosphate-buffered saline (PBS), and finally suspended in PBS. *E. coli* was injected to the tail veins of mice at 2.5×10^7 CFU/g body weight. Infected mice were euthanized 24 h later by pentobarbital over dose. Blood was collected through cardiac puncture, and coagulated blood was centrifuged to obtain serum. The liver of each mouse was excised with a small piece preserved in formalin for histological assessment and the rest of the liver snap-frozen in liquid nitrogen prior to storage at −80 °C.

4.2. Biochemical Assessment of Liver Lipid

Hepatic total lipid was extracted and determined gravimetrically as described [62]. Extracted liver lipid was solubilized to determine triglyceride and cholesterol spectrophotometrically using a triglyceride or cholesterol measuring kit (Pointe Scientific, Canton, MI, USA), according to the manufacturer's instructions.

4.3. Histological Assessment of Liver Lipid Content

Hematoxylin and eosin (H&E) staining was conducted on paraffin-embedded liver sections (5 μm). The slides were evaluated blindly by a veterinary pathologist for histologically abnormalities. Images of ten randomly selected fields were captured (400× magnification) to assess liver lipid content using the established Brunt scoring system for assessing liver steatosis [63]. In brief, the liver lipid levels were scored as: grade 0 for <5% hepatocytes without lipid droplets; grade 1 for 5–33% of hepatocytes containing visible lipid droplets; grade 2 for fatty hepatocytes occupying 33–66% of the hepatic parenchyma; or grade 3 for >66% hepatocytes containing lipid droplets.

4.4. Liver RNA Extraction

Total RNA was extracted from frozen liver tissues using Trizol reagent (Thermo Fisher Scientific, Waltham, MA, USA), solubilized in UltraPure RNase/DNase-free water (Thermo Fisher Scientific), and quantified by using NanoDrop ND-1000 spectrophotometer (Marshall Scientific, Hampton, NH, USA).

4.5. RNA-Seq

For RNA-seq analyses, 1 µg total RNA was used as starting material. First, cytoplasmic and mitochondrial ribosomal RNAs were depleted using a NEBNext rRNA Depletion Kit (New England Biolabs, Ipswich, MA, USA), according to the manufacture's recommendations. The samples were then digested with DNase I to remove contaminated genomic DNA, and purified using NEBNext RNA Sample Purification Beads (New England Biolabs). The RNA library was prepared using a NEBNext Ultra Directional RNA Library Prep Kit for Illumina (New England Biolabs). Briefly, RNA was fragmented, and then cDNAs were synthesized using random primers. The resulting double-strand cDNA was then subject to end-repair, adapter ligation, and PCR amplification to generate the library. The indexed RNA-seq libraries were quantitated by quantitative PCR, pooled with equimolar amounts and sequenced on an illumina HiSeq 3000 sequencer using a 2 × 125 cycle run, as previously described [64,65].

Following computational de-multiplexing, single end reads (50 bp) in the FASTQ format were generated. Quality control and adapter trimming were accomplished using the FastQC (version 0.11.3) and Trim Galore (version 0.4.0) software packages. Trimmed reads were mapped to the Genome Reference Consortium GRCm38 (mm10) murine genome assembly using TopHat2 (version 2.1.0), and feature counts were generated using HTSeq (version 0.6.1). Statistical analysis for differential expression was performed using the DESeq2 package (version 1.16.1) in R, with the default Benjamini-Hochberg *p* value adjustment method. Statistically significant differential expression thresholds included an adjusted *p* value <0.05 and an absolute value linear fold change of 2 or greater. Overrepresentation analysis for select gene sets comprising five or more members was determined using hypergeometric statistical testing (hyper function in R Documentation). Additionally, significantly impacted pathways were analyzed using Advaita Bio's iPathwayGuide (https://www.advaitabio.com/ipathwayguide).

4.6. qRT-PCR

To confirm the result of RNA-seq, qRT-PCR was performed as previously described [62] with minor modifications. Briefly, genomic DNA was removed by digesting the total RNA with RQ1 RNase-Free DNase (Promega, Madison, WI, USA). Liver RNA was then reverse transcribed on PTC-200 DNA Engine Cycler (Bio-Rad, Hercule, CA, USA) with High-Capacity cDNA Reverse Transcription Kit (Applied Biosystems, Foster City, CA, USA). qRT-PCR was performed using PowerUp SYBR Green PCR Master Mix (Applied Biosystems) on a Realplex^2 Mastercycler (Eppendorf, Hauppauge, NY, USA). All primers were synthesized by Integrated DNA Technologies (Coralville, IA, USA). Table 1 listed the primer sequences for the mRNAs of the following proteins, including proteins involved in fatty acid uptake, synthesis, oxidation, or their regulation: Cd36, Cpt1a/2, Acox1, Pck1, and Mtor. Hepatic mRNA expression of the genes of interest was calculated relative to 18s using the $2^{-\Delta\Delta Ct}$ method [66].

Table 1. Primers used for qRT-PCR reactions.

Gene	Forward Primer	Reverse Primer
18s	GTAACCCGTTGAACCCCATT	CCATCCAATCGGTAGTAGCG
Acox1	CAGGAAGAGCAAGGAAGTGG	CCTTTCTGGCTGATCCCATA
Cd36	ATGGGCTGTGATCGGAACTG	TTTGCCACGTCATCTGGGTTT
Cpt1a	CAGAGGATGGACACTGTAAAGG	CGGCACTTCTTGATCAAGCC
Cpt2	GGATAAACAGAATAAGCACACCA	GAAGGAACAAAGCGGATGAG
Mtor	ATTCAATCCATAGCCCCGTC	TGCATCACTCGTTCATCCTG
Pck1	TCTCTGATCCAGACCTTCCAA	GAAGTCCAGACCGTTATGCAG
Pparg	CCAGAGTCTGCTGATCTGCG	GCCACCTCTTTGCTCTGATC

Int. J. Mol. Sci. **2018**, *19*, 3904

4.7. Western Blotting

Frozen liver tissues were homogenized in lysis buffer (20 mM HEPES, pH7.4, 50 mM β-glycerol phosphate, 2 mM EGTA, 1 mM DTT, 10 mM NaF, 1 mM sodium orthovanadate, 10% glycerol, 1 mM PMSF, 2 μM leupeptin, 1.5 μM pepstatin, 0.3 μM aprotinin, and 50 nM microcystin-LR), using a Bullet Blender (Next Advance, Troy, NY, USA). Triton X-100 was then added to the homogenates to a final concentration of 1%, and the homogenates were incubated at 4 °C on a rotator at 300 rpm for 30 min. The homogenates were then centrifuged at 14,000× *g* for 10 min to collect the supernatants, and the protein concentrations were measured using a Protein Assay Kit (Bio-Rad). Extracted liver proteins were then separated on a NuPage 10% Bis-Tris gel (Invitrogen, Carlsbad, CA, USA), and transferred to nitrocellulose membranes. Membranes were probed for 1 h at room temperature or overnight at 4 °C with a primary antibody against a protein of interest. After washing three times with Tris-buffered saline with 0.1% Tween-20, membranes were incubated with a horseradish peroxidase-conjugated anti-rabbit or anti-mouse secondary antibody (GE Healthcare, Piscataway, NJ, USA). Mouse monoclonal antibodies against Cpt1a, Dgat2, Fasn, Pck1, and Scd1 as well as the rabbit polyclonal antibody against total p38 were purchased from Santa Cruz Biotechnology (Santa Cruz, CA, USA). Mouse monoclonal antibody against β-actin was purchased from Sigma-Aldrich (St. Louis, MO, USA). The rabbit polyclonal antibody against phospho-p38 was purchased from Cell Signaling (Danvers, MA, USA). Immunoreactive bands were developed using enhanced chemiluminescence reagent (Millipore, Burlington, MA, USA). The Western blot images were acquired using Epson Perfection 4990 PHOTO scanner (Epson, Long Beach, CA, USA) and intensities of the immunoreactive bands were measured by densitometry using the UVP Vison Works LS software (Upland, CA, USA).

4.8. Statistical Analysis

Data were analyzed using GraphPad Prism 7 (GraphPad Software; La Jolla, CA, USA). The main effects and their interaction were evaluated by two-way ANOVA with Tukey post-hoc test to detect group differences. In addition, two-tail student's *t*-test was conducted to detect statistical difference of the biomarkers measured in only two groups. Variables with unequal variance were log-transformed to achieve a normal distribution. Differences with $p < 0.05$ were considered significant.

4.9. Data Availability

The main data supporting the findings of this study are available from the NCBI GENE Expression Omnibus GES122741. For additional information contact the corresponding author.

Author Contributions: Y.L. and J.L. designed the study; J.L., X.W. (Xanxi Wang), X.W. (Xiantao Wang), A.J.B., W.M.W., and S.G.K. conducted experiments; J.L., A.J.B., S.G.K., W.E.A.I., D.A., L.D.N., L.S., I.B., M.H., and Y.L. analyzed the data. J.L. and Y.L. wrote the manuscript. All authors read, provided substantial intellectual input, and approved the final manuscript.

Funding: This research was funded by grants from NIH (R01 AI68956 and R01 AI124029 to Y.L., and R01 HL113508 to L.S.).

Acknowledgments: We are grateful to Bristol-Myers Squibb Pharmaceutical Research Institute for proving *Mkp-1* knockout mice. We thank Tiffany Jenkins for assistance with histological analysis. We also gratefully acknowledge Xiaomei Meng for technical assistance. We thank Harshan Pisharath, Laura Goodchild, and James Cooper for vivarium expertise.

Conflicts of Interest: The authors declare no conflict of interest.

Abbreviations

Acaa1	Acetyl-CoA acyltransferase 1
Acaca	Acetyl-CoA carboxylase α
Acacb	Acetyl-CoA carboxylase β
Acsl1	Long chain fatty acid-CoA ligase 1
Acot	Acyl-CoA thioesterase
Acox1	Acyl-CoA oxidase 1
ANOVA	Analysis of variance
Apo	Apolipoprotein
Cd36	Cluster of differentiation 36
CFU	Colony forming units
Cebp	CCAAT/enhancer-binding protein
Cpt	Carnitine palmitoyltransferase
Dgat2	Diglyceride acyltransferase 2
Fabp1	Fatty acid-binding protein 1
Fasn	Fatty acid synthase
Foxo1	Forkhead box protein O1
JNK	c-Jun N-terminal kinase
LPS	Lipopolysaccharide
Mkp-1	MAP kinase phosphatase-1
Mtor	Mammalian target of rapamycin
NF-κB	Nuclear factor-κB
PBS	Phosphate-buffered saline
Pck1/PEPCK-c	Phosphoenolpyruvate carboxykinase, cytosolic isoform
Plin	Perilipin
Pparg/PPARγ	Peroxisome proliferator-activated receptor γ
Ppargc1/PGC-1	PPARγ coactivator 1
Scd1	Stearoyl-CoA desaturase 1
qRT-PCR	Quantitative reverse transcription PCR
Srebf1/SREBP-1	Sterol regulatory element-binding transcription factor 1
TLR	Toll-like receptor
TNF-α	Tumor necrosis factor-α
VLDL	Very low-density lipoprotein

References

1. Angus, D.C.; Linde-Zwirble, W.T.; Lidicker, J.; Clermont, G.; Carcillo, J.; Pinsky, M.R. Epidemiology of Severe Sepsis in the United States: Analysis of Incidence, Outcome, and Associated Costs of Care. *Crit. Care Med.* **2001**, *29*, 1303–1310. [CrossRef] [PubMed]
2. Mayr, F.B.; Yende, S.; Angus, D.C. Epidemiology of Severe Sepsis. *Virulence* **2014**, *5*, 4–11. [CrossRef] [PubMed]
3. Trager, K.; DeBacker, D.; Radermacher, P. Metabolic Alterations in Sepsis and Vasoactive Drug-Related Metabolic Effects. *Curr. Opin. Crit Care.* **2003**, *9*, 271–278. [CrossRef] [PubMed]
4. Miller, S.I.; Wallace, R.J., Jr.; Musher, D.M.; Septimus, E.J.; Kohl, S.; Baughn, R.E. Hypoglycemia As a Manifestation of Sepsis. *Am. J. Med.* **1980**, *68*, 649–654. [CrossRef]
5. Chiolero, R.; Revelly, J.P.; Tappy, L. Energy Metabolism in Sepsis and Injury. *Nutrition* **1997**, *13*, 45S–51S. [CrossRef]
6. Cetinkaya, A.; Erden, A.; Avci, D.; Karagoz, H.; Karahan, S.; Basak, M.; Bulut, K.; Gencer, V.; Mutlu, H. Is Hypertriglyceridemia a Prognostic Factor in Sepsis? *Ther. Clin. Risk Manag.* **2014**, *10*, 147–150. [CrossRef] [PubMed]
7. Naylor, J.M.; Kronfeld, D.S. In Vivo Studies of Hypoglycemia and Lactic Acidosis in Endotoxic Shock. *Am. J. Physiol.* **1985**, *248*, E309–E316. [CrossRef]

8.   Berk, J.L.; Hagen, J.F.; Beyer, W.H.; Gerber, M.J. Hypoglycemia of Shock. *Ann. Surg.* **1970**, *171*, 400–408. [CrossRef]

9.   Woodske, M.E.; Yokoe, T.; Zou, B.; Romano, L.C.; Rosa, T.C.; Garcia-Ocana, A.; Alonso, L.C.; O'Donnell, C.P.; McVerry, B.J. Hyperinsulinemia Predicts Survival in a Hyperglycemic Mouse Model of Critical Illness. *Crit Care Med.* **2009**, *37*, 2596–2603. [CrossRef]

10. Frazier, W.J.; Wang, X.; Wancket, L.M.; Li, X.A.; Meng, X.; Nelin, L.D.; Cato, A.C.; Liu, Y. Increased Inflammation, Impaired Bacterial Clearance, and Metabolic Disruption after Gram-Negative Sepsis in Mkp-1-Deficient Mice. *J. Immunol.* **2009**, *183*, 7411–7419. [CrossRef]

11. Kaufmann, R.L.; Matson, C.F.; Rowberg, A.H.; Beisel, W.R. Defective Lipid Disposal Mechanisms During Bacterial Infection in Rhesus Monkeys. *Metabolism* **1976**, *25*, 615–624. [CrossRef]

12. Griffiths, J.; Groves, A.C.; Leung, F.Y. Hypertriglyceridemia and Hypoglycemia in Gram-Negative Sepsis in the Dog. *Surg. Gynecol. Obstet.* **1973**, *136*, 897–903. [PubMed]

13. Gallin, J.I.; Kaye, D.; O'Leary, W.M. Serum Lipids in Infection. *N. Engl. J. Med.* **1969**, *281*, 1081–1086. [CrossRef] [PubMed]

14. Harris, H.W.; Gosnell, J.E.; Kumwenda, Z.L. The Lipemia of Sepsis: Triglyceride-Rich Lipoproteins As Agents of Innate Immunity. *J. Endotoxin. Res.* **2000**, *6*, 421–430. [PubMed]

15. Wendel, M.; Paul, R.; Heller, A.R. Lipoproteins in Inflammation and Sepsis. II. Clinical Aspects. *Intensive Care Med.* **2007**, *33*, 25–35. [CrossRef]

16. Lefer, A.M. Significance of Lipid Mediators in Shock States. *Circ. Shock* **1989**, *27*, 3–12. [PubMed]

17. Kawakami, M.; Cerami, A. Studies of Endotoxin-Induced Decrease in Lipoprotein Lipase Activity. *J. Exp. Med.* **1981**, *154*, 631–639. [CrossRef]

18. Lanza-Jacoby, S.; Lansey, S.C.; Cleary, M.P.; Rosato, F.E. Alterations in Lipogenic Enzymes and Lipoprotein Lipase Activity During Gram-Negative Sepsis in the Rat. *Arch. Surg.* **1982**, *117*, 144–147. [CrossRef]

19. Beutler, B.; Greenwald, D.; Hulmes, J.D.; Chang, M.; Pan, Y.C.; Mathison, J.; Ulevitch, R.; Cerami, A. Identity of Tumour Necrosis Factor and the Macrophage-Secreted Factor Cachectin. *Nature* **1985**, *316*, 552–554. [CrossRef]

20. Beutler, B.A.; Cerami, A. Recombinant Interleukin 1 Suppresses Lipoprotein Lipase Activity in 3T3-L1 Cells. *J. Immunol.* **1985**, *135*, 3969–3971.

21. Tall, A.R.; Yvan-Charvet, L. Cholesterol, Inflammation and Innate Immunity. *Nat. Rev. Immunol.* **2015**, *15*, 104–116. [CrossRef] [PubMed]

22. Aspichueta, P.; Perez-Agote, B.; Perez, S.; Ochoa, B.; Fresnedo, O. Impaired Response of VLDL Lipid and ApoB Secretion to Endotoxin in the Fasted Rat Liver. *J. Endotoxin. Res.* **2006**, *12*, 181–192. [CrossRef] [PubMed]

23. Endo, M.; Masaki, T.; Seike, M.; Yoshimatsu, H. TNF-Alpha Induces Hepatic Steatosis in Mice by Enhancing Gene Expression of Sterol Regulatory Element Binding Protein-1c (SREBP-1c). *Exp. Biol. Med.* **2007**, *232*, 614–621.

24. Ohhira, M.; Motomura, W.; Fukuda, M.; Yoshizaki, T.; Takahashi, N.; Tanno, S.; Wakamiya, N.; Kohgo, Y.; Kumei, S.; Okumura, T. Lipopolysaccharide Induces Adipose Differentiation-Related Protein Expression and Lipid Accumulation in the Liver Through Inhibition of Fatty Acid Oxidation in Mice. *J. Gastroenterol.* **2007**, *42*, 969–978. [CrossRef]

25. Liu, Y.; Shepherd, E.G.; Nelin, L.D. MAPK Phosphatases—Regulating the Immune Response. *Nat. Rev. Immunol.* **2007**, *7*, 202–212. [CrossRef]

26. Wang, X.; Nelin, L.D.; Kuhlman, J.R.; Meng, X.; Welty, S.E.; Liu, Y. The Role of MAP Kinase Phosphatase-1 in the Protective Mechanism of Dexamethasone Against Endotoxemia. *Life Sci.* **2008**, *83*, 671–680. [CrossRef]

27. Wang, X.; Zhao, Q.; Matta, R.; Meng, X.; Liu, X.; Liu, C.G.; Nelin, L.D.; Liu, Y. Inducible Nitric-Oxide Synthase Expression Is Regulated by Mitogen-Activated Protein Kinase Phosphatase-1. *J. Biol. Chem.* **2009**, *284*, 27123–27134. [CrossRef] [PubMed]

28. Zhao, Q.; Shepherd, E.G.; Manson, M.E.; Nelin, L.D.; Sorokin, A.; Liu, Y. The Role of Mitogen-Activated Protein Kinase Phosphatase-1 in the Response of Alveolar Macrophages to Lipopolysaccharide: Attenuation of Proinflammatory Cytokine Biosynthesis Via Feedback Control of P38. *J. Biol. Chem.* **2005**, *280*, 8101–8108. [CrossRef]

29. Zhao, Q.; Wang, X.; Nelin, L.D.; Yao, Y.; Matta, R.; Manson, M.E.; Baliga, R.S.; Meng, X.; Smith, C.V.; Bauer, J.A.; et al. MAP Kinase Phosphatase 1 Controls Innate Immune Responses and Suppresses Endotoxic Shock. *J. Exp. Med.* **2006**, *203*, 131–140. [CrossRef]

30. Lang, R.; Hammer, M.; Mages, J. DUSP Meet Immunology: Dual Specificity MAPK Phosphatases in Control of the Inflammatory Response. *J. Immunol.* **2006**, *177*, 7497–7504. [CrossRef]

31. Franklin, C.C.; Kraft, A.S. Conditional Expression of the Mitogen-Activated Protein Kinase (MAPK) Phosphatase MKP-1 Preferentially Inhibits P38 MAPK and Stress-Activated Protein Kinase in U937 Cells. *J. Biol. Chem.* **1997**, *272*, 16917–16923. [CrossRef] [PubMed]

32. Chi, H.; Barry, S.P.; Roth, R.J.; Wu, J.J.; Jones, E.A.; Bennett, A.M.; Flavell, R.A. Dynamic Regulation of Pro- and Anti-Inflammatory Cytokines by MAPK Phosphatase 1 (MKP-1) in Innate Immune Responses. *Proc. Natl. Acad. Sci. USA* **2006**, *103*, 2274–2279. [CrossRef] [PubMed]

33. Salojin, K.V.; Owusu, I.B.; Millerchip, K.A.; Potter, M.; Platt, K.A.; Oravecz, T. Essential Role of MAPK Phosphatase-1 in the Negative Control of Innate Immune Responses. *J. Immunol.* **2006**, *176*, 1899–1907. [CrossRef] [PubMed]

34. Wu, J.J.; Roth, R.J.; Anderson, E.J.; Hong, E.G.; Lee, M.K.; Choi, C.S.; Neufer, P.D.; Shulman, G.I.; Kim, J.K.; Bennett, A.M. Mice Lacking MAP Kinase Phosphatase-1 Have Enhanced MAP Kinase Activity and Resistance to Diet-Induced Obesity. *Cell Metab.* **2006**, *4*, 61–73. [CrossRef] [PubMed]

35. Laplante, M.; Sabatini, D.M. An Emerging Role of MTOR in Lipid Biosynthesis. *Curr. Biol.* **2009**, *19*, R1046–R1052. [CrossRef] [PubMed]

36. Cases, S.; Stone, S.J.; Zhou, P.; Yen, E.; Tow, B.; Lardizabal, K.D.; Voelker, T.; Farese, R.V., Jr. Cloning of DGAT2, a Second Mammalian Diacylglycerol Acyltransferase, and Related Family Members. *J. Biol. Chem.* **2001**, *276*, 38870–38876. [CrossRef] [PubMed]

37. Lardizabal, K.D.; Mai, J.T.; Wagner, N.W.; Wyrick, A.; Voelker, T.; Hawkins, D.J. DGAT2 Is a New Diacylglycerol Acyltransferase Gene Family: Purification, Cloning, and Expression in Insect Cells of Two Polypeptides From Mortierella Ramanniana With Diacylglycerol Acyltransferase Activity. *J. Biol. Chem.* **2001**, *276*, 38862–38869. [CrossRef]

38. Zhou, J.; Febbraio, M.; Wada, T.; Zhai, Y.; Kuruba, R.; He, J.; Lee, J.H.; Khadem, S.; Ren, S.; Li, S.; et al. Hepatic Fatty Acid Transporter Cd36 Is a Common Target of LXR, PXR, and PPARgamma in Promoting Steatosis. *Gastroenterology* **2008**, *134*, 556–567. [CrossRef]

39. Nakamura, M.T.; Yudell, B.E.; Loor, J.J. Regulation of Energy Metabolism by Long-Chain Fatty Acids. *Prog. Lipid Res.* **2014**, *53*, 124–144. [CrossRef]

40. Bonnefont, J.P.; Djouadi, F.; Prip-Buus, C.; Gobin, S.; Munnich, A.; Bastin, J. Carnitine Palmitoyltransferases 1 and 2: Biochemical, Molecular and Medical Aspects. *Mol. Asp. Med.* **2004**, *25*, 495–520. [CrossRef]

41. Moreno-Fernandez, M.E.; Giles, D.A.; Stankiewicz, T.E.; Sheridan, R.; Karns, R.; Cappelletti, M.; Lampe, K.; Mukherjee, R.; Sina, C.; Sallese, A.; et al. Peroxisomal Beta-Oxidation Regulates Whole Body Metabolism, Inflammatory Vigor, and Pathogenesis of Nonalcoholic Fatty Liver Disease. *JCI Insight* **2018**, *3*, 93626. [CrossRef] [PubMed]

42. Hakimi, P.; Johnson, M.T.; Yang, J.; Lepage, D.F.; Conlon, R.A.; Kalhan, S.C.; Reshef, L.; Tilghman, S.M.; Hanson, R.W. Phosphoenolpyruvate Carboxykinase and the Critical Role of Cataplerosis in the Control of Hepatic Metabolism. *Nutr. Metab.* **2005**, *2*, 33. [CrossRef] [PubMed]

43. Horike, N.; Sakoda, H.; Kushiyama, A.; Ono, H.; Fujishiro, M.; Kamata, H.; Nishiyama, K.; Uchijima, Y.; Kurihara, Y.; Kurihara, H.; et al. AMP-Activated Protein Kinase Activation Increases Phosphorylation of Glycogen Synthase Kinase 3beta and Thereby Reduces CAMP-Responsive Element Transcriptional Activity and Phosphoenolpyruvate Carboxykinase C Gene Expression in the Liver. *J. Biol. Chem.* **2008**, *283*, 33902–33910. [CrossRef]

44. Gomez-Valades, A.G.; Mendez-Lucas, A.; Vidal-Alabro, A.; Blasco, F.X.; Chillon, M.; Bartrons, R.; Bermudez, J.; Perales, J.C. Pck1 Gene Silencing in the Liver Improves Glycemia Control, Insulin Sensitivity, and Dyslipidemia in Db/Db Mice. *Diabetes* **2008**, *57*, 2199–2210. [CrossRef] [PubMed]

45. Xiong, Y.; Lei, Q.Y.; Zhao, S.; Guan, K.L. Regulation of Glycolysis and Gluconeogenesis by Acetylation of PKM and PEPCK. *Cold Spring Harb. Symp. Quant. Biol.* **2011**, *76*, 285–289. [CrossRef] [PubMed]

46. Lawan, A.; Zhang, L.; Gatzke, F.; Min, K.; Jurczak, M.J.; Al-Mutairi, M.; Richter, P.; Camporez, J.P.; Couvillon, A.; Pesta, D.; et al. Hepatic Mitogen-Activated Protein Kinase Phosphatase 1 Selectively Regulates Glucose Metabolism and Energy Homeostasis. *Mol. Cell Biol.* **2015**, *35*, 26–40. [CrossRef] [PubMed]

47. Sweeney, S.E.; Firestein, G.S. Mitogen Activated Protein Kinase Inhibitors: Where are we now and where are we going? *Ann. Rheum. Dis.* **2006**, *65* (Suppl. 3), iii83–iii88. [CrossRef]

48. Chen, P.; Li, J.; Barnes, J.; Kokkonen, G.C.; Lee, J.C.; Liu, Y. Restraint of Proinflammatory Cytokine Biosynthesis by Mitogen-Activated Protein Kinase Phosphatase-1 in Lipopolysaccharide-Stimulated Macrophages. *J. Immunol.* **2002**, *169*, 6408–6416. [CrossRef]

49. Hammer, M.; Mages, J.; Dietrich, H.; Servatius, A.; Howells, N.; Cato, A.C.; Lang, R. Dual Specificity Phosphatase 1 (DUSP1) Regulates a Subset of LPS-Induced Genes and Protects Mice From Lethal Endotoxin Shock. *J. Exp. Med.* **2006**, *203*, 15–20. [CrossRef]

50. Lee, J.C.; Kassis, S.; Kumar, S.; Badger, A.; Adams, J.L. P38 Mitogen-Activated Protein Kinase Inhibitors—Mechanisms and Therapeutic Potentials. *Pharmacol. Ther.* **1999**, *82*, 389–397. [CrossRef]

51. Bagby, G.J.; Spitzer, J.A. Lipoprotein Lipase Activity in Rat Heart and Adipose Tissue During Endotoxic Shock. *Am. J. Physiol.* **1980**, *238*, H325–H330. [CrossRef] [PubMed]

52. Harris, H.W.; Grunfeld, C.; Feingold, K.R.; Rapp, J.H. Human Very Low Density Lipoproteins and Chylomicrons Can Protect Against Endotoxin-Induced Death in Mice. *J. Clin. Investig.* **1990**, *86*, 696–702. [CrossRef] [PubMed]

53. Bechmann, L.P.; Hannivoort, R.A.; Gerken, G.; Hotamisligil, G.S.; Trauner, M.; Canbay, A. The Interaction of Hepatic Lipid and Glucose Metabolism in Liver Diseases. *J. Hepatol.* **2012**, *56*, 952–964. [CrossRef] [PubMed]

54. Beutler, B.; Mahoney, J.; Le, T.N.; Pekala, P.; Cerami, A. Purification of Cachectin, a Lipoprotein Lipase-Suppressing Hormone Secreted by Endotoxin-Induced RAW 264.7 Cells. *J. Exp. Med.* **1985**, *161*, 984–995. [CrossRef] [PubMed]

55. Beutler, B.A.; Milsark, I.W.; Cerami, A. Cachectin/Tumor Necrosis Factor: Production, Distribution, and Metabolic Fate in Vivo. *J. Immunol.* **1985**, *135*, 3972–3977. [PubMed]

56. Chen, H.C.; Farese, R.V., Jr. DGAT and Triglyceride Synthesis: A New Target for Obesity Treatment? *Trends Cardiovasc. Med.* **2000**, *10*, 188–192. [CrossRef]

57. Vluggens, A.; Andreoletti, P.; Viswakarma, N.; Jia, Y.; Matsumoto, K.; Kulik, W.; Khan, M.; Huang, J.; Guo, D.; Yu, S.; et al. Reversal of Mouse Acyl-CoA Oxidase 1 (ACOX1) Null Phenotype by Human ACOX1b Isoform. *Lab Investig.* **2010**, *90*, 696–708. [CrossRef]

58. Xiao, Y.; Wang, J.; Yan, W.; Zhou, K.; Cao, Y.; Cai, W. P38alpha MAPK Antagonizing JNK to Control the Hepatic Fat Accumulation in Pediatric Patients Onset Intestinal Failure. *Cell Death Dis.* **2017**, *8*, e3110. [CrossRef]

59. Lawan, A.; Shi, H.; Gatzke, F.; Bennett, A.M. Diversity and Specificity of the Mitogen-Activated Protein Kinase Phosphatase-1 Functions. *Cell Mol. Life Sci.* **2013**, *70*, 223–237. [CrossRef]

60. Lawan, A.; Bennett, A.M. Mitogen-Activated Protein Kinase Regulation in Hepatic Metabolism. *Trends Endocrinol. Metab.* **2017**, *28*, 868–878. [CrossRef]

61. Dorfman, K.; Carrasco, D.; Gruda, M.; Ryan, C.; Lira, S.A.; Bravo, R. Disruption of the Erp/Mkp-1 Gene Does Not Affect Mouse Development: Normal MAP Kinase Activity in ERP/MKP-1-Deficient Fibroblasts. *Oncogene* **1996**, *13*, 925–931. [PubMed]

62. Li, J.; Sapper, T.N.; Mah, E.; Rudraiah, S.; Schill, K.E.; Chitchumroonchokchai, C.; Moller, M.V.; McDonald, J.D.; Rohrer, P.R.; Manautou, J.E.; et al. Green Tea Extract Provides Extensive Nrf2-Independent Protection Against Lipid Accumulation and NFkappaB Pro-Inflammatory Responses During Nonalcoholic Steatohepatitis in Mice Fed a High-Fat Diet. *Mol. Nutr. Food Res.* **2016**, *60*, 858–870. [CrossRef] [PubMed]

63. Brunt, E.M.; Kleiner, D.E.; Wilson, L.A.; Belt, P.; Neuschwander-Tetri, B.A. Nonalcoholic Fatty Liver Disease (NAFLD) Activity Score and the Histopathologic Diagnosis in NAFLD: Distinct Clinicopathologic Meanings. *Hepatology* **2011**, *53*, 810–820. [CrossRef] [PubMed]

64. Greer, Y.E.; Porat-Shliom, N.; Nagashima, K.; Stuelten, C.; Crooks, D.; Koparde, V.N.; Gilbert, S.F.; Islam, C.; Ubaldini, A.; Ji, Y.; et al. ONC201 Kills Breast Cancer Cells in Vitro by Targeting Mitochondria. *Oncotarget* **2018**, *9*, 18454–18479. [CrossRef] [PubMed]

Int. J. Mol. Sci. **2018**, *19*, 3904

65. Benhalevy, D.; Gupta, S.K.; Danan, C.H.; Ghosal, S.; Sun, H.W.; Kazemier, H.G.; Paeschke, K.; Hafner, M.; Juranek, S.A. The Human CCHC-Type Zinc Finger Nucleic Acid-Binding Protein Binds G-Rich Elements in Target MRNA Coding Sequences and Promotes Translation. *Cell Rep.* **2017**, *18*, 2979–2990. [CrossRef] [PubMed]

66. Livak, K.J.; Schmittgen, T.D. Analysis of Relative Gene Expression Data Using Real-Time Quantitative PCR and the $2^{-\Delta\Delta Ct}$ Method. *Methods* **2001**, *25*, 402–408. [CrossRef] [PubMed]

International Journal of
Molecular Sciences

MDPI

Article

Whole Genome Microarray Analysis of DUSP4-Deletion Reveals A Novel Role for MAP Kinase Phosphatase-2 (MKP-2) in Macrophage Gene Expression and Function

Thikryat Neamatallah [1],*[iD], Shilan Jabbar [2], Rothwelle Tate [2], Juliane Schroeder [2,†], Muhannad Shweash [2], James Alexander [2] and Robin Plevin [2]

1   Department of Pharmacology and Toxicology, Faculty of Pharmacy, King Abdulaziz University, P.O. Box 80260, Jeddah 21589, Saudi Arabia
2   Strathclyde Institute for Pharmacy & Biomedical Sciences, University of Strathclyde, 161 Cathedral Street, Glasgow G4 0RE, UK
*   Correspondence: taneamatallah@kau.edu.sa; Tel.: +966555651066
†   Current Address: Institute of Infection, Immunity and Inflammation, University of Glasgow, Glasgow G12 8TA, UK.

Received: 29 March 2019; Accepted: 9 July 2019; Published: 12 July 2019

Abstract: Background: Mitogen-activated protein kinase phosphatase-2 (MKP-2) is a type 1 nuclear dual specific phosphatase (DUSP-4). It plays an important role in macrophage inflammatory responses through the negative regulation of Mitogen activated protein kinase (MAPK) signalling. However, information on the effect of MKP-2 on other aspect of macrophage function is limited. Methods: We investigated the impact of MKP-2 in the regulation of several genes that are involved in function while using comparative whole genome microarray analysis in macrophages from MKP-2 wild type (wt) and knock out (ko) mice. Results: Our data showed that the lack of MKP-2 caused a significant down-regulation of colony-stimulating factor-2 (*Csf2*) and monocyte to macrophage-associated differentiation (*Mmd*) genes, suggesting a role of MKP-2 in macrophage development. When treated with macrophage colony stimulating factor (M-CSF), *Mmd* and *Csf2* mRNA levels increased but significantly reduced in ko cells in comparison to wt counterparts. This effect of MKP-2 deletion on macrophage function was also observed by cell counting and DNA measurements. On the signalling level, M-CSF stimulation induced extracellular signal-regulated kinases (ERK) phosphorylation, which was significantly enhanced in the absence of MKP-2. Pharmacological inhibition of ERK reduced both *Csf2* and *Mmd* genes in both wild type and ko cultures, which suggested that enhanced ERK activation in ko cultures may not explain effects on gene expression. Interestingly other functional markers were also shown to be reduced in ko macrophages in comparison to wt mice; the expression of CD115, which is a receptor for M-CSF, and CD34, a stem/progenitor cell marker, suggesting global regulation of gene expression by MKP-2. Conclusions: Transcriptome profiling reveals that MKP-2 regulates macrophage development showing candidate targets from monocyte-to-macrophage differentiation and macrophage proliferation. However, it is unclear whether effects upon ERK signalling are able to explain the effects of DUSP-4 deletion on macrophage function.

Keywords: MAP Kinase Phosphatase-2; DUSP-4; macrophages; proliferation; differentiation

1. Introduction

It is now well recognized that the mitogen-activated protein (MAP) kinases are the essential regulators of a diverse range of immune responses that are linked to normal physiology and disease [1]. The MAP kinase phosphatases (MKPs) finely regulate MAP kinase activity in cells [2]. This group consists of a ten membered family of dual specific phosphatases (DUSPs) that function to terminate MAP kinase signaling within a defined subcellular location. There are three classes I, II, and III defined by subcellular location, specificity for each MAP kinase and the modes of regulation [3]. For example, MKP-1 is a nuclear DUSP of the type 1 class and is selective for all three major MAP kinases in vitro, whilst MKP-3, which is a type II DUSP, is a cytosolic phosphatase selective solely for ERK over the other kinases.

Previous evidence indicates the potential of DUSPs to play an important role in the regulation of immune function [4]. It has been shown that the deletion of MKP-1 results in severe exacerbation of septic shock and arthritis via enhanced activity of p38 and JNK [5,6]. In contrast, PAC1 (DUSP-2) deletion mice have reduced inflammatory responses in arthritis [7], whilst the deletion of MKP-5 (DUSP-10) results in marked changes in T- cell proliferation and in CD4$^+$ and CD8$^+$ function [8].

MAP Kinase phosphatase-2 (MKP-2) is a type 1 DUSP that is nuclear located due to the expression of two nuclear localization sequences (NLS) [9]. MKP-2 is selective for ERK and JNK in vitro [10] and is induced by multiple agents, including serum, growth factors, UV-light, and oxidative stress (see review by [11]). In comparison to other DUSPs, in particular, MKP-1, the function of MKP-2 has not been as comprehensively studied and its role within the immune system has not been well defined. Using a novel DUSP-4 deletion mouse model we have demonstrated a key role for MKP-2 in regulating infection mediated by either *Leishmania mexicana* [12], *Leishmania donovani* [13], or *Toxoplasma gondii* [14]. The responses are underpinned by an enhanced macrophage expression of Arginase-1 and reduced expression of iNOS in two of these models [12]. Other studies have implicated a role for DUSP-4 in mediating sepsis via the regulation of cytokine production [15] and in T- cell proliferation [16]. However, the exact role of MKP-2 in regulating macrophage gene expression and function is not fully investigated.

We conducted a comparative microarray gene expression analysis on MKP-2 wt and ko macrophages following lipopolysaccharide (LPS) activation in this study. We examined the impact of MKP-2 in the regulation of several genes that are involved in macrophage development and function for the first time. In particular, we identified a role for MKP-2 in the expression of two genes that are related to macrophage development, *Csf2*, and *Mmd*. We further confirmed the correlation of both genes and the defect in macrophage development linked to MKP-2 loss. Additionally, we further confirmed the effect of MKP-2 deletion on the expression of progenitor markers CD115 and CD34, suggesting a more global effect for DUSP-4 in macrophage function.

2. Results

2.1. The Effect of MKP-2 in the Gene Expression Pattern of LPS-Stimulated Macrophages

The relative importance of MKP-2 in macrophage function was studied using microarray gene expression analysis. A four-hour LPS stimulation was selected to cover early and late innate immune response genes in macrophages since most of the genes are induced or repressed in a temporal pattern. Generally, LPS induced the expression of a large number of cytokines, chemokines, and growth factors with disparate expressions between wt and MKP-2-ko macrophages. Table 1 shows the top 10-upregulated Agilent SurePrint G3 (8× 60K) array probes in response to LPS. Gene ontology analysis showed that highly enriched biological processes following LPS stimulation in both MKP-2-wt and ko macrophages are related to immune function (Table 2). Genes, such as *Il19*, *Cxcl1*, *Cxcl2 Prst1*, and *Il1a*, were more enhanced in MKP-2$^{+/+}$ when compared to MKP-2$^{-/-}$ counterparts. However, a slight increase was recorded for *Il1b*, *Ifng*, *Cxcl10*, and *Ccl22* in MKP-2-deficient macrophages in relation to wt cells. In addition, the gene encoding CD14 was also upregulated in MKP-2$^{+/+}$ over MKP-2$^{-/-}$ macrophages (27.9 vs. 10.1, respectively). CD14 is a surface antigen that is expressed on monocytes and macrophages to mediate the innate immune response to bacterial lipopolysaccharide). The decrease in *CD14* upregulation by MKP-2$^{-/-}$ macrophages could explain the attenuation in the inflammatory response in these macrophages, as CD14 is critical for the recognition of LPS by macrophages to initiate the innate immunity and release of cytokines and chemokines.

Table 1. Top 10 up-regulated Agilent (8× 60K) SurePrint G3 Mouse Gene Expression Array probe sets by LPS.

No.	FC-MKP-2$^{+/+}$	FC-MKP-2$^{-/-}$	Gene Symbol	Gene Name
1	4717.09	3292.84	*Il19*	interleukin 19
2	2975.03	1604.87	*Cxcl1*	chemokine (C-X-C motif) ligand 1
3	1850.59	1746.34	*Ptgs2*	prostaglandin-endoperoxide synthase 2
4	1758.98	1369.49	*Il1α*	interleukin 1 alpha
5	1574.15	663.96	*Gfi1*	growth factor independent 1
6	1094.3	5176.78	*Edn1*	endothelin 1
7	769.02	742.88	*Il6*	interleukin 6
8	710.97	446.13	*Cxcl2*	chemokine (C-X-C motif) ligand 2
9	434.82	425.43	*Dusp2*	dual specificity phosphatase 2
10	423.49	494.97	*Il1β*	interleukin 1 beta

A fold change (FC) of 2 or more was used for significantly regulated probes using one-way ANOVA (p-value < 0.05). Expression responses were analyzed in LPS treated cells and compared to that of untreated cells. The list is sorted according to WT macrophages.

Table 2. GO Analysis of the genes expressed between MKP-2$^{+/+}$ and MKP-2$^{-/-}$ macrophages with ≥2.0-fold change.

GO Term	Description	p-Value
GO:0009611	response to wounding	4.10E−14
GO:0050727	regulation of inflammatory response	1.97E−13
GO:0001817	regulation of cytokine production	2.75E−13
GO:0051239	regulation of multicellular organismal process	2.07E−12
GO:0032101	regulation of response to external stimulus	1.42E−11
GO:0048518	positive regulation of biological process	4.44E−11
GO:0023051	regulation of signalling	6.00E−11
GO:0010646	regulation of cell communication	8.51E−11
GO:0008284	positive regulation of cell proliferation	9.07E−11
GO:0050865	regulation of cell activation	1.63E−10
GO:0010941	regulation of cell death	1.32E−09
GO:0008009	chemokine activity	8.31E−08
GO:0032496	response to lipopolysaccharide	1.09E−08
GO:0001818	negative regulation of cytokine production	1.50E−08
GO:0045595	regulation of cell differentiation	7.88E−08
GO:0032680	regulation of tumor necrosis factor production	1.06E−07
GO:0042625	ATPase activity, coupled to transmembrane movement of ions	5.32E−05
GO:0017017	MAP kinase tyrosine/serine/threonine phosphatase activity	3.86E−04

p-value is the enrichment p-value computed according to the hypergeometric distribution model.

Among the highly induced genes, *Gfi1* was also differently expressed in MKP-2 wt versus ko LPS-induced macrophages (1574.1 vs. 663.9, respectively). Gfi-1 is a transcriptional repressor that regulates the Toll-like receptor (TLR) inflammatory response by antagonising the NF-κB pathway [17]. Endothelin-1 (*Edn1*), a gene that is encoded for a vasoconstrictor protein EDN-1, significantly differed between LPS-stimulated MKP-2 wt and ko macrophages (1094 vs. 5176, respectively, p-value < 0.05), with *Edn1* showing the highest fold change of any gene in the induced MKP-2 ko group.

Interestingly, genes for macrophage development, such as monocyte to macrophage differentiation associated (*Mmd*) and colony-stimulating factor-2 (*Csf2*), were also upregulated following LPS stimulation (FC= 33.7- and 24.5- for wt, respectively, vs. 17.6 and 7.8 for ko, respectively. *Csf2* encodes for CSF-2 cytokine or GM-CSF that controls the production, differentiation and growth of granulocytes and macrophages [18]. *Mmd* is mainly upregulated upon monocyte differentiation and it is expressed in mature, in vitro differentiated macrophages whilst missing in monocytes [19]. Figure 1 illustrates the expression differences of significantly regulated genes in the LPS-stimulated macrophages.

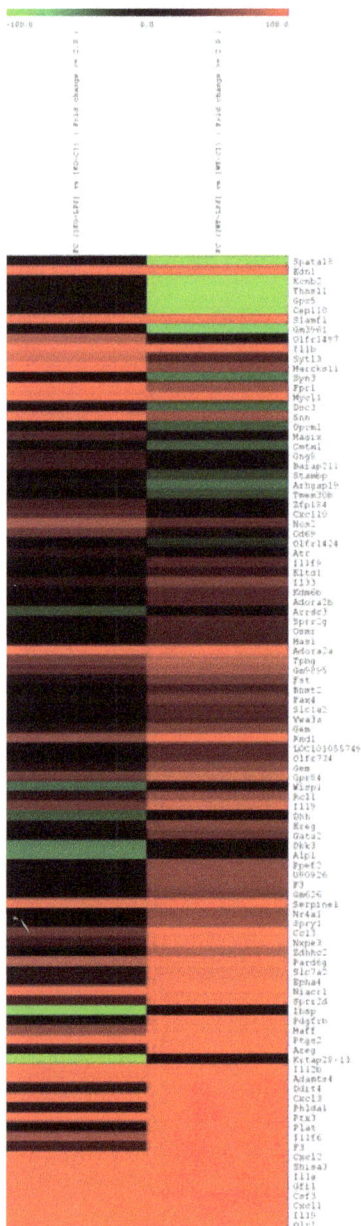

Figure 1. Heat maps representing hierarchical clustering of the differentially regulation, plotted using the log2 fold change (FC) values of the genes with *p*-value ≤0.05 (unpaired *t*-test) between Mitogen-activated protein kinase phosphatase-2 (MKP-2)$^{+/+}$ vs. MKP-2$^{-/-}$. Each column represents a sample and each row in the heat map represents a gene that is differentially regulated in that particular comparison of samples. Signal intensities were normalized to the expression data of unstimulated macrophages. The color scale represents the degree of expression of the gene, green being the downregulated (below −100) and red being (above +100) the upregulated genes in the sample sets with black as the center of the scale at '0'.

2.2. The Role of MKP-2 in the Expression of M-CSF-Induced Macrophage Differentiation and Proliferation Genes

Having identified the differential expression of macrophage genes following LPS stimulation, changes in *Edn1* mRNA, and protein levels were further confirmed (Supplementary Figure S1). This validated, while using RT-qPCR, the microarray data as a bonafide measure of gene expression. Thus, using RT-qPCR, we then examined the differential expression of important macrophage development genes, *Csf2* and *Mmd*. Macrophage colony stimulating factor (M-CSF), which is a specific growth factor for macrophages development, was used to stimulate the cells. As shown in Figure 2A, the induction of *Csf2* in MKP-2$^{+/+}$ macrophages by M-CSF was increased at 2 h, reaching a peak by 8 h, and decreased again by 24 h (fold stimulation at 8 h = 2.45 ± 0.85). Surprisingly, in MKP-2$^{-/-}$ macrophages, the expression of *Csf2* was reduced between 2 and 8 h, reaching a significant difference at 4 and 8 h ($p < 0.05$). However, the level of *Csf2* was slightly but not significantly enhanced in MKP-2$^{-/-}$ macrophages at 12 h (Figure 2A). During the same time course and, as demonstrated in Figure 2B, *Mmd* mRNA levels were increased in MKP-2$^{+/+}$ macrophages at 2 h, reaching a peak by 4 h and decreased by 24 h (fold stimulation at 4 h = 2.53 ± 0.5). Similar to *Csf2*, *Mmd* expression in MKP-2$^{-/-}$ macrophages was reduced at 2, 4, and 8h, reaching a significant difference at 4 h ($p < 0.05$). This demonstrates that MKP-2 can be an important factor in the maximal regulation of *Mmd* and *Csf2* gene expression in the early stages of stimulation. The increase of *Mmd* and *Csf2* mRNA at 12 h in MKP-2$^{-/-}$ mice suggests a late response from this group to induce both genes (Figure 2A,B).

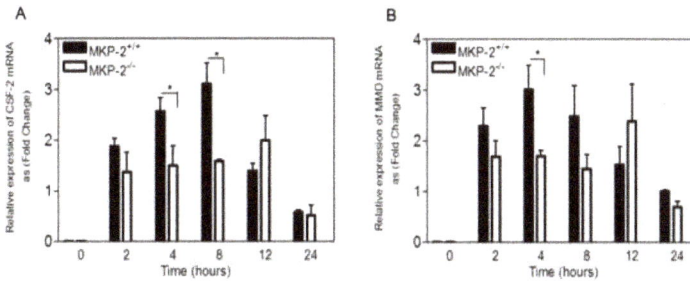

Figure 2. MKP-2 deletion negatively affects *Csf2* and *Mmd* gene expression in bone marrow-derived macrophages. Cells from day 3 of culture, were rendered quiescent in macrophage colony stimulating factor (M-CSF) free medium and then stimulated with recombinant M-CSF (10 ng/ml) for the indicated periods of time. Control cells (0) were left untreated. Total RNA was prepared from the cells. After reverse transcription, quantitative PCR analysis was performed on cDNA while using primers designed to detect *Csf2* (**A**) and *Mmd* (**B**). Expression levels of *Mmd* and *Csf-2* mRNA transcripts were normalized to the reference gene *QARS* using the delta-delta Ct method [20]. Error bars represent the mean ± SEM from three individual experiments. * $p < 0.05$, one-tailed unpaired t test comparing MKP-2$^{-/-}$ to MKP-2$^{+/+}$ macrophages.

2.3. MKP-2 Deletion Negatively Effects M-CSF Stimulated Macrophage Proliferation

The proliferative potential of macrophages following MKP-2 deletion was examined over a period of ten days using L929-conditioned medium, having established the negative regulation of *Csf2* and *Mmd* genes following MKP-2 deletion. Macrophages were harvested on days 5, 7, and 10 and counted in a haemocytometer. In the absence of MKP-2, macrophage proliferation was reduced by 20.04%, 21.1%, and 20.1% at days 5, 7, and 10, respectively, in comparison to wild type cells (Figure 3A). We assessed the proliferative capacity of macrophages over 72 h to further confirm our data. In MKP-2$^{+/+}$ macrophages, M-CSF was able to increase the number of MKP-2$^{+/+}$ cells 13-fold over 72 h (Figure 3B,C). In contrast, M-CSF-stimulated proliferation was significantly slower at 48 and 72 h when compared to wild type ($p < 0.05$) in MKP-2$^{-/-}$ macrophages. Next, we studied macrophage proliferation and confirmed itusing a BrdU cell proliferation assay. Cells from both MKP-2$^{+/+}$ and MKP-2$^{-/-}$ mice were harvested on day three and the proliferative rate was tracked for up to 72 h, which would be day seven, the time point when the macrophages are expected to be fully differentiated. The results in Figure 4A indicate a significant increase in the macrophage proliferation rate in both MKP-2$^{+/+}$ and MKP-2$^{-/-}$ after 24 h of M-CSF stimulation. This was approximately 3–4 fold in both populations with a slight increase in the wild type MKP-2 population over MKP-2 deleted macrophages. After 48 h (Figure 4B), proliferation further increased in the wt MKP-2 macrophages, however in MKP-2 ko macrophages, cell growth significantly lagged. By 72 h, the proliferation rates for both cultures had fallen and they were similar (Figure 4C). This provided additional evidence that the deletion of MKP-2 significantly impairs the growth of macrophages from precursors and that MKP-2 is essential for appropriate macrophage proliferation.

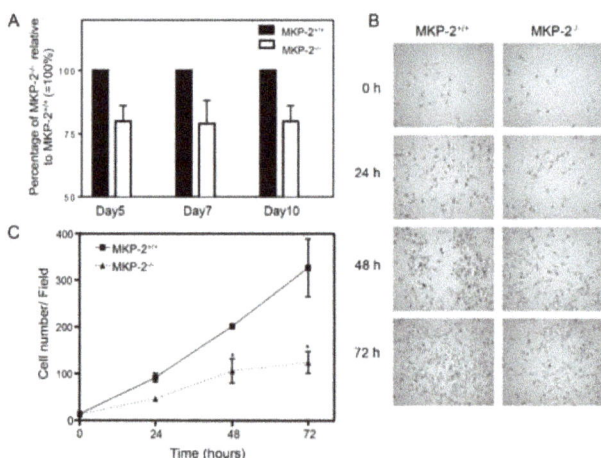

Figure 3. MKP-2 deletion reduces proliferative of bone marrow macrophages. (**A**) An equal number of cells (3.5×10^6) from the bone marrow cell suspension were cultured for 10 days in media supplemented with 30% L-929 conditioned medium (LCM) as a source for M-CSF. Macrophages were harvested from petri dishes at days 5, 7, and 10 following isolation. (**B,C**) Cells from day 3 of culture were rendered quiescent in cover slips for 24 h in M-CSF free medium and then stimulated with recombinant M-CSF (10 ng/mL) for the times indicated or left unstimulated (Controls, 0 h). The number of cells was determined, as described in Section 4.7.1. Control values, (0 h) MKP-2$^{+/+}$ = 20, MKP-2$^{-/-}$ = 18. Each value represents the mean ± SEM from three individual experiments. * $p < 0.05$, two-tailed unpaired *t*-test comparing MKP-2$^{-/-}$ to MKP-2$^{+/+}$ macrophages. Images were taken using a 20× objective, scale bar = 50 μm.

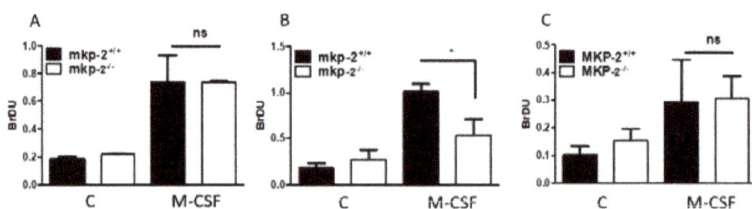

Figure 4. Effect of MKP-2 deletion on macrophage proliferation. Macrophages from both MKP-2$^{+/+}$ and MKP-2$^{-/-}$ were harvested on day 3 of culture, seeded at 4×10^4 cells/well in a 96-well plate and incubated overnight. Cells were stimulated then with 10 ng/ml M-CSF for (**A**) 24 h, (**B**) 48, and (**C**) 72 h. 10μM BrDU was added to each well. The cells were incubated for 6 h. Incorporated BrDU was measured at absorbance of 450 nM. Data represents mean ± SEM of three independent experiments. * $p < 0.05$ two-tailed unpaired *t*-test comparing MKP-2$^{-/-}$ to MKP-2$^{+/+}$ macrophages.

2.4. MKP-2 Deletion Enhances ERK Kinase Signaling

We studied the relationship between MKP-2 expression, MAP kinase signaling, and whether MAPK activation was necessary for the proliferative response of macrophages to M-CSF. M-CSF stimulated a rapid increase in MKP-2 expression, which was detected as early as 15 minutes, and was maximal, by 1 h (Figure 5A). No MKP-2 expression was observed in equivalently stimulated MKP-2$^{-/-}$ macrophages (see Figure 5A insert).

In the absence of MKP-2, ERK phosphorylation by M-CSF was enhanced relative to wild type controls (Figure 5B). Studying a time course between 5 to 60 minutes revealed a significant difference at 15 min. (fold stimulation at 15 minutes: MKP-2$^{+/+}$ = 16.26 ± 8.04, MKP-2$^{-/-}$ = 29.59 ± 2.82, $p < 0.05$). This response declined at 2 h in both genotypes. In contrast, there were no measurable differences in the phosphorylation of JNK and p38 MAPK between MKP-2$^{+/+}$ and MKP-2$^{-/-}$ macrophages over the same time course (Figure 5C,D). These data indicate that M-CSF-induced ERK activation was selectively enhanced in MKP-2 deficient macrophages in comparison to wild type.

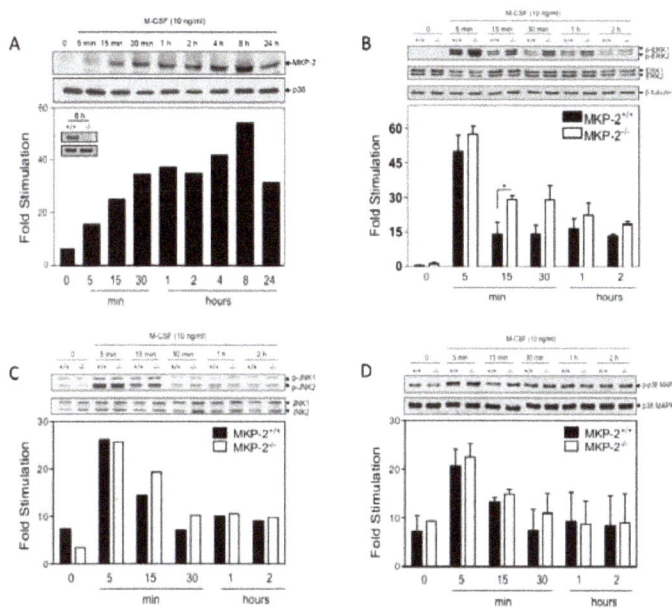

Figure 5. MKP-2 deletion enhances M-CSF-mediated ERK phosphorylation in macrophages. Cells harvested on day 3 of culture, were rendered quiescent in M-CSF free medium then stimulated with recombinant M-CSF (10 ng/mL) for the times indicated. Control cells (0) were left untreated. Whole cell lysates were prepared, and assessed for (**A**) MKP-2 (43 kDa) (**B**) p-ERK (42/44 kDa), (**C**) p-JNK (46/65 kDa), and (**D**) p-p38 (38 kDa) or total controls by Western blotting, as described in the Methods. MKP-2 was not induced in cells that were derived from MKP-2$^{-/-}$ mice (upper left insert of Panel A). Blots were semi-quantified for fold stimulation by scanning densitometry relative to the background signal. The blot represents the data from 2–3 individual experiments. * $p < 0.05$, one-tailed unpaired *t* test comparing MKP-2$^{-/-}$ to MKP-2$^{+/+}$ macrophages.

2.5. The Effect of ERK Inhibition on M-CSF Induced Proliferation and Mmd and Csf2 Expression

Differences in the activation of ERK following MKP-2 deletion suggest the potential involvement of ERK activation in macrophage proliferation. The effect of ERK inhibition on M-CSF-induced ERK phosphorylation was examined to study this correlation. The results that are presented in Figure 6A show that the pre-treatment of cells with U0126 completely inhibited ERK phosphorylation at 5 min in both genotypes relative to their respective control.

An additional ERK inhibition experiment was carried out to investigate the definitive role of ERK in macrophage proliferation, since MKP-2 deletion was found to negatively affect proliferation (Figure 3). This was achieved by studying the effect of MEK1,2 inhibitor, U0126, on macrophage proliferation using cell counting using hematoxylin staining. M-CSF was generally able to increase the number of cells at 48 h, but to a greater extent in wild type cultures (Figure 6B), a finding that is consistent with previous experiments. Inhibiting ERK activity while using U0126 significantly impaired cell growth, but to an equivalent extent in each genotype (Figure 6B). This experiment provided additional evidence that ERK significantly regulates the proliferation of bone marrow derived macrophages from precursors and that ERK activation by M-CSF is essential for appropriate macrophage development.

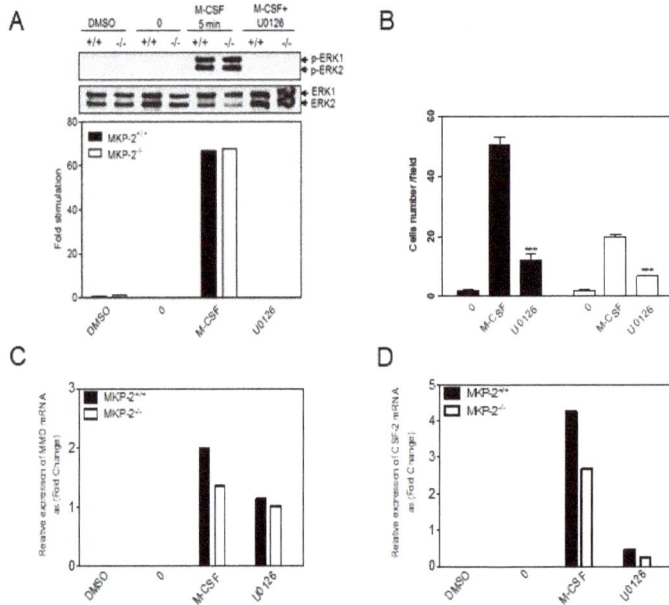

Figure 6. The effect of MAPK inhibition on M-CSF induced MAPK phosphorylation and *Csf2/Mmd* expression. (**A**) Cells from day 3 of culture were rendered quiescent in M-CSF free medium and then pre-incubated with DMSO (vehicle control) or 10 µM of U0126 for 1 h where indicated. The cells were either left unstimulated (0), or stimulated with M-CSF (10 ng/mL) for 5 min. Whole cell lysates were prepared, and assessed for p-ERK (42/44 kDa) by Western blotting. The blot represents the data from two individual experiments. (**B**) Cells that were harvested on day 3 of culture (5×10^3) were rendered quiescent in coverslips for 24 h in M-CSF free medium and then pre-treated with (U0126) for 1 h prior to stimulation with recombinant M-CSF (10 ng/ml) for 48 h, stimulated with M-CSF alone or left unstimulated (0 h). Each value represents the mean ± SEM from three individual experiments. *** $p < 0.001$, one-tailed unpaired t test relative to 48 h M-CSF treated macrophages. (**C,D**) Relative expression levels of *Mmd* mRNA transcripts were normalized to the reference gene *QARS* while using the delta-delta Ct method [17]. The figure represents the data from three individual experiments.

The effect of ERK inhibition on *Mmd* and *Csf2* expression was also examined (Figure 6C,D). The results presented in Figure 6C show that, relative to stimulated controls, the pre-treatment of cells with U0126 reduced M-CSF-induced *Mmd* expression by ~52.2% and ~33% for wild type and knock-out cells, respectively (Figure 6C). In contrast, ERK inhibition substantially reduced the *Csf2* mRNA levels in both wild type and knock-out cells, where inhibition was approximately 85% for each condition.

(Figure 6D). This shows for the first time that the expression of both genes is positively regulated through ERK activation. However, taking Figure 5; Figure 6 together, the data show that enhanced ERK signaling in MKP-2 deficient macrophages does not translate into enhanced *Mmd* or *Csf2* gene expression nor increased proliferation, but a reduction in all three parameters.

2.6. Differential Expression of CD34 and CD115 on the Surface of Bone Marrow Cells

We next examined the effect of MKP-2 loss on the expression of macrophage surface proteins CD115 and CD34. CD34 is an important progenitor cell marker [21], whereas (CD115) is a receptor for M-CSF that has been used to identify cells that belong to the mononuclear phagocyte system, including macrophages [19].

The analysis of bone marrow cells showed a significant reduction in CD115 expression on cells from MKP-2$^{-/-}$ mice in comparison to wild type; 5.6–8.2% and 9.1–13.8%, respectively, $p < 0.05$ (Figure 7A). Based on this, the bone marrow cell suspension from MKP-2$^{-/-}$ mice contained less macrophage precursors within the population. Analysis of the cell suspension showed that approximately 11.3–16.5% and 2.3–7.8% of the cells were CD34$^+$ for MKP-2$^{+/+}$ and MKP-2$^{-/-}$, respectively. The percentage of CD34 positive cells was significantly lower in MKP-2$^{-/-}$ when compared to the wild type counterparts ($p < 0.05$) (Figure 7B). This indicated that MKP-2$^{-/-}$ bone marrow consisted of fewer myeloid progenitors. It is noteworthy that the percentage of CD34$^+$CD115$^+$ cells was also reduced in MKP-2$^{-/-}$ bone marrow with 6.3%±0.9 when compared to 3.75 ± 0.45 in MKP-2$^{+/+}$ bone marrow.

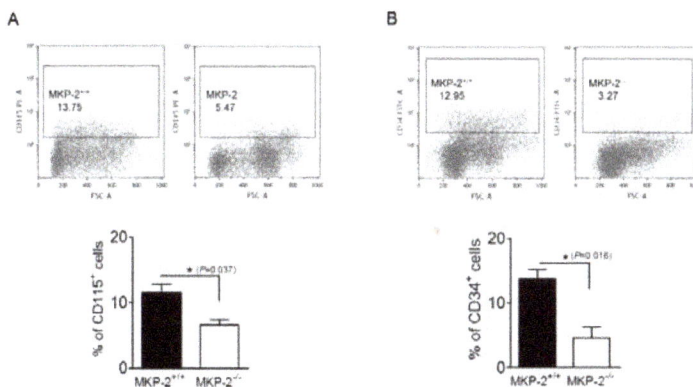

Figure 7. MKP-2 deletion negatively affects CD115 and CD34 expression. Native bone marrow cells were isolated and subsequently analyzed for CD115 and CD34, as described in Section 4.6. The gates represent the CD115 (**A**) or CD34 (**B**) positive population as percent within the native cell suspension. Gating was performed according to negative isotype-matched controls. Flow cytometry data represent the mean ± SEM of three separate experiments. * $p < 0.05$, two-tailed unpaired *t* test comparing MKP-2$^{-/-}$ to MKP-2$^{+/+}$ macrophages.

3. Discussion

For the first time we conducted a whole genome array to investigate the role of MKP-2 in the macrophage function, building on the initial finding using a DUSP-4 ko mouse model that showed effects upon inflammatory gene expression and resistance to leishmania and toxoplasma infection [12,14]. The initial results are consistent with previously published studies demonstrating LPS-induced gene programs in murine macrophages [22,23]. We and others have also shown that MKP-2 deletion has an impact on number of cytokines, such as IL-6, IL-12, TNFα, and COX-2 in LPS-stimulated macrophages [12], and these findings were replicated in the whole genome study. However, differences in other genes were also observed, such as CD14 and End1, and in preliminary RT-qPCR experiments, we used *End1* to validate the gene array recapitulating the results at the mRNA and protein level. Endothelin-1 has the potential to regulate the macrophage function [24–26], but more importantly smooth muscle contractility and proliferation [27,28], and this is currently being studied.

However, the study also showed that MKP-2 deletion reduced the expression of *Mmd* and *Csf2* genes, both of which are believed to be critical in macrophage differentiation and proliferation. We further showed, for the first time, that these two key genes, *Csf2* and *Mmd* were substantially reduced during the initial phase of stimulation in MKP-2$^{-/-}$ macrophages relative to wild type macrophages following stimulation with M-CSF, which is a key driver of macrophage proliferation and differentiation. The Monocyte to Macrophage Differentiation (*Mmd*) gene was first identified in 1995 and it is up-regulated during the differentiation of monocytes to macrophages [29]. The protein that is encoded by the *Csf2* is a cytokine that controls the production, differentiation, and function of macrophages [18]. Studies examining the regulation and function of these genes are limited, however Liu, Q., et al. has shown that LPS significantly up-regulated *Mmd* expression in macrophages and Mmdover-expression enhanced LPS-stimulated production of TNF-α and NO via ERK and Akt phosphorylation [30]. Similarly, an early study showed that, total cell numbers of blastocysts from CSF-2$^{-/-}$ mice were reduced to 18% compared with wild-type controls [31].

Thus, MKP-2 deletion adversely affected the ability of cells to differentiate into adult macrophages over the 10-day period. Our data also showed a significant reduction in macrophage proliferation in response to M-CSF following to loss of MKP-2. This phenomenon is consistent with studies in fibroblasts that showed a potential effect on G2/M-Phase transition [32]. In those studies, serum or PDGF was used as a stimulant suggesting the involvement of pathways, such as ERK or PI3K/PKB. In this study, we utilized M-CSF [33,34], which interacts with colony stimulating factor-1 receptor (CSFR1) and couples to MAPKs and PI3K/Akt [35,36]. In response to M-CSF, we consistently found that the deletion of MKP-2 enhanced ERK activation, but it had no significant effect upon JNK and p38 activation. This profile is different to that observed in adult macrophages that were stimulated with LPS in which JNK and p38 MAP kinase phosphorylation was increased but ERK activation was not affected [12]. In another MKP-2 deletion model, an increase in ERK phosphorylation in response to LPS has been demonstrated [15]. Significantly, we found that M-CSF stimulated a very rapid expression of MKP-2 relative to other studies, including our own, however it is recognized that ERK activation controls MKP-2 expression [32]. Thus, the extremely rapid activation of ERK by M-CSF would coincide with this profile. It should also be noted that ERK-mediated phosphorylation is also able to promote the stabilization of MKP-2 [37] and this may contribute to the rapid cellular expression in the cells that were stimulated by M-CSF.

A key issue is determining whether enhanced ERK signaling following MKP-2 deletion plays a role in the functional effects that were observed following MKP-2 deletion. To this end, we treated cells with the MEK1 inhibitor U0126. For proliferation, it was found that ERK inhibition significantly reduced the proliferation in both set of cultures. This is consistent with a number of studies that show that ERK is important in M-CSF-dependent macrophage proliferation [38,39] and is able to regulate a number of early genes and is essential for cell cycle progression [40]. This might suggest that the effect of MKP-2 deletion to enhance ERK is not related to the overall negative effect on proliferation. However, several studies have indicated that the kinetics of ERK phosphorylation is significant in determining whether macrophages are stimulated to proliferate or activated immunologically; a rapid transient signal mediates proliferation, a more prolonged signal mediates activation [38,41,42], thus MKP-2 deletion may be steering macrophages towards activation. Consistent with our results, it has been shown that the knock down of MKP-1 in macrophages also leads to prolonged ERK activation and the inhibition of proliferation [43]. A similar argument may be posited for the consistent and inhibition of both *Mmd* and *Csf2* expression following pretreatment with U0126, which again we show for the first time.

However, it is still very possible that the enhancement of ERK signaling is not linked to the negative effect of MKP-2 on the proliferation and expression of *Mmd* and *Csf2*, despite the arguments above. Thus, another potential mechanism might be involved. Recently, Joeng et al. has demonstrated the direct binding of MKP-2 to VRK1, an effect that is independent of phosphatase activity [44]. Given that MKP-2 is nuclear located, it has the potential to positively regulate *Mmd* and *Csf2* gene transcription potential via the effect on Histone H3. However, we also show, for the first time, that a number of key macrophage markers are reduced following DUSP-4 deletion. This includes the expression of the M-CSF receptor, CD115 [45,46], and this finding would explain the consistent inhibitory effect of DUSP-4 deletion on the proliferation and differentiation of macrophages in response to M-CSF. However, we also show that the hematopoietic stem cell progenitors are also significantly reduced. CD34 is an important stem/progenitor cell marker [21] and CD34+ myeloid progenitors differentiate into monocytes and further to macrophages in the presence of M-CSF [47]. Thus, macrophages may lack responsiveness due to a number of factors as a result of MKP-2 deficiency but not directly related to kinase activation in the early stages of development. Interestingly, we found that adult macrophage markers where not altered including F4/80, CD14, CD11b and MHC-II (results not shown), thus the deletion is very specific to the early stages of the differentiation process. This may be related to a reduced of progenitor cells to initiate the differentiation with and less receptors responding to M-CSF. Therefore, proliferation will be correspondingly reduced.

A future analysis would therefore help to uncover the potential role of MKP-2 in the early stages of myeloid differentiation. Macrophages are derived from myeloid progenitor cells differentiated from hematopoietic stem cells (HSCs) in the bone marrow [48]. Studies have identified common CD115+ CD135+ CX3CR1+ markers that are associated with the myeloid stem cells committed toward macrophage/DC lineage [49,50]. These and other myeloid precursors generating cells such as neutrophils, eosinophils, basophils and mast cells would have to be analysed. Furthermore, it would be important to determine whether the effect of DUSP4 deletion can generally attenuate other lineages originate from HSCs, such as lymphoid progenitors (LPs), the source of CD19/CD45 B cells and CD4/CD8 T cells. We have already shown the reduced frequency of CD4+ and CD8+ T cells in spleens upon MKP-2 deletion, which suggests that this lineage is also compromised [51]. However, if DUSP-4 deletion is not affecting the number macrophage progenitors, the function of these cells, including the potential to differentiate, would have to be analysed across a host of cellular parameters, including the expression of differentiation markers, proliferation, and the effect of ERK signaling. A recent study has shown that MAP kinase is linked to macrophage progenitor development [39].

In conclusion, using a whole genome microarray of macrophages from wt and DUSP-4 deletion mice for the first time, we uncover the potential for MKP-2 to regulate a significant number of genes. Two of these, *Mmd* and *Csf-2*, are implicated in the proliferation of macrophages. Furthermore, whilst MKP-2 deletion enhanced ERK signaling, this did not readily explain the effect upon macrophage function, which is profound. Finally, it highlights the importance to carry out further studies on the role of MKP-2 in macrophages and other myeloid cells.

4. Materials and Methods

4.1. Reagents

The antibodies were purchased as follows: Rabbit polyclonal anti-JNK-1 (S-18), anti-ERK-1 (K-23) and anti-p38 (N-20) antibodies were all purchased from Santa Cruz (Dallas, TX, USA). Anti-phospho-ERK (T202/Y204) and anti-phospho-JNK (T183/Y185) were purchased from Cell Signaling Technology (Hitchin, UK). Anti-phospho-p38 (44684-G) was purchased from Applied Biosystems, Life Technologies (Paisley, UK). HRP- conjugated anti-mouse, and conjugated anti-rabbit antibodies were purchased from Jackson Immunoresearch laboratories (Luton, Beds, UK). Anti-Mouse CD34-FITC conjugated, anti-Mouse CD115 (c-fms)-PE conjugated, and recombinant M-CSF were all purchased from eBioscience (Hatfield, UK). SP600125 JNK inhibitor, SB 203580 p38 inhibitor, and U2106 MEK1,2 inhibitor was purchased from Tocris Bioscience (Bristol, UK). BrdU Cell Proliferation kit was purchased from Cell Signaling (New England Biolabs, Hitchin, UK). SuperScript® III First-Strand Kit (cDNA synthesis system for RT-PCR), oligonucleotide PCR Primers, SYBR® Select real-time PCR Master Mix were purchased from Applied Biosystems, Life Technologies (Paisley, UK). RNeasy Mini Kit (RNA extraction kit), and RNase-Free DNase Set were obtained from QIAGEN (Manchester, UK). All other materials used were of the highest commercial grade available and they were obtained from Sigma Aldrich (Poole, Dorset, UK).

4.2. DUSP-4 Deletion Mice

The DUSP-4 deletion mice were generated in collaboration with Geno-way, France, as described previously [12].

4.3. Cell Culture

Bone marrow derived macrophages were isolated from femurs and tibia of 6–8 weeks old MKP-2$^{-/-}$ or $^{+/+}$ mice and that were grown in DMEM (Sigma-Aldrich), containing 10% (v/v) heat-inactivated FCS (Gibco Life Technologies, paisley) and supplemented with 30% L929 cell-conditioned medium as a source of M-CSF supplemented with 5 mM L-Glutamine, 100 U/mL penicillin, 1 μg/mL streptomycin. After culture, the adherent cells were harvested and then seeded in either 12 well (1×10^6 cells/mL) or 6 well (2×10^6 cells/2 mL) plates with RPMI 1641 supplemented with 10% FCS. The cells at this stage were deprived of M-CSF for 18 h and then treated with M-CSF.

4.4. Microarray Expression Analysis

Total RNA was extracted with the RNase Mini Kit, following the manufacturer's recommendations (QIAGEN, Manchester, UK). RNA concentrations and A260/280 ratios were obtained while using a Nanodrop 2000c spectrophotometer (Thermo Scientific, Paisley, UK). All of the RNA samples had rations of around 2.0. The integrity of the RNA isolated was tested while using an Experion automated electrophoresis system (Bio-Rad, Watford, UK). All of the samples in the study had RNA quality indexes (RQI) of >9.4. 100 ng of RNA from each sample was amplified and labeled with cyanine 3-CTP while using the one-color Low RNA Input Linear Amplification Kit according to the manufacturer's instructions (Agilent, Stockport, UK). Labeled cRNA with a specific activity of at least 6.0 pmol Cy3 per μg cRNA were fragmented and hybridized on SurePrint G3 8× 60K Mouse gene expression microarrays (Agilent, Stockport, UK). Tab-delimited data files in text format that were obtained from the feature extraction were analysed using GeneSpring GX 12.0.1 software (Agilent, Stockport, UK) in order to compare the gene expression profiles between the samples. Data for a given gene were normalized to the median expression level of that gene across all samples. Fold change of 2 or more was used for significantly regulated genes using One-way ANOVA (p-value < 0.05). Pathway analysis from Pathvisio (http://www.pathvisio.org/)was used to set out the groups of differently expressed genes [52]. Hierarchical clustering was used to classify genes that were co-expressed across samples in groups according their functional relationship based on the fold change while using MeV software (http://www.tm4.org/mev.html) [53]. Gorilla carried out gene ontology (GO) analysis, which is the Gene Ontology enrichment analysis and visualization tool [54], available at http://cbl-gorilla.cs.technion.ac.il/.

4.5. Quantitative Real-Time Polymerase Chain Reaction Analysis

The cells were washed once with PBS and total RNA was extracted with the RNase Mini Kit following the manufacturer's instructions, which included an on-column DNase treatment (QIAGEN, UK). For cDNA synthesis, 2 μg RNA was subjected to reverse transcription while using SuperScript III cDNA Synthesis system (Invitrogen, Paisley, UK), following the manufacturer's manual with Oligo-dT as the primer. Real-time PCR (RT-qPCR) was performed while using SYBR select Master Mix chemistry (Applied Biosystems), following the manufacturer's instructions. The sequence of primers that were used throughout this study were specifically designed to only bind the selected target using Gene Runner v3.1 Software (Hastings Software, Hastings, NY, US) and they were checked for specificity while using the National Centre for Biotechnology Information (NCBI) Primer-BLAST web tool [55]. Table 3 displays the sequence of primers. The thermal cycling and detection was carried out on a StepOne Plus real-time PCR system (Applied Biosystems, UK). The thermal cycle consisted of an initial uracil-DNA glycosylase activation of 2 min. at 50 °C to control PCR product carryover contamination, then a DNA polymerase activation of 2 min. at 95 °C, being followed by 40 cycles of 15 s at 95 °C, 60 s at 60 °C. A final extension of 5 min. at 72 °C followed and the reaction was stopped by extended incubation and cooling down to 4 °C. Melt curve analysis was conducted on the reactions to assay specificity. The expression levels of *Mmd* and *Csf2* mRNA transcripts were normalized to the housekeeping gene *QARS* while using the delta-delta Ct method [20].

Table 3. Nucleotide sequences of the primers used for the analysis of gene expression by Quantitative Real Time Polymerase Chain Reaction amplification (qRT-PCR).

Primer Name	Sequence	PCR Product Size (bp)
Qars	Forward: 5′-GGACTCCAGCTGAGCGCTGCTC-3′ Reverse: 5′-GGTGGACTCCACAGCTTCCTCAAT-3′	138
Csf-2	Forward: 5′-TGCCTGTCACGTTGAATGAAGAGG-3′ Reverse: 5′-TGTCTGGTAGTAGCTGGCTGTCATGTTC-3′	164
Mmd	Forward: 5′-TGGCCGCTACAAACCAACGTG-3′ Reverse: 5′-CAAAGGCCCATCCCGTAGATCC-3′	156
Edn1	Forward: 5′- ACA CTC CCG AGC GCG TCG TA -3′ Reverse: 5′- TCT TGT CTT TTT GGT GAG CGC ACT G -3′	142

4.6. Flow Cytometry Assay

Macrophages (1×10^6) were harvested, passed through a nitex mesh to remove clumps, washed with PBS, incubated with Fc-Block (5 µg/mL αCD16/CD32, BD Bioscience, 1% mouse serum in RPMI) for 10 min. at room temperature, and the cell surface markers were then stained with conjugated antibodies for CD11b, CD115, MHCII, F4/80, or CD34 for 1 h at 4 °C. After several washing steps, the cells were resuspended in 200 mL PBS and subsequently run and analyzed on the FACS Canto flow cytometer (BD Bioscience) while using FACS Diva software and data analyzed using Kaluza software (Beckman Coulter, High Wycombe, UK). All of the values were normalized to their respective isotype control.

4.7. Proliferation Assay

4.7.1. Cell Counting by Hematoxylin

Confluent macrophages were detached with cold PBS, seeded onto coverslips in 24-well plates (5000 cells/well) in 10% FCS-DMEM, and then allowed to attach for 24 h. Cells were starved in serum free media for 24 h and then stimulated for either 24, 48, 72 h, with 10 ng/mL of recombinant M-CSF. The cultures were washed with PBS and stained with hematoxylin. The number of cells was determined by counting in 10 random fields per each coverslip.

4.7.2. BrdU Cell Proliferation Assay

The cells were harvested on the third day of the culture, seeded at a density of 4×10^4 per well of 96-well plate and allowed to attach overnight in 10% FCS-DMEM. The next day, macrophages were treated with 10 ng/mL of M-CSF for 24, 48, 72 h, or left untreated as controls. Macrophage proliferation was assessed while using an anti-BrDU antibody detection assay kit, according to manufacturer's instructions (Cell signalling, New England Biolabs, Hitchin, UK).

4.8. Western Blotting

The macrophages were lysed in SDS-PAGE sample buffer (63 mM Tris-HCL, pH 6.8, 2 mM $Na_4P_2O_7$, 5 mM EDTA, 10% (*v/v*) glycerol, 2% (*w/v*) SDS, 50 mM DTT, 0.007% (*w/v*) bromophenol blue). The lysates were resolved by SDS-PAGE and transferred to nitrocellulose membranes. The membranes were blocked for non-specific binding for 2 h in 2% BSA (*w/v*), diluted in NATT buffer 50 mM Tris-HCl, 150 mM NaCl, 0.2% (*v/v*) Tween-20. The blots were then incubated overnight with 50 ng/mL primary antibody that was diluted in 0.2% BSA (*w/v*) in NATT buffer. The blots were washed with NATT buffer for 90 min. and incubated with HRP-conjugated secondary antibody (20 ng/mL in 0.2% BSA (*w/v*) diluted in NaTT buffer) for 2 h. After a further 90 minutes wash, the blots were subjected to ECL reagent and then exposed to Kodak X-ray film.

4.9. Data Analysis

Each figure represents one of at least three separate experiments, unless otherwise stated. The Western blots were scanned while using an Epson perfection 1640SU scanner using Adobe Photoshop 5.0.2 software. For gels, densitometry measurement was performed while using the Scion Image program. Data were expressed as mean ± SEM. Statistical analysis was performed while using GraphPad Prism (Version 5.0, GraphPad Software, San Diego, CA, USA). Statistically significant differences were determined using students *t*-test. P values equal or below 0.05 were considered to be significant.

5. Conclusions

We conduct a whole genome array in macrophages from wt and ko DUSP-4 deletion mice for the first time. The study reveals major effects on a number of genes, including those that are involved in macrophages differentiation. We go on to show that loss in MKP-2 has profound effects upon the proliferation and differentiation of macrophages. However, it is unclear whether these effects are directly linked to effects upon ERK signaling, but is related to the very early stage of macrophage formation. In addition, novel findings from this study highlight the significant role of MKP-2 in the immune response via regulating macrophage functions and adds new findings, which help in understanding the role of MKP-2 in immune biology.

Supplementary Materials: Supplementary materials can be found at http://www.mdpi.com/1422-0067/20/14/3434/s1.

Author Contributions: T.N., R.P. and J.A. conceived the original idea, T.N., R.T., J.S. and S.J. co-designed the study. T.N., R.T., S.J., M.S. performed the experiments. T.N., R.P. and J.A. co-wrote the manuscript. R.P. supervised the project.

Funding: This work was supported by a Wellcome Trust grant to R.P. and J.A. (089273/Z/09/Z.). T.N is a recipient of a King Abdulaziz University scholarship program. S.J. is sponsored by Kirkuk University scholarship program.

Acknowledgments: The authors would like to thank Carol Whitehouse for her help with the animal studies. We also thank our colleagues from Plevin's lab for providing insight and expertise that greatly assisted the research.

Conflicts of Interest: The authors declare no conflict of interest.

References

1. Rincon, M.; Pedraza-Alva, G. JNK and p38 MAP kinases in CD4+ and CD8+ T cells. *Immunol. Rev.* **2003**, *192*, 131–142. [CrossRef] [PubMed]
2. Caunt, C.J.; Keyse, S.M. Dual-specificity MAP kinase phosphatases (MKPs): Shaping the outcome of MAP kinase signalling. *FEBS J.* **2013**, *280*, 489–504. [CrossRef] [PubMed]
3. Keyse, S.M. Dual-specificity MAP kinase phosphatases (MKPs) and cancer. *Cancer Metastasis Rev.* **2008**, *27*, 253–261. [CrossRef] [PubMed]
4. Jeffrey, K.L.; Camps, M.; Rommel, C.; Mackay, C.R. Targeting dual-specificity phosphatases: Manipulating MAP kinase signalling and immune responses. *Nat. Rev. Drug Discov.* **2007**, *6*, 391–403. [CrossRef] [PubMed]
5. Hammer, M.; Mages, J.; Dietrich, H.; Servatius, A.; Howells, N.; Cato, A.C.; Lang, R. Dual specificity phosphatase 1 (DUSP1) regulates a subset of LPS-induced genes and protects mice from lethal endotoxin shock. *J. Exp. Med.* **2006**, *203*, 15–20. [CrossRef] [PubMed]

6. Salojin, K.V.; Owusu, I.B.; Millerchip, K.A.; Potter, M.; Platt, K.A.; Oravecz, T. Essential role of MAPK phosphatase-1 in the negative control of innate immune responses. *J. Immunol.* **2006**, *176*, 1899–1907. [CrossRef]

7. Jeffrey, K.L.; Brummer, T.; Rolph, M.S.; Liu, S.M.; Callejas, N.A.; Grumont, R.J.; Gillieron, C.; Mackay, F.; Grey, S.; Camps, M.; et al. Positive regulation of immune cell function and inflammatory responses by phosphatase PAC-1. *Nat. Immunol.* **2006**, *7*, 274–283. [CrossRef] [PubMed]

8. Zhang, Y.; Blattman, J.N.; Kennedy, N.J.; Duong, J.; Nguyen, T.; Wang, Y.; Davis, R.J.; Greenberg, P.D.; Flavell, R.A.; Dong, C. Regulation of innate and adaptive immune responses by MAP kinase phosphatase 5. *Nature* **2004**, *430*, 793–797. [CrossRef]

9. Sloss, C.M.; Cadalbert, L.; Finn, S.G.; Fuller, S.J.; Plevin, R. Disruption of two putative nuclear localization sequences is required for cytosolic localization of mitogen-activated protein kinase phosphatase-2. *Cell Signal* **2005**, *17*, 709–716. [CrossRef]

10. Chen, P.; Hutter, D.; Yang, X.; Gorospe, M.; Davis, R.J.; Liu, Y. Discordance between the binding affinity of mitogen-activated protein kinase subfamily members for MAP kinase phosphatase-2 and their ability to activate the phosphatase catalytically. *J. Biol. Chem.* **2001**, *276*, 29440–29449. [CrossRef]

11. Lawan, A.; Torrance, E.; Al-Harthi, S.; Al-Harthi, M.; Alnasser, S.; Neamatallah, T.; Schroeder, J.; Plevin, R. MKP-2: Out of the DUSP-bin and back into the limelight. *Biochem. Soc. Trans.* **2012**, *40*, 235–239. [CrossRef] [PubMed]

12. Al-Mutairi, M.S.; Cadalbert, L.C.; McGachy, H.A.; Shweash, M.; Schroeder, J.; Kurnik, M.; Sloss, C.M.; Bryant, C.E.; Alexander, J.; Plevin, R. MAP kinase phosphatase-2 plays a critical role in response to infection by Leishmania mexicana. *PLoS Pathog.* **2010**, *6*, e1001192. [CrossRef] [PubMed]

13. Schroeder, J.; McGachy, H.A.; Woods, S.; Plevin, R.; Alexander, J. T cell hypo-responsiveness against Leishmania major in MAP kinase phosphatase (MKP) 2 deficient C57BL/6 mice does not alter the healer disease phenotype. *PLoS Negl. Trop. Dis.* **2013**, *7*, e2064. [CrossRef] [PubMed]

14. Woods, S.; Schroeder, J.; McGachy, H.A.; Plevin, R.; Roberts, C.W.; Alexander, J. MAP kinase phosphatase-2 plays a key role in the control of infection with Toxoplasma gondii by modulating iNOS and arginase-1 activities in mice. *PLoS Pathog.* **2013**, *9*, e1003535. [CrossRef]

15. Cornell, T.T.; Rodenhouse, P.; Cai, Q.; Sun, L.; Shanley, T.P. Mitogen-activated protein kinase phosphatase 2 regulates the inflammatory response in sepsis. *Infect. Immun.* **2010**, *78*, 2868–2876. [CrossRef] [PubMed]

16. Huang, C.Y.; Lin, Y.C.; Hsiao, W.Y.; Liao, F.H.; Huang, P.Y.; Tan, T.H. DUSP4 deficiency enhances CD25 expression and CD4+ T-cell proliferation without impeding T-cell development. *Eur. J. Immunol.* **2012**, *42*, 476–488. [CrossRef] [PubMed]

17. Sharif-Askari, E.; Vassen, L.; Kosan, C.; Khandanpour, C.; Gaudreau, M.C.; Heyd, F. Zinc finger protein Gfi1 controls the endotoxin-mediated Toll-like receptor inflammatory response by antagonizing NF-kappaB p65. *Mol. Cell. Biol.* **2010**, *30*, 3929–3942. [CrossRef] [PubMed]

18. Morstyn, G.; Burgess, A.W. Hemopoietic growth factors: A review. *Cancer Res.* **1988**, *48*, 5624–5637. [PubMed]

19. Francke, A.; Herold, J.; Weinert, S.; Strasser, R.H.; Braun-Dullaeus, R.C. Generation of mature murine monocytes from heterogeneous bone marrow and description of their properties. *J. Histochem. Cytochem.* **2011**, *59*, 813–825. [CrossRef]

20. Schmittgen, T.D.; Livak, K.J. Analyzing real-time PCR data by the comparative C(T) method. *Nat. Protoc.* **2008**, *3*, 1101–1108. [CrossRef]

21. Matsuoka, S.; Ebihara, Y.; Xu, M.J.; Ishii, T.; Sugiyama, D.; Yoshino, H.; Nakahata, T. CD34 expression on long-term repopulating hematopoietic stem cells changes during developmental stages. *Blood* **2001**, *97*, 419–425. [CrossRef]

22. Björkbacka, H.; Fitzgerald, K.A.; Huet, F.; Li, X.; Gregory, J.A.; Lee, M.A.; Freeman, M.W. The induction of macrophage gene expression by LPS predominantly utilizes Myd88-independent signaling cascades. *Physiol. Genom.* **2004**, *19*, 319–330. [CrossRef]

23. Shuyi, Z.; Kim, C.C.; Batra, S.; McKerrow, J.H.; Loke, P.N. Delineation of Diverse Macrophage Activation Programs in Response to Intracellular Parasites and Cytokines. *PLoS Negl. Trop. Dis.* **2010**, *4*, e648.

24. Sikkeland, L.I.B.; Dahl, C.P.; Ueland, T.; Andreassen, A.K.; Gude, E.; Edvardsen, T.; Aukrust, P. Increased levels of inflammatory cytokines and endothelin-1 in alveolar macrophages from patients with chronic heart failure. *PLoS ONE* **2012**, *7*, e36815. [CrossRef] [PubMed]

25. Wahl, J.R.; Goetsch, N.J.; Young, H.J.; Maanen, R.J.V.; Johnson, J.D.; Pea, A.S.; Brittingham, A. Murine macrophages produce endothelin-1 after microbial stimulation. *Exp. Biol. Med. (Maywood)* **2005**, *230*, 652–658. [CrossRef] [PubMed]

26. Javeshghani, D.; Barhoumi, T.; Idris-Khodja, N.; Paradis, P.; Schiffrin, E.L. Reduced Macrophage-Dependent Inflammation Improves Endothelin-1–Induced Vascular Injury. *Hypertension* **2013**, *62*, 112–117. [CrossRef] [PubMed]

27. Elisa, T.; Antonio, P.; Giuseppe, P.; Alessandro, B.; Giuseppe, A.; Federico, C.; Daniela, R. Endothelin Receptors Expressed by Immune Cells Are Involved in Modulation of Inflammation and in Fibrosis: Relevance to the Pathogenesis of Systemic Sclerosis. *J. Immunol. Res.* **2015**, *2015*, 147616. [CrossRef]

28. Freeman, B.D.; Machado, F.S.; Tanowitz, H.B.; Desruisseaux, M.S. Endothelin-1 and its role in the pathogenesis of infectious diseases. *Life Sci.* **2014**, *118*, 110–119. [CrossRef] [PubMed]

29. Rehli, M.; Krause, S.W.; Schwarzfischer, L.; Kreutz, M.; Andreesen, R. Molecular cloning of a novel macrophage maturation-associated transcript encoding a protein with several potential transmembrane domains. *Biochem. Biophys. Res. Commun.* **1995**, *217*, 661–667. [CrossRef]

30. Liu, Q.; Zheng, J.; Yin, D.D.; Xiang, J.; He, F.; Wang, Y.C.; Han, H. Monocyte to macrophage differentiation-associated (MMD) positively regulates ERK and Akt activation and TNF-α and NO production in macrophages. *Mol. Biol. Rep.* **2012**, *39*, 5643–5650. [CrossRef]

31. Robertson, S.A.; Sjoblom, C.; Jasper, M.J.; Norman, R.J.; Seamark, R.F. Granulocyte-Macrophage Colony-Stimulating Factor Promotes Glucose Transport and Blastomere Viability in Murine Preimplantation Embryos. *Biol. Reprod.* **2001**, *64*, 1206–1215. [CrossRef] [PubMed]

32. Lawan, A.; Al-Harthi, S.; Cadalbert, L.; McCluskey, A.G.; Shweash, M.; Grassia, G.; Plevin, R. Deletion of the dual specific phosphatase-4 (DUSP-4) gene reveals an essential non-redundant role for MAP kinase phosphatase-2 (MKP-2) in proliferation and cell survival. *J. Biol. Chem.* **2011**, *286*, 12933–12943. [CrossRef] [PubMed]

33. Jaworowski, A.; Christy, E.; Yusoff, P.; Byrne, R.; Hamilton, J.A. Differences in the kinetics of activation of protein kinases and extracellular signal-related protein kinase 1 in colony-stimulating factor 1-stimulated and lipopolysaccharide-stimulated macrophages. *Biochem. J.* **1996**, *320*, 1011–1016. [CrossRef] [PubMed]

34. Jaworowski, A.; Wilson, N.J.; Christy, E.; Byrne, R.; Hamilton, J.A. Roles of the mitogen-activated protein kinase family in macrophage responses to colony stimulating factor-1 addition and withdrawal. *J. Biol. Chem.* **1999**, *274*, 15127–15133. [CrossRef] [PubMed]

35. Hamilton, J.A. CSF-1 signal transduction. *J. Leukoc. Biol.* **1997**, *62*, 145–155. [CrossRef] [PubMed]

36. Murray, J.T.; Craggs, G.; Wilson, L.; Kellie, S. Mechanism of phosphatidylinositol 3-kinase-dependent increases in BAC1.2F5 macrophage-like cell density in response to M-CSF: Phosphatidylinositol 3-kinase inhibitors increase the rate of apoptosis rather than inhibit DNA synthesis. *Inflamm. Res.* **2000**, *49*, 610–618. [CrossRef] [PubMed]

37. Torres, C.; Francis, M.K.; Lorenzini, A.; Tresini, M.; Cristofalo, V.J. Metabolic stabilization of MAP kinase phosphatase-2 in senescence of human fibroblasts. *Exp. Cell Res.* **2003**, *290*, 195–206. [CrossRef]

38. Valledor, A.F.; Comalada, M.; Xaus, J.; Celada, A. The Differential Time-course of Extracellular-regulated Kinase Activity Correlates with the Macrophage Response toward Proliferation or Activation. *J. Biol. Chem.* **2000**, *275*, 7403–7409. [CrossRef] [PubMed]

39. Richardson, E.T.; Shukla, S.; Nagy, N.; Boom, W.H.; Beck, R.C.; Zhou, L.; Harding, C.V. ERK Signaling Is Essential for Macrophage Development. *PLoS ONE* **2015**, *10*, e0140064. [CrossRef] [PubMed]

40. Torii, S.; Yamamoto, T.; Tsuchiya, Y.; Nishida, E. ERK MAP kinase in G cell cycle progression and cancer. *Cancer Sci.* **2006**, *97*, 697–702. [CrossRef] [PubMed]

41. Sanchez-Tillo, E.; Comalada, M.; Farrera, C.; Valledor, A.F.; Lloberas, J.; Celada, A. Macrophage-colony-stimulating factor-induced proliferation and lipopolysaccharide-dependent activation of macrophages requires Raf-1 phosphorylation to induce mitogen kinase phosphatase-1 expression. *J. Immunol.* **2006**, *176*, 6594–6602. [CrossRef] [PubMed]

42. Valledor, A.F.; Xaus, J.; Marquès, L.; Celada, A. Macrophage colony-stimulating factor induces the expression of mitogen-activated protein kinase phosphatase-1 through a protein kinase C-dependent pathway. *J. Immunol.* **1999**, *163*, 2452–2462. [PubMed]

43. Valledor, A.F.; Arpa, L.; Sánchez-Tilló, E.; Comalada, M.; Casals, C.; Xaus, J.; Celada, A. IFN-{gamma}-mediated inhibition of MAPK phosphatase expression results in prolonged MAPK activity in response to M-CSF and inhibition of proliferation. *Blood* **2008**, *112*, 3274–3282. [CrossRef] [PubMed]

44. Jeong, M.-W.; Kang, T.H.; Kim, W.; Choi, Y.H.; Kim, K.T. Mitogen-activated protein kinase phosphatase 2 regulates histone H3 phosphorylation via interaction with vaccinia-related kinase 1. *Mol. Biol. Cell* **2013**, *24*, 373–384. [CrossRef] [PubMed]

45. Fend, L.; Accart, N.; Kintz, J.; Cochin, S.; Reymann, C.; Le Pogam, F.; Fournel, S. Therapeutic effects of anti-CD115 monoclonal antibody in mouse cancer models through dual inhibition of tumor-associated macrophages and osteoclasts. *PLoS ONE* **2013**, *8*, e73310. [CrossRef] [PubMed]

46. Sherr, C.J. Colony-stimulating factor-1 receptor. *Blood* **1990**, *75*, 1–12. [PubMed]

47. Martinez, F.O.; Gordon, S.; Locati, M.; Mantovani, A. Transcriptional Profiling of the Human Monocyte-to-Macrophage Differentiation and Polarization: New Molecules and Patterns of Gene Expression. *J. Immunol.* **2006**, *177*, 7303–7311. [CrossRef] [PubMed]

48. Chaplin, D.D. Overview of the immune response. *J. Allergy Clin. Immunol.* **2010**, *125* (Suppl. 2), S3–S23. [CrossRef]

49. Auffray, C.; Fogg, D.K.; Narni-Mancinelli, E.; Senechal, B.; Trouillet, C.; Saederup, N.; Molina, T. CX3CR1+ CD115+ CD135+ common macrophage/DC precursors and the role of CX3CR1 in their response to inflammation. *J. Exp. Med.* **2009**, *206*, 595–606. [CrossRef]

50. Jacome-Galarza, C.E.; Lee, S.K.; Lorenzo, J.A.; Aguila, H.L. Identification, characterization, and isolation of a common progenitor for osteoclasts, macrophages, and dendritic cells from murine bone marrow and periphery. *J. Bone Miner. Res.* **2013**, *28*, 1203–1213. [CrossRef]

51. Barbour, M.; Plevin, R.; Jiang, H.-R. MAP kinase phosphatase 2 deficient mice develop attenuated experimental autoimmune encephalomyelitis through regulating dendritic cells and T cells. *Sci. Rep.* **2016**, *6*, 38999. [CrossRef] [PubMed]

52. Van Iersel, M.P.; Kelder, T.; Pico, A.R.; Hanspers, K.; Coort, S.; Conklin, B.R.; Evelo, C. Presenting and exploring biological pathways with PathVisio. *BMC Bioinf.* **2008**, *9*, 399. [CrossRef] [PubMed]

53. Saeed, A.I.; Sharov, V.; White, J.; Li, J.; Liang, W.; Bhagabati, N.; Sturn, A. TM4: A free, open-source system for microarray data management and analysis. *Biotechniques* **2003**, *34*, 374–378. [CrossRef] [PubMed]

54. Eden, E.; Navon, R.; Steinfeld, I.; Lipson, D.; Yakhini, Z. GOrilla: A tool for discovery and visualization of enriched GO terms in ranked gene lists. *BMC Bioinf.* **2009**, *10*, 48. [CrossRef] [PubMed]

55. Ye, J.; Coulouris, G.; Zaretskaya, I.; Cutcutache, I.; Rozen, S.; Madden, T.L. Primer-BLAST: A tool to design target-specific primers for polymerase chain reaction. *BMC Bioinf.* **2012**, *13*, 134. [CrossRef] [PubMed]

International Journal of
Molecular Sciences

MDPI

Article

Nuclear-Biased DUSP6 Expression is Associated with Cancer Spreading Including Brain Metastasis in Triple-Negative Breast Cancer

Fan Wu [1], Robert D. McCuaig [1], Christopher R. Sutton [1], Abel H. Y. Tan [1], Yoshni Jeelall [1],
Elaine G. Bean [2], Jin Dai [1], Thiru Prasanna [1,3], Jacob Batham [1], Laeeq Malik [3,4],
Desmond Yip [3,4], Jane E. Dahlstrom [2,4] and Sudha Rao [1,*]

[1] Melanie Swan Memorial Translational Centre, Faculty of Sci-Tech, University of Canberra, Bruce ACT 2617,
     Australia; Fan.Wu@canberra.edu.au (F.W.); Robert.McCuaig@canberra.edu.au (R.D.M.);
     Chris.Sutton@canberra.edu.au (C.R.S.); Abel.Tan@canberra.edu.au (A.H.Y.T.);
     Yoshni.Jeelall@canberra.edu.au (Y.J.); Jin.dai@canberra.edu.au (J.D.); Thiru.Prasanna@canberra.edu.au (T.P.);
     Jacob.batham@canberra.edu.au (J.B.)
[2] Anatomical Pathology, ACT Pathology, The Canberra Hospital, Canberra health Services, Garran ACT 2606,
     Australia; Elaine.Bean@act.gov.au (E.G.B.); Jane.Dahlstrom@act.gov.au (J.E.D.)
[3] Department of Medical Oncology, The Canberra Hospital, Canberra health Services, Garran ACT 2606,
     Australia; Laeeq.Malik@act.gov.au (L.M.); Desmond.Yip@anu.edu.au (D.Y.)
[4] ANU Medical School, College of Health and Medicine, The Australian National University,
     Canberra ACT 0200, Australia
*   Correspondence: Sudha.Rao@Canberra.edu.au

Received: 30 April 2019; Accepted: 22 June 2019; Published: 24 June 2019

Abstract: DUSP6 is a dual-specificity phosphatase (DUSP) involved in breast cancer progression, recurrence, and metastasis. DUSP6 is predominantly cytoplasmic in HER2+ primary breast cancer cells, but the expression and subcellular localization of DUSPs, especially DUSP6, in HER2-positive circulating tumor cells (CTCs) is unknown. Here we used the DEPArray system to identify and isolate CTCs from metastatic triple negative breast cancer (TNBC) patients and performed single-cell NanoString analysis to quantify cancer pathway gene expression in HER2-positive and HER2-negative CTC populations. All TNBC patients contained HER2-positive CTCs. HER2-positive CTCs were associated with increased ERK1/ERK2 expression, which are direct DUSP6 targets. DUSP6 protein expression was predominantly nuclear in breast CTCs and the brain metastases but not pleura or lung metastases of TNBC patients. Therefore, nuclear DUSP6 may play a role in the association with cancer spreading in TNBC patients, including brain metastasis.

Keywords: circulating tumor cells (CTCs); DEPArray; dual-specificity phosphatase; HER2; brain metastasis; single cell analysis; triple-negative breast cancer (TNBC)

---

1. Introduction

Breast cancer is the most common malignant disease in women worldwide. While there has been significant progress in breast cancer treatment over recent decades through new chemotherapies, hormone therapies, targeted therapies, and more recently immunotherapies, 25–40% of breast cancer patients still develop primary resistance and metastasis to eventually die from their disease [1,2]. Identification of predictive resistance biomarkers has so far remained elusive, especially in triple negative breast cancer (TNBC), a breast cancer molecular subtype with currently few treatment options and an often-aggressive disease course.

Epithelial to mesenchymal transition (EMT) is a biological program in which epithelial cells lose cell–cell junctions, apical–basal polarity, and acquire an invasive mesenchymal phenotype.

EMT is induced via several interacting signaling pathways including the WNT, TGF-β, Notch, Hedgehog, PI3-kinase/AKT, PKC-theta, and mitogen-activated protein kinase (MAPK) pathways [3–5]. EMT is implicated in cancer initiation, progression, metastasis, resistance to conventional therapies, and recurrence by inducing cancer stem cells (CSCs), tumor-initiating cells (TICs), and circulating tumor cells (CTCs) [3,5,6].

Breast CTCs are defined as cells circulating in the blood that lack CD45 expression but express cytokeratins 7, 8, 18, and 19 [7,8]. Breast CTCs are associated with progressive cancer and a higher risk of recurrence and metastasis [6,9]. Both HER2-positive CTCs and HER2-negative CTCs have been detected in patients with HER2-negative primary tumors (e.g., TNBCs) after adjuvant chemotherapy for the management of their primary breast cancer [9–12]. HER2-positive CTCs are associated with progression following multiple courses of systemic therapy in HER2-negative metastatic breast cancer patients [13].

Dual-specificity phosphatases (DUSPs) are a family of proteins responsible for dephosphorylating threonine/serine and tyrosine residues on their substrates, mainly MAP kinases. Several DUSPs are implicated in breast cancer metastasis, for example, DUSP1, DUSP4, and DUSP6 [4,14]. DUSP1 and DUSP4 are predominantly expressed in the nucleus, while DUSP6 is predominantly cytoplasmic [4]. DUSP6 has an interdomain linker region that contains a binding motif of ERK1/2 and nuclear export signal (NES) [15]. NES mediated nuclear-cytoplasmic shuttling of DUSP6 undergoes a highly specific interaction with ERK1 and ERK2 at dual threonine and tyrosine residues of the Thr-Glu-Tyr (TEY) motif, thereby inactivating DUSP6 [16]. Caspase-3 cleavage of DUSP6 mediates the subcellular localization and activation of ERK1/2 [15]. DUSP6 phosphorylation and ERK1/2 dephosphorylation creates a negative feedback loop to control ERK activity and indirectly the MAPK signal, which is vital for cellular proliferation and differentiation [15–22]. DUSP6 is enriched in HER2-positive breast cancer tumor [23], the overexpression of which confers resistance to tamoxifen endocrine therapy [4,24], and we previously showed that DUSP6 knockdown by RNA interference increased CSC formation but inhibited cell proliferation in TNBC cell lines in vitro [4,21]. DUSP6 is also upregulated in the cytoplasm of primary malignant breast cancer cells in HER2-positive breast cancer patients regardless of estrogen receptor/progesterone receptor (ER/PR) status [4]. Importantly, DUSP6 is cytoplasmic in ER⁻PR⁻HER2⁺ tissue but not in TNBCs [4]. The nuclear role of DUSP6 in EMT and metastasis in TNBC remains elusive.

HER2-positive CTCs were recently identified in breast cancer patients with HER2-negative primary tumors [8]. However, it is unknown how the mRNA profiles of HER2-positive CTCs and HER2-negative CTCs differ. We therefore analyzed mRNA profiles in HER2-positive CTCs and HER2-negative CTCs isolated from five metastatic breast cancer patients with TNBC. At least one HER2-positive CTC was detected in every patient. mRNA expression profiles were distinct in HER2-positive and HER2-negative CTCs in the same patient. Pathway analysis revealed differential expression of genes involved in ERK1/2-MAPK signaling between HER2-positive and HER2-negative CTCs and involvement of DUSP6. We report for the first time that nuclear DUSP6 was present in CTCs from metastatic breast cancer patients regardless of CTC HER2 status. Importantly, nuclear-biased DUSP6 was also present in brain metastases but not pleura or lung metastases in TNBC patients, suggesting that the nuclear DUSP6-associated pathway participates in cancer spreading in TNBC, including brain metastasis.

2. Results

2.1. HER2-Positive and HER2-Negative CTCs Detected and Isolated from Primary TNBC Patients

CD45⁻/pan-CK⁺ CTCs along with HER2-positive and HER2-negative CTC sub-populations were identified, enumerated and isolated from five metastatic TNBC patients using the DEPArray system (Silicon Biosystems, Castel Maggiore, Italy). There were 5 to 150 CTCs per 7.5 mL blood (in average of 103 CTCs per 7.5 mL blood) (Figure 1A,B). All TNBC patients in this group had HER2-positive CTCs detected ranging from 10% to 100% of total (Figure 1B–D). Post standard chemotherapy in combination with an inhibitor targeting Lysine-specific histone demethylase 1 (LSD1), HER2-positive CTCs were more frequent in patients with disease progression with increased total CTC number and less frequent in patients with stable disease (Figure 1D).

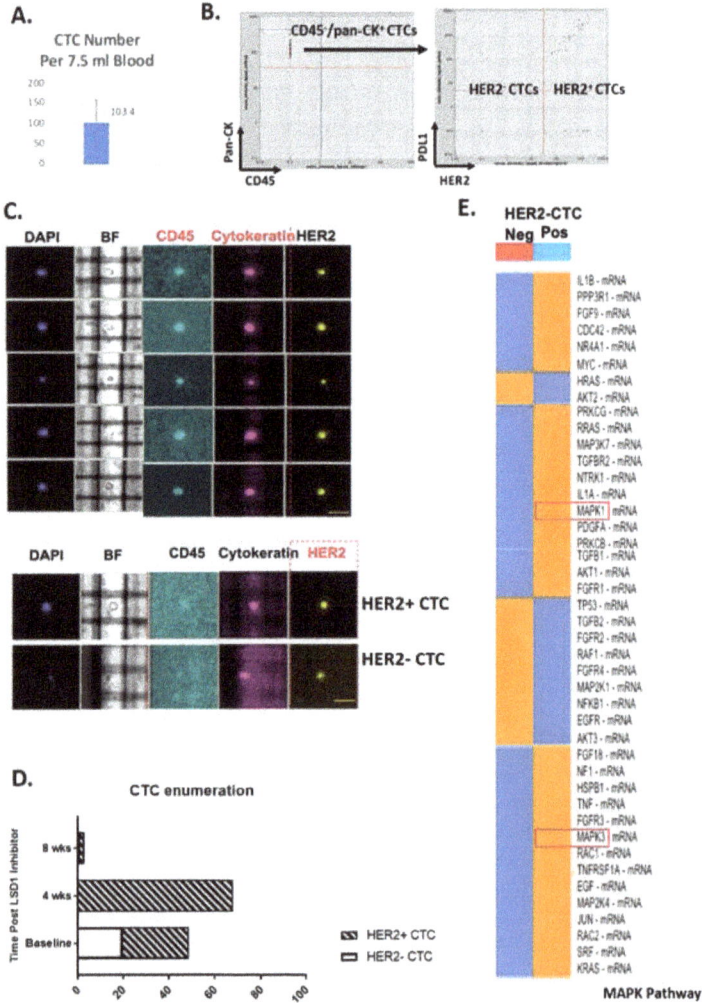

Figure 1. HER2-positive and negative circulating tumor cells (CTCs) have distinct mRNA profiles in metastatic triple negative breast cancer (TNBC) patients. CD45-negative enriched PBMCs were labelled with anti-human CD45-APC and pan-cytokeratin-FITC for DEPArray CTC enumeration and isolation. (**A**) CTCs were detected in metastatic TNBC patients. CTCs were enumerated in five metastatic TNBC patients using the DEPArray platform. Data represent the mean of CTC counts ± standard deviation in 7.5 mL equivalent whole blood. (**B**) Identification and isolation of HER2-positive and negative CTCs in the same patients.

Plot shows the gating strategy for identifying HER2-positive and negative CTCs using DEPArray. Example images are indicated with scale bars (10 μm). (**C**) HER2-positive CTCs and HER2-negative CTCs using DEPArray. BF: brightfield. (**D**) CTC HER2 proportions over the course of chemotherapy in combination of Lysine-specific histone demethylase 1 (LSD1) inhibitor. Plot represents the number of HER2-positive and negative CTCs from 7.5 mL equivalent whole blood. (**E**) HER2-positive CTCs show distinct mRNA profiles in genes involved in mitogen-activated protein kinase (MAPK) signaling. NanoString analysis was performed using their pan-cancer progression panel to investigate differential gene expression in HER2-positive and negative CTCs isolated from the same patient. Data generated by pathway analysis using nSolver 4.0 software (NanoString). Orange indicates higher mRNA expression and blue indicates lower expression.

2.2. Distinct ERK1/2-MAPK Pathway Gene Profiles in HER2-Positive CTCs and HER2-Negative CTCs

To assess the mRNA profiles of HER2-positive and negative CTCs, both populations were harvested from the same patient using the DEPArray system and, after cDNA synthesis and un-biased pre-amplification, expression of 739 mRNAs from the NanoString PanCancer Progression panel assessed. ERK1/ERK2-MAPK signaling pathway analysis showed that HER2-positive CTCs had decreased AKT1/2 mRNA expression and increased MAPK1 and MAPK3 expression, which are the direct substrates of DUSP6 (Figure 1E).

2.3. Nuclear Distribution of DUSP6 in CTCs and Brain Metastases from TNBC Patients

Since DUSP6 directly dephosphorylates ERK1/ERK2, we reasoned that DUSP6 may participate in HER2-positive CTCs in metastatic breast cancer patients. DUSP6 showed nuclear and cytoplasmic expression and was biased towards nuclear expression in the CTCs as measured by the ratio of nuclear to cytoplasmic (Fn/c), (Figure 2A,B and Figure S1C). Moreover, nuclear DUSP6 was also co-expressed with cell surface vimentin (CSV) and ABCB5 known mesenchymal markers in CTCs (Figure 2A–C). We also examined the distribution of DUSP6 protein expression in brain metastasis tissue in TNBC patients (Figure 2C,D and Figure S1A,B). In 19 out of 21 patients, approximately 40% of cells were triple-positive for DUSP6, CSV, and ABCB5 (Figure 2C). Surprisingly, DUSP6 nuclear expression was also observed in the brain tissue, with highest DUSP6 expression observed in $CSV^+ABCB5^+DUSP6^+$ triple-positive cells, (Figure 2C). To further understand whether nuclear expression of DUSP6 is exclusively presented in brain metastasis, we examined five lung and pleura metastases tissues from TNBC patients. Interestingly, virtually no neoplastic cell had nuclear DUSP6 expression in all five samples, although large numbers of CSV^+ABCB5^+ cells were detected (Figure 2E,F).

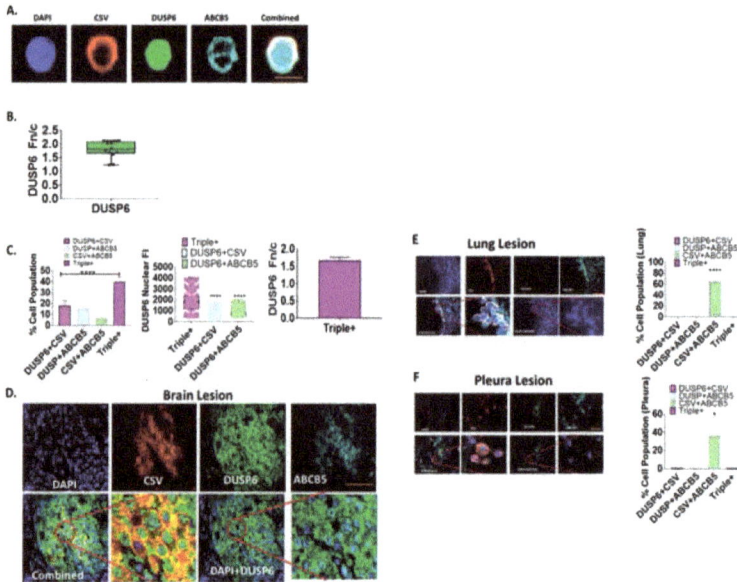

Figure 2. Nuclear-biased dual-specificity phosphatase 6 (DUSP6) in CTCs and brain metastases from metastatic TNBC patients. (**A**) Nuclear DUSP6 presented in CTCs from metastatic breast cancer patients. CTCs were labelled with anti-human DUSP6, cell surface vimentin (CSV), ABCB5, and DAPI for microscopy. (**B**) The nuclear bias of DUSP6 in CTCs from stage IV metastatic breast cancer patients (*n* = 5 patients). Nuclear/cytoplasmic intensity of DUSP6 was measured by microscopy. (**C**) DUSP6 expression in breast cancer metastases to the brain. Fluorescence labelling of DUSP6, CSV, and ABCB5 was performed for ASI microscopy analysis. Bar graph represents cells double-positive for either DUSP6/ABCB5/CSV or triple-positive for all three markers. The second bar graph displays the nuclear fluorescence intensity in double- or triple-positive cells for DUSP6. >500 cells counted per patient. Representative image of nuclear DUSP6 expression in (**D**) brain metastasis tissue (*n* = 21 patients). (**E**) DUSP6 expression in breast cancer metastases to the lung and (**F**) pleura (*n* = 5 patients) from metastatic TNBC patients. Bar graph represents cells double-positive for either DUSP6/ABCB5/CSV or triple-positive for all three markers. Scale bars = 10 μm and are shown in orange. Data represent mean ± SE. Unpaired *t*-test. *$p < 0.05$, ****$p < 0.0001$ denote significant differences.

2.4. Nuclear Distribution of DUSP6 in Primary Tumor Sections in the Chemotherapy or Immunotherapy Treated 4T1 Metastatic Mouse Model

To understand whether nuclear DUSP6 is associated with drug resistance, we next examined DUSP6 expression in the 4T1 TNBC metastatic mouse breast cancer model treated with control vehicle, Abraxane (nab-paclitaxel, 30 mg/kg), or anti-PD1 (RMP1–14, 10 mg/kg) using the ASI digital pathology system. We first examined the changes in tumor volume caused by treatment with either Abraxane or anti-PD1 immunotherapy (Figure 3A) reduced overall tumor volume equally (Figure 3A). In surviving 4T1 tumor cells in Abraxane or anti-PD1-treated mice, nuclear DUSP6 was enriched in both triple-positive cells (CSV$^+$ABCB5$^+$DUSP6$^+$) and the double-positive ABCB5$^+$DUSP6$^+$ tumor cell pool (Figure 3B). We next examined the Fn/c of DUSP6 expression that revealed that DUSP6 is predominantly cytoplasmic in control tumors but becomes nuclear biased in surviving tumor cells following treatment with Abraxane or anti-PD1 (Figure 3C). DUSP6 expression correlated with tumor cell survival in response to either chemotherapy (Abraxane) or immunotherapy (anti-PD1) (Figure S2).

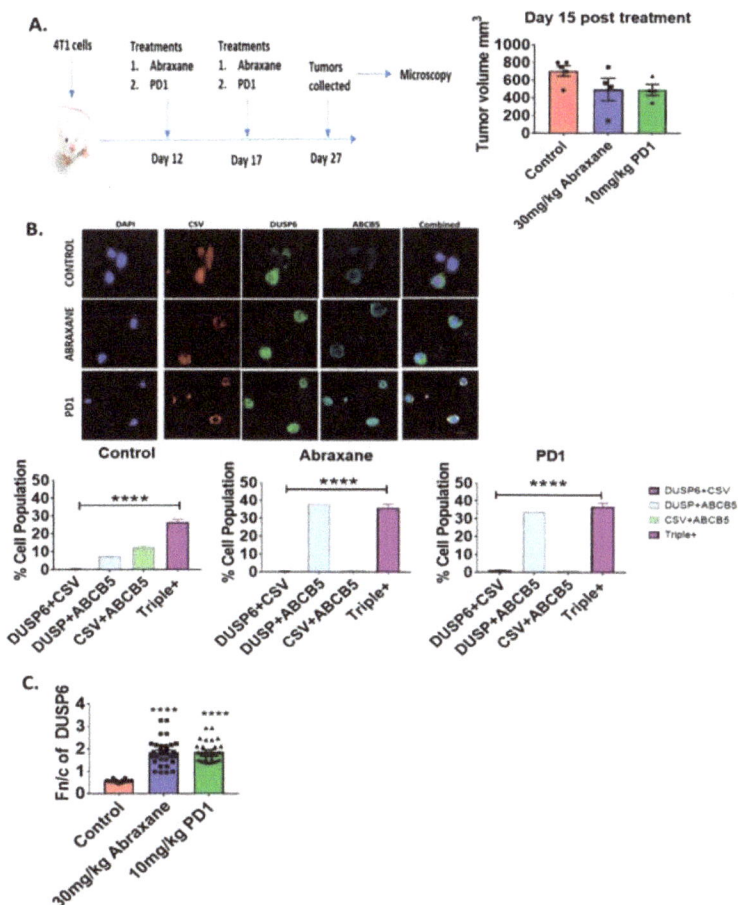

Figure 3. Nuclear-biased DUSP6 in the 4T1 metastatic mouse cancer model. (**A**) Treatment regimen using the BALB/c 4T1 breast cancer model and day 15 tumor volumes of mice treated with vehicle control, Abraxane, or PD-1 (*n* = 4). (**B**) DUSP6 expression in the primary 4T1 tumor model. Fluorescence labelling of DUSP6, CSV, and ABCB5 was performed for ASI microscopy analysis. Bar graph represents cells double-positive for either DUSP6/ABCB5/CSV or triple-positive for these three markers. *n* = 3 mice per group (>500 cells per patient counted). Data represent mean ± SE. Unpaired *t*-test. ****p < 0.0001 denote significant differences. Example images of each group are indicated with scale bars (10 μm). (**C**) DUSP6 Fn/c (ratio of nuclear to cytoplasmic staining) was quantified in the primary 4T1 tumor model. Fn/c was determined by performing ASI microcopy analysis and using ImageJ-Fiji to determine the ratio. Bar graphs represent fn/c mean ± SE. Unpaired *t*-test., ****p < 0.0001 denote significant differences. Example images are in 3B.

2.5. Novel Inhibition of Nuclear DUSP6 Mediated the Expression of ABCB5 and CSV In Vitro

To understand the effect of inhibition of nuclear DUSP6 in TNBC, we designed a novel competitive peptide inhibitor targeting a bipartite nuclear localization signal (NLS) of DUSP6 that was identified using the NLS Mapper predictive tool. In MDA-MB-231 TNBC cells, treatment of Paclitaxel induced nuclear translocation of DUSP6. In comparison, the DUSP6 NLS peptide inhibitor significantly reduced nuclear DUSP6 expression (Figure 4A) and inhibited the expression of CSV and ABCB5 in vitro suggesting alteration of protein function (Figure 4A,B). These results suggest that inhibition of nuclear expression of DUSP6 subsequently impacts expression of the mesenchymal, stem-like markers such as CSV or ABCB5.

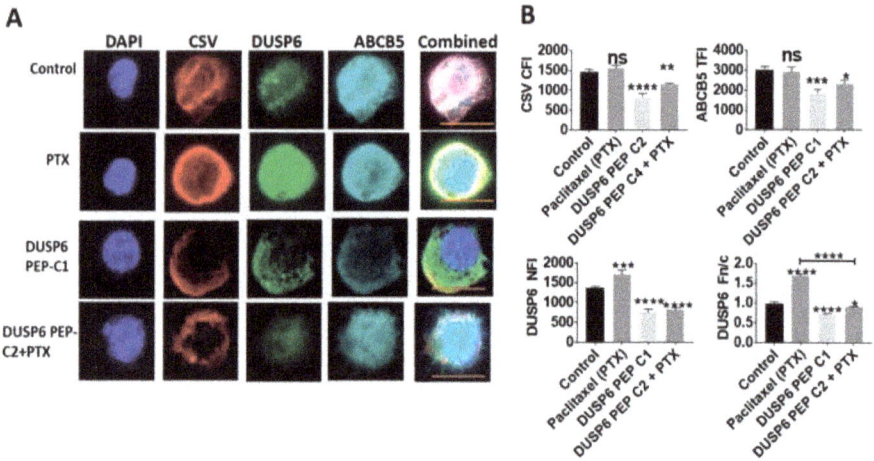

Figure 4. Inhibition of nuclear DUSP6 with novel peptide inhibitor. DUSP6, CSV and ABCB5 were labelled in TNBC cell line, MDA-MB-231 and treated with either vehicle alone, Paclitaxel (PTX) or DUSP6 Nuclear Peptide inhibitor (DUSP6 PEP, C1 = 60 mg/mL) or a combination (PTX + C2 40 mg/mL) with $n > 20$ individual cells. Representative images for each dataset are shown (**A**). Graphs represent the Fn/c, nuclear fluorescence intensity (NFI), cytoplasmic fluorescence intensity (CFI) or total fluorescence intensity (TFI) measured with the ASI automated digital pathology system (**B**). Scale bar is in orange and equals 10 μM. $*p < 0.05$, $**p < 0.01$, $***p < 0.001$, $****p < 0.0001$ denote significant differences. ns denotes non-significant.

2.6. Nuclear Association of DUSP6 with P300 and H3K9me2

Nucleosomes are enriched in highly compacted chromatin structures which are transcriptionally silent. An important mechanism of regulating gene expression is chromatin remodeling which can be orchestrated via a number of mechanisms including post-translational modifications (PTMs) of histone proteins such as the repressive mark histone H3K9me2 as well co-activator proteins such as the histone acetyltransferase P300 that mediate active gene transcription and we have previously show to interact with DUSP family members [4].

Our results of H3k9me2 and DUSP6 co-expression revealed that while there was some positive co-localization the score was low (Figure 5). Whereas analysis revealed that P300 positively co-localized with DUSP6 in the nucleus of these CTCs with a stronger positive score (Figure 5). Collectively our data suggest that DUSP6 associates with P300 akin to our previous results for DUSP4 [4]. This suggests a potential role for DUSP6 to mediate active gene transcription programs in CTCs from metastatic breast cancer patients.

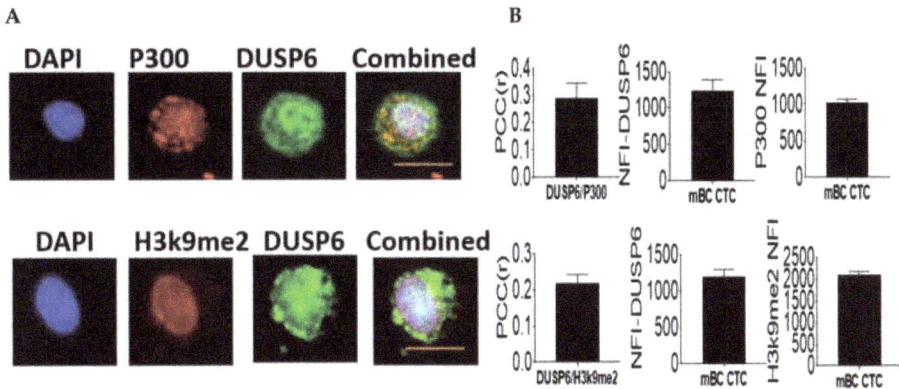

Figure 5. Nuclear DUSP6 interaction with P300 or H3k9me2. Confocal laser scanning microscopy was performed on CTCs (circulating tumor cells) isolated from metastatic breast cancer patients (Stage IV $n = 10$). Cells were fixed and probed with primary rabbit antibodies to DUSP6 and primary mouse antibodies to P300 or H3k9me2 followed by the corresponding secondary antibody conjugated to Alexa-Fluor 568 or Alexa-Fluor 488. Representative images for each antibody dataset pair are shown: red = p300 or H3k9me2; green = DUSP6. (**A**) The The Pearson correlation coefficient (PCC) was determined as described in methods. PCC indicates the strength of relation between the two fluorochrome signals for at least 20 individual cells ± SE. Bar graphs (**B**) indicate the total NFI of DUSP6, p300, or H3K9me2 as measured using ASI software to select the nucleus of each cell and measure the total NFI signal minus background for at least 20 individual cells ± SE. Scale bar is in orange and equals 10 μM.

3. Discussion

TNBC remains the most difficult group of metastatic breast cancers to manage due to a lack of targetable proteins expressed in these tumor cells [1]. Despite therapeutic advances, the overall response and survival rates for patients with TNBC are low using conventional chemotherapy and novel immunotherapy compared to other solid organ malignancies. Increased CTC numbers are associated with rapid cancer progression and a higher risk of recurrence and metastasis. Here we confirmed the presence of both HER2-positive and negative CTCs in patients with a primary tissue diagnosis of TNBC, supporting previous findings [9,11,12]. For the first time, we compared the mRNA profiles of these two populations directly in the same patient using a panel of cancer progression-related genes, and found differential MAPK pathway gene expression. Specifically, there was increased expression of *MAPK1/2*, encoding ERK1/2, which are the direct dephosphorylation targets of DUSP6, in HER2-positive CTCs.

DUSP6 is predominately expressed in the cytoplasm of breast cancer cells such as the commonly used cell lines MCF-7 and MDA-MB-231 [4]. DUSP6 nuclear expression increases in tumor cells that have transformed into CSCs via EMT [4]. Dual inhibition of DUSP6 and DUSP1 significantly reduces CSC formation in vitro. We previously showed that DUSP6 was exclusively cytoplasmic in primary HER2$^+$ER$^-$PR$^-$ breast cancers from metastatic breast cancer patients, with no nuclear DUSP6 expression in primary TNBCs [4]. Interestingly, in CTCs, DUSP6 was nuclear in all vimentin-positive mesenchymal CTCs from stage IV metastatic patients evaluated. In 19 out of 21 patients TNBC metastatic breast cancer patient brain metastases, DUSP6 had nuclear expression, suggesting that the brain metastasis may form from CTCs via a nuclear DUSP6-dependent pathway. In comparison, our previous study showed that TNBC patients had undetectable levels of nuclear DUSP6 expression [4]. Overall, this suggests that distinct mechanisms are likely at play in the aggressive cancer cells in primary tissue, lung and pleura metastases tissue compared to brain metastases events. Future studies, particularly larger blind studies, will be required to fully validate the results of this study.

These results have a number of potential important implications for patient care. First, a subset of TNBC patients may in fact benefit from trastuzumab (Herceptin) therapy if they harbor significant numbers of HER2-positive CTCs and that HER2-CTC testing may be required in routine practice. Second, the presence of HER2-positive CTCs in primary TNBC may be a risk factor for metastatic disease, especially to the brain. Third, the nuclear distribution of DUSP6 in CTCs and brain metastases may be prognostic, the majority of metastatic breast cancer patients with brain metastases had cancer cells positive for DUSP6 and all CTCs examined were also DUSP6 positive. Four, the novel nuclear DUSP6 peptide inhibitor can potentially reduce the risk of brain metastasis. Therefore, DUSP6 may be a therapeutic target to prevent or treat brain metastasis. We also examined DUSP6 expression in surviving tumor cells in a mouse model of breast cancer metastasis treated with Abraxane or anti-PD1. DUSP6-positive cells were enriched in surviving tumor cells, which may indicate a role in resistance to chemotherapy or immunotherapy.

Although nuclear DUSP6 expression was present in CTCs and in the brain metastases of 19 out of 21 patients TNBC patients, this result requires further confirmation in a larger cohort and the samples need to be stratified based on the original hormonal status of the primary tumor in future experiments. Nuclear DUSP6 was not expressed in pleura or lung metastases tissue of TNBC patients, and it remains uncertain whether nuclear-biased DUSP6 in CTCs is exclusively associated with brain metastasis. DUSP6 expression should be analyzed in metastases at other sites as well as obtaining matched CTCs and metastatic lesion tissue samples from different sites to compare for DUSP6 expression. In addition, future studies should examine the nuclear interaction partners of DUSP6 along with the epigenetic machinery that DUSP6 may regulate in cancer biology as well as the underpinning gene signatures associated with DUSP6 in brain metastases. Also, further work should investigate if MAPKs such as ERK1/2 and AKT1/2 are directly mediated by nDUSP6 enriched HER2-postive CTCs. Overall these findings suggest the potential importance of developing nuclear targeting DUSP6 therapeutics for targeting aggressive TNBC cancer spreading, including brain metastasis.

4. Materials and Methods

4.1. Patients

Peripheral blood (25 to 40 mL) was collected into 3 or 4 EDTA tubes from five patients with radiologically and histologically confirmed metastatic TNBC before starting a new chemotherapy regimen. Samples were maintained at room temperature and processed within four hours after blood collection. Samples were anonymized, and operators were blinded to clinical outcome.

4.2. Tissues

Tissue from routinely processed archival primary and metastatic breast cancers from the five patients were retrieved from the Department of Anatomical Pathology, ACT Pathology, Canberra Hospital. The tissue diagnosis and receptor status were reviewed by a specialized breast pathologist (JED), who excluded areas of tumor necrosis.

4.3. Animal Studies

Five-week-old female BALB/c mice were obtained from the Animal Resources Centre (ARC), Perth, and allowed to acclimatize for one week in the containment suites at The John Curtin School of Medical Research (JCSMR). All experimental procedures were performed in accordance with the guidelines and regulations approved by the Australian National University Animal Experimentation Ethics Committee (ANU AEEC). Mice were shaved at the site of inoculation the day before subcutaneous injection with 2×10^5 4T1 cells in 50 μL PBS into the right mammary gland. Treatment was started on day 12 post inoculation, when tumors reached approximately 50 mm^3. Tumors were measured using external calipers and volume calculated using a modified ellipsoidal formula $\frac{1}{2}$ (a/b^2), where a = longest diameter and b = shortest diameter. Mice were treated with the chemotherapy Abraxane (30 mg/kg) and anti-PD1 immunotherapy (10 mg/kg) every 5 days (twice). All treatments were given intraperitoneally in PBS. Tumors were collected on day 27 post inoculation of 4T1 cells for immunofluorescence microscopy.

4.4. Ethics

All subjects gave their informed consent for inclusion before study participation. The study was conducted in accordance with the Declaration of Helsinki, and the ACT Heath Human Research Ethics Committee (ID code: ETH.11.15.217) approved the protocol on the 4 December 2017.

4.5. CTC Enrichment

CD45-positive cell depletion was performed to enrich CD45-negative expressing cells using the RosettSep CD45 depletion kit (StemCell Technologies, Vancouver, Canada) according to the manufacturer's instructions and as described previously [5].

4.6. CTC Detection, Imaging, and Isolation

CD45-negative enriched cells were stained with antibodies targeting human CD45, pan-cytokeratin, and HER2/vimentin/DUSP6 prior to CTC enumeration and sorting by DEPArray V2 or NXT (Silicon Biosystems, Italy). CTCs were defined as nucleated cells lacking CD45 and expressing pan-cytokeratin (CK7, CK8, CK18, and CK19) as described previously [5]. CTCs for mRNA analysis were lysed using the Single Cell Lysis Kit (Thermo Fisher Scientific, Waltham, MA, USA) and stored at −80 °C prior to further analysis. CTCs for immunofluorescence microscopy analysis were cytospun onto coverslips and fixed prior to fluorescence labeling and imaging.

4.7. RNA Isolation, cDNA Synthesis and Pre-Amplification from CTCs

Lysed single CTCs were treated with DNAse I (Thermo Fisher Scientific), and the RNA was converted to cDNA using Superscript VILO (Thermo Fisher Scientific). Pre-amplification was performed using the Nanostring Pre-Amp kit (NanoString Technologies, Seattle, WA).

4.8. Nanostring Analysis

The nCounter XT protocol was used for hybridization, and the data were captured using a Nanostring Sprint instrument (NanoString Technologies). Data analysis was performed using nSolver 4.0 (NanoString Technologies).

4.9. ASI Microscopy

For high-throughput microscopy in human tissue samples, protein targets were localized by confocal laser scanning microscopy. Single 0.5 μm sections were obtained using an Olympus-ASI automated microscope with a 60× oil immersion lens running ASI software. The final image was obtained by employing a high-throughput automated stage with ASI spectral capture software. Digital images were analyzed using automated ASI software (Applied Spectral Imaging, Carlsbad, CA) to determine the distribution and intensities automatically with automatic thresholding and background correction of the average nuclear fluorescence intensity (NFI), allowing for the specific targeting of nuclear expression of proteins of interest (in this case DUSP6).

4.10. Nuclear to Cytoplasmic Fluorescence Ratio (Fn/c) Analysis

Digital confocal images were analyzed using Fiji-ImageJ software (Schindelin et al., 2012) to determine the TNFI, TCFI, TFI, or the nuclear to cytoplasmic fluorescence ratio (Fn/c) using the equation: $Fn/c = (Fn − Fb)/(Fc − Fb)$, where Fn is nuclear fluorescence, Fc is cytoplasmic fluorescence, and Fb is background fluorescence. A minimum of $n = 20$ cells were analyzed for each sample set. The Mann–Whitney non-parametric test (GraphPad Prism, GraphPad Software, San Diego, CA) was used to determine significant differences between datasets.

4.11. Statistical Analysis

PRISM-Graph pad was used to calculate the significant differences using the Mann–Whitney Unpaired *t*-test. Significance was plotted as indicated: * $p < 0.05$, ** $p < 0.01$, *** $p < 0.001$, **** $p < 0.0001$

5. Conclusions

We showed for the first time that nuclear-biased DUSP6 was present in CTCs from TNBC patients. Importantly, nuclear DUSP6 was also observed in brain metastases but not lung or pleura metastases from metastatic TNBC patients, while DUSP6 was always cytoplasmic in the primary tumor tissue. These results indicate that a nuclear-biased DUSP6-dependent pathway may be involved in cancer spreading in TNBC patients, including brain metastasis. Inhibition of DUSP6 nuclear translocation reduced the expression of aggressive mesenchymal protein and may be a good candidate therapeutic target to reduce or prevent metastases via CTC in metastatic breast cancer and TNBC patients. Future studies are required to address the precise transcriptional programs underpinning nuclear DUSP6 in cancer spreading in TNBC, especially in brain metastases.

Supplementary Materials: Supplementary materials can be found at http://www.mdpi.com/1422-0067/20/12/3080/s1.

Author Contributions: F.W. wrote the manuscript together with S.R. and carried out the DEPArray and Nanostring experiments. R.D.M. carried out all the ASI imaging and analysis and wrote those sections of the paper. C.R.S. carried out the Nanostring experiments together with F.W. A.H.Y.T. carried out the 4T1 mouse experiments. J.E.D., Y.J., J.D. and J.B. assisted R.M. with ASI Tissue imaging. E.G.B provided and prepared all tissue sectioning. T.P., L.M. and D.Y. provided the patient samples and clinical expertise on metastatic breast cancer. J.E.D. provided the tissue blocks and expertise on the pathology of metastatic breast cancer. S.R. developed the concept, oversaw the project and co-wrote the manuscript.

Funding: This work was funded by the UC Cancer foundation through donations to the Melanie Swan Memorial Translation Centre.

Acknowledgments: We acknowledge all of the donors, patients, medical practitioners, nurses, and coordinators who supported this research. This study was supported by UC Foundation funds, a University of Canberra PhD Scholarship, and a Deep Water Scholarship awarded to Thiru Prasanna. We also acknowledge ACT Pathology staff who originally processed the tissue and the pathologists who initially reported on the samples.

Conflicts of Interest: In accordance with NHMRC guidelines and our ethical obligations as researchers, we report that the University of Canberra, Rao, McCuaig, Dahlstrom, and Tan have a financial interest in EpiAxis Therapeutics Pty Ltd. and Fan Wu's salary was paid in part by EpiAxis Therapeutics. Rao is also Chief Scientific Officer of EpiAxis Therapeutics Pty Ltd. We have in place a plan for managing any potential conflicts arising from that involvement. The authors declare no conflict of interest.

Abbreviations

MDPI	Multidisciplinary Digital Publishing Institute
DOAJ	Directory of open access journals
TLA	Three letter acronym
LD	linear dichroism
LSD1	Lysine-specific histone demethylase 1

References

1. Lin, N.U.; Vanderplas, A.; Hughes, M.E.; Theriault, R.L.; Edge, S.B.; Wong, Y.N.; Blayney, D.W.; Niland, J.C.; Winer, E.P.; Weeks, J.C. Clinicopathologic features, patterns of recurrence, and survival among women with triple-negative breast cancer in the national comprehensive cancer network. *Cancer* **2012**, *118*, 5463–5472. [CrossRef] [PubMed]

2. Prasanna, T.; Wu, F.; Khanna, K.K.; Yip, D.; Malik, L.; Dahlstrom, J.E.; Rao, S. Optimizing poly (adp-ribose) polymerase inhibition through combined epigenetic and immunotherapy. *Cancer Sci.* **2018**, *109*, 3383–3392. [CrossRef] [PubMed]

3. Zafar, A.; Wu, F.; Hardy, K.; Li, J.; Tu, W.J.; McCuaig, R.; Harris, J.; Khanna, K.K.; Attema, J.; Gregory, P.A.; et al. Chromatinized protein kinase c-theta directly regulates inducible genes in epithelial to mesenchymal transition and breast cancer stem cells. *Mol. Cell Biol.* **2014**, *34*, 2961–2980. [CrossRef] [PubMed]

4. Boulding, T.; Wu, F.; McCuaig, R.; Dunn, J.; Sutton, C.R.; Hardy, K.; Tu, W.; Bullman, A.; Yip, D.; Dahlstrom, J.E.; et al. Differential roles for dusp family members in epithelial-to-mesenchymal transition and cancer stem cell regulation in breast cancer. *PLoS ONE* **2016**, *11*, e0148065. [CrossRef] [PubMed]
5. Boulding, T.; McCuaig, R.D.; Tan, A.; Hardy, K.; Wu, F.; Dunn, J.; Kalimutho, M.; Sutton, C.R.; Forwood, J.K.; Bert, A.G.; et al. Lsd1 activation promotes inducible emt programs and modulates the tumour microenvironment in breast cancer. *Sci. Rep.* **2018**, *8*, 73. [CrossRef]
6. Cristofanilli, M.; Budd, G.T.; Ellis, M.J.; Stopeck, A.; Matera, J.; Miller, M.C.; Reuben, J.M.; Doyle, G.V.; Allard, W.J.; Terstappen, L.W.; et al. Circulating tumor cells, disease progression, and survival in metastatic breast cancer. *N Engl. J. Med.* **2004**, *351*, 781–791. [CrossRef] [PubMed]
7. Gradilone, A.; Raimondi, C.; Nicolazzo, C.; Petracca, A.; Gandini, O.; Vincenzi, B.; Naso, G.; Agliano, A.M.; Cortesi, E.; Gazzaniga, P. Circulating tumour cells lacking cytokeratin in breast cancer: The importance of being mesenchymal. *J. Cell Mol. Med.* **2011**, *15*, 1066–1070. [CrossRef]
8. Krishnamurthy, S.; Bischoff, F.; Ann Mayer, J.; Wong, K.; Pham, T.; Kuerer, H.; Lodhi, A.; Bhattacharyya, A.; Hall, C.; Lucci, A. Discordance in her2 gene amplification in circulating and disseminated tumor cells in patients with operable breast cancer. *Cancer Med.* **2013**, *2*, 226–233. [CrossRef]
9. Azim, H.A., Jr.; Rothe, F.; Aura, C.M.; Bavington, M.; Maetens, M.; Rouas, G.; Gebhart, G.; Gamez, C.; Eidtmann, H.; Baselga, J.; et al. Circulating tumor cells and response to neoadjuvant paclitaxel and her2-targeted therapy: A sub-study from the neoaltto phase iii trial. *Breast* **2013**, *22*, 1060–1065. [CrossRef]
10. Yu, M.; Bardia, A.; Wittner, B.S.; Stott, S.L.; Smas, M.E.; Ting, D.T.; Isakoff, S.J.; Ciciliano, J.C.; Wells, M.N.; Shah, A.M.; et al. Circulating breast tumor cells exhibit dynamic changes in epithelial and mesenchymal composition. *Science* **2013**, *339*, 580–584. [CrossRef]
11. Jordan, N.V.; Bardia, A.; Wittner, B.S.; Benes, C.; Ligorio, M.; Zheng, Y.; Yu, M.; Sundaresan, T.K.; Licausi, J.A.; Desai, R.; et al. Her2 expression identifies dynamic functional states within circulating breast cancer cells. *Nature* **2016**, *537*, 102–106. [CrossRef] [PubMed]
12. Jaeger, B.A.S.; Neugebauer, J.; Andergassen, U.; Melcher, C.; Schochter, F.; Mouarrawy, D.; Ziemendorff, G.; Clemens, M.; Abel, E.V.; Heinrich, G.; et al. The her2 phenotype of circulating tumor cells in her2-positive early breast cancer: A translational research project of a prospective randomized phase iii trial. *PLoS ONE* **2017**, *12*, e0173593. [CrossRef] [PubMed]
13. Levitan, D. Breast cancer ctcs can flip between her2 negativity and positivity. *Cancer Netw.* **2016**. Available online: https://www.cancernetwork.com/her2-positive-breast-cancer/breast-cancer-ctcs-can-flip-between-her2-negativity-and-positivity (accessed on 20 September 2016).
14. Buiga, P.; Elson, A.; Tabernero, L.; Schwartz, J.M. Regulation of dual specificity phosphatases in breast cancer during initial treatment with herceptin: A boolean model analysis. *BMC Syst. Biol.* **2018**, *12*, 11. [CrossRef] [PubMed]
15. Cejudo-Marin, R.; Tarrega, C.; Nunes-Xavier, C.E.; Pulido, R. Caspase-3 cleavage of dusp6/mkp3 at the interdomain region generates active mkp3 fragments that regulate erk1/2 subcellular localization and function. *J. Mol. Biol.* **2012**, *420*, 128–138. [CrossRef] [PubMed]
16. Karlsson, M.; Mathers, J.; Dickinson, R.J.; Mandl, M.; Keyse, S.M. Both nuclear-cytoplasmic shuttling of the dual specificity phosphatase mkp-3 and its ability to anchor map kinase in the cytoplasm are mediated by a conserved nuclear export signal. *J. Biol Chem.* **2004**, *279*, 41882–41891. [CrossRef] [PubMed]
17. Ahmad, M.K.; Abdollah, N.A.; Shafie, N.H.; Yusof, N.M.; Razak, S.R.A. Dual-specificity phosphatase 6 (dusp6): A review of its molecular characteristics and clinical relevance in cancer. *Cancer Biol. Med.* **2018**, *15*, 14–28. [PubMed]
18. Arkell, R.S.; Dickinson, R.J.; Squires, M.; Hayat, S.; Keyse, S.M.; Cook, S.J. Dusp6/mkp-3 inactivates erk1/2 but fails to bind and inactivate erk5. *Cell Signal.* **2008**, *20*, 836–843. [CrossRef]
19. Lonne, G.K.; Masoumi, K.C.; Lennartsson, J.; Larsson, C. Protein kinase cdelta supports survival of mda-mb-231 breast cancer cells by suppressing the erk1/2 pathway. *J. Biol. Chem.* **2009**, *284*, 33456–33465. [CrossRef]
20. Nunes-Xavier, C.E.; Tarrega, C.; Cejudo-Marin, R.; Frijhoff, J.; Sandin, A.; Ostman, A.; Pulido, R. Differential up-regulation of map kinase phosphatases mkp3/dusp6 and dusp5 by ets2 and c-jun converge in the control of the growth arrest versus proliferation response of mcf-7 breast cancer cells to phorbol ester. *J. Biol. Chem.* **2010**, *285*, 26417–26430. [CrossRef]

21. Song, H.; Wu, C.; Wei, C.; Li, D.; Hua, K.; Song, J.; Xu, H.; Chen, L.; Fang, L. Silencing of dusp6 gene by rnai-mediation inhibits proliferation and growth in mda-mb-231 breast cancer cells: An in vitro study. *Int J. Clin. Exp. Med.* **2015**, *8*, 10481–10490. [PubMed]

22. Hagan, C.R.; Knutson, T.P.; Lange, C.A. A common docking domain in progesterone receptor-b links dusp6 and ck2 signaling to proliferative transcriptional programs in breast cancer cells. *Nucleic Acids Res.* **2013**, *41*, 8926–8942. [CrossRef] [PubMed]

23. Lucci, M.A.; Orlandi, R.; Triulzi, T.; Tagliabue, E.; Balsari, A.; Villa-Moruzzi, E. Expression profile of tyrosine phosphatases in her2 breast cancer cells and tumors. *Cell Oncol.* **2010**, *32*, 361–372. [PubMed]

24. Cui, Y.; Parra, I.; Zhang, M.; Hilsenbeck, S.G.; Tsimelzon, A.; Furukawa, T.; Horii, A.; Zhang, Z.Y.; Nicholson, R.I.; Fuqua, S.A. Elevated expression of mitogen-activated protein kinase phosphatase 3 in breast tumors: A mechanism of tamoxifen resistance. *Cancer Res.* **2006**, *66*, 5950–5959. [CrossRef] [PubMed]

International Journal of
Molecular Sciences

MDPI

Article

Pseudophosphatase MK-STYX Alters Histone Deacetylase 6 Cytoplasmic Localization, Decreases Its Phosphorylation, and Increases Detyrosination of Tubulin

Yuming Cao [1,†], Dallas A. Banks [1,2,†], Andrew M. Mattei [1], Alexys T. Riddick [1], Kirstin M. Reed [1], Ashley M. Zhang [1], Emily S. Pickering [1] and Shantá D. Hinton [1,*]

[1] Department of Biology, Integrated Science Center, College of William and Mary, Williamsburg, VA 23185, USA; yuming.cao@umassmed.edu (Y.C.); dbanks13@umd.edu (D.A.B.); ammattei@email.wm.edu (A.M.M.); atriddic@aggies.ncat.edu (A.T.R.); kmreed01@email.wm.edu (K.M.R.); amzhang@email.wm.edu (A.M.Z.); espickering@email.wm.edu (E.S.P.)

[2] Department of Cell Biology and Molecular Genetics, University of Maryland, College Park, MD 20742, USA

* Correspondence: sdhinton@wm.edu

† These authors contributed equally to this work.

Received: 29 January 2019; Accepted: 19 March 2019; Published: 22 March 2019

Abstract: The catalytically inactive mitogen-activated protein (MAP) kinase phosphatase, MK-STYX (MAPK (mitogen-activated protein kinase) phosphoserine/threonine/tyrosine-binding protein) interacts with the stress granule nucleator G3BP-1 (Ras-GAP (GTPase-activating protein) SH3 (Src homology 3) domain-binding protein-1), and decreases stress granule (stalled mRNA) formation. Histone deacetylase isoform 6 (HDAC6) also binds G3BP-1 and serves as a major component of stress granules. The discovery that MK-STYX and HDAC6 both interact with G3BP-1 led us to investigate the effects of MK-STYX on HDAC6 dynamics. In control HEK/293 cells, HDAC6 was cytosolic, as expected, and formed aggregates under conditions of stress. In contrast, in cells overexpressing MK-STYX, HDAC6 was both nuclear and cytosolic and the number of stress-induced aggregates significantly decreased. Immunoblots showed that MK-STYX decreases HDAC6 serine phosphorylation, protein tyrosine phosphorylation, and lysine acetylation. HDAC6 is known to regulate microtubule dynamics to form aggregates. MK-STYX did not affect the organization of microtubules, but did affect their post-translational modification. Tubulin acetylation was increased in the presence of MK-STYX. In addition, the detyrosination of tubulin was significantly increased in the presence of MK-STYX. These findings show that MK-STYX decreases the number of HDAC6-containing aggregates and alters their localization, sustains microtubule acetylation, and increases detyrosination of microtubules, implicating MK-STYX as a signaling molecule in HDAC6 activity.

Keywords: pseudophosphatase; MK-STYX (MAPK (mitogen-activated protein kinase) phosphoserine/threonine/tyrosine-binding protein); HDAC6 (histone deacetylase isoform 6); post-translational modification; microtubules

1. Introduction

MK-STYX (MAPK (mitogen-activated protein kinase) phosphoserine/threonine/tyrosine-binding protein) is a catalytically inactive member of the protein tyrosine phosphatase (PTP) superfamily, classified as a pseudophosphatase [1–4]. In MK-STYX, critical histidine and cysteine residues in its PTP active site signature motif (HCX$_5$R), which are essential for phosphatase activity, are replaced by phenylalanine and serine residues (FSTQGISR) [1–3]. Despite being catalytically inactive, MK-STYX plays an important regulatory role in a number of cellular pathways, including apoptosis, neurite formation, and stress granule assembly [3–10]. Elucidating the molecular mechanisms by which MK-STYX functions as a signaling molecule is a subject of much interest. For example, MK-STYX has been shown to promote stress-activated mitochondrial-dependent apoptosis and release of cytochrome c, by sequestering an inhibitor of apoptosis, the mitochondrial phosphatase, PTPM1 (PTP localized to the mitochondrion 1) [8,9]. Similarly, MK-STYX binds G3BP-1 (Ras-GTPase activating protein SH3 domain binding protein-1) [3,4,10], a nucleator of stress granules [11], thereby inhibiting stress granule formation. Stress granules are cytoplasmic aggregates of stalled mRNA [11], which accumulate as an immediate protective response to a stressful environment [12]. Interestingly, the inhibitory effects of MK-STYX are independent of the phosphorylation status of G3BP-1 at Ser149, the critical residue that regulates its activity as a stress granule nucleator [4,10,11].

In addition to MK-STYX and G3BP1, there have been numerous other proteins implicated in the stress granule life cycle [13]. Among them is histone deacetylase isoform 6 (HDAC6). HDAC6 is a member of the HDAC superfamily, which, along with the histone acetyltransferase (HAT) superfamily, regulates gene expression and chromatin remodeling [14]. Important antagonistic actions of HATs and HDACs are accomplished by post-translational modification of histones [15,16], which tightly pack DNA into chromatin. In particular, HATs acetylate lysine residues of histones, loosening the packed DNA and making it available for transcription [16]. HDACs remove the acetyl (CH_3CO_2) group from histones, repressing gene expression [17,18]. HATs and HDACs are primarily localized in the nucleus [19]; however, there are non-canonical classes of HDACs that are localized in the cytoplasm [14,19]. The substrates of cytoplasmic HDACs are non-histone proteins such as heat shock proteins, tubulin, and cortactin [20]; they regulate various important cellular processes such as cilia formation, ubiquitination, and autophagy [14,21–23]. HDAC6, the prototypical cytoplasmic HDAC [24], is involved in a variety of cellular processes including cytoskeleton-associated stress responses and cell signaling [14,20,23].

HDAC6 contains two catalytic domains, DD1 and DD2, both of which are indispensable for its deacetylase activity. A dynein-binding domain (DMB) separates the two catalytic domains and is required for the recruitment of HDAC6 onto the microtubule motor protein dynein [19,25,26]. Within the HDAC6 N-terminus, there is a nuclear export signal (NES) and a SE14 domain, which ensure the stable anchoring of HDAC6 in the cytoplasm. The C-terminus of HDAC6 contains a cysteine and histidine-rich ZnF-UBP (zinc-finger ubiquitin binding protein) domain that binds mono- and poly-ubiquitin chains with high affinity [14]. In addition, this domain plays a critical role in the function of HDAC6 in the cellular stress response pathway [22,27]. HEK/293 (human embryonic kidney cells) cells treated with HDAC6 catalytic inhibitors, such as trichostatin A (TSA), have fewer or no HDAC6-positive stress granules [28], indicating that the deacetylase activity of HDAC6 regulates stress granule formation. Furthermore, HDAC6 binds the stress granule nucleator G3BP-1, acting as a crucial component in stress granule formation [28]. Immunoprecipitation assays showed that an HDAC6-G3BP-1 complex is dependent on the catalytic domain of HDAC6 and the acidic-rich domain of G3BP-1 [28]. Stress induces dephosphorylation of G3BP-1 at phosphoserine 149, which resides in the acidic-rich domain, resulting in stress granule assembly [29]. Intriguingly, HDAC6 binds with higher affinity to dephosphorylated G3BP-1 than phosphorylated G3BP-1 [28].

Given that both HDAC6 and MK-STYX interact with G3BP-1, we sought to ascertain whether MK-STYX has an impact on the dynamics of HDAC6. Here, we demonstrate that HDAC6 is cytosolic and nuclear in the presence of MK-STYX, instead of solely cytosolic. We also show that MK-STYX inhibits HDAC6 aggregate formation under stress conditions and decreases HDAC6 phosphorylation at Ser22, tyrosine phosphorylation, and lysine acetylation. Microtubule organization was not disrupted by MK-STYX. However, when cells were treated with nocodazole, which normally depolymerizes microtubules and decreases the acetylation and detyrosination of tubulin in certain cells [30–32], MK-STYX induced an increase in microtubule detyrosination.

2. Results

2.1. MK-STYX Causes HDAC6 Localization to Become Whole Cell (Nuclear and Cytosolic) and Decreases the Number of HDAC6 Aggregates

HDACs are grouped into four classes, I, II, IV (Zn^+ dependent proteases), and III (sirtuins; NAD^+ mechanism) according to their structural homologies [33,34]. Class II consists of HDAC6, which is unique because it is a cytosolic protein with a nuclear export signal (NES) and cytoplasmic anchoring domain [20,26]. To determine whether MK-STYX affects the subcellular localization of HDAC6, HEK/293 cells were transfected with expression plasmids for mCherry (vehicle control) or mCherry-MK-STYX. Using immunofluorescence microscopy, we examined the localization of endogenous HDAC6 in the absence or presence of stressful (serum starved) conditions. In control cells overexpressing mCherry under non-stressful conditions, HDAC6 was cytosolic, as expected (Figure 1A,B). However, HDAC6 was both cytosolic and nuclear in the presence of mCherry-MK-STYX under non-stressful and stressful conditions (Figure 1A–C). MK-STYX significantly decreased the cytosolic localization of HDAC6 (Figure 1B); 62% of control cells showed HDAC6 in the cytosol, whereas 47.7% of cells showed a whole cell distribution of HDAC6 in the presence of mCherry-MK-STYX (Figure 1B; $p < 0.05$). HDAC6 shifted towards a whole cell distribution under stressful conditions (serum starvation) (Figure 1C). However, serum starvation did not significantly affect the subcellular localization of HDAC6 between control and MK-STYX-expressing cells (Figure 1C); HDAC6 was localized throughout the cell in both control cells and cells overexpressing mCherry-MK-STYX (Figure 1A,C). Furthermore, HDAC6 formed aggregates in 39% of cells under stress conditions (Figure 1A,D). However, MK-STYX significantly decreased the number of cells with these aggregates (Figure 1A,D); only 8.3% of MK-STYX-expressing cells formed HDAC aggregates (pairwise *t*-test; $p < 0.05$).

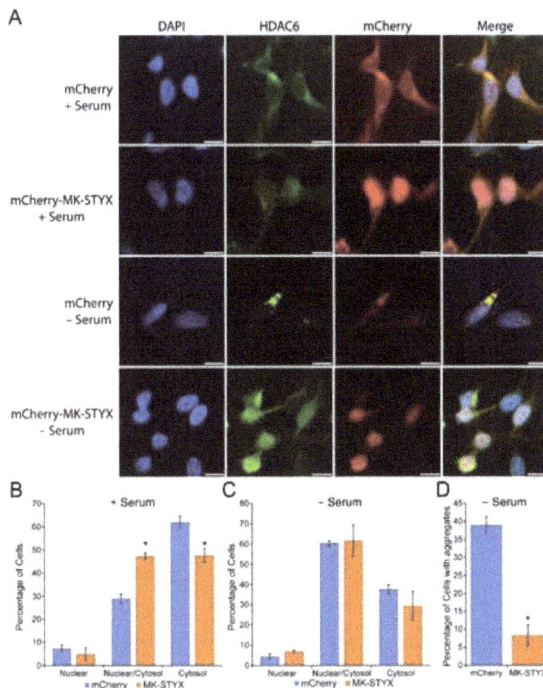

Figure 1. MAPK (mitogen-activated protein kinase) phosphoserine/threonine/tyrosine-binding

protein (MK-STYX) causes histone deacetylase isoform 6 (HDAC6) to become partly nuclear and decreases HDAC6 aggregates. (**A**) Representative examples of the subcellular distribution of endogenous HDAC6, detected with an anti-HDAC6 antibody (anti-HDAC6) and anti-rabbit conjugated to FITC, in HEK/293 cells transfected with mCherry (control vector) or mCherry-MK-STYX. We scored the localization of HDAC6 as cytosolic or whole cell (cytosolic and nuclear) in the (**B**) presence or (**C**) absence of serum, in HEK/293 cells overexpressing mCherry-MK-STYX or mCherry control plasmid. (**D**) We scored the number of cells with HDAC6 aggregates that formed in the absence of serum. Paired *t*-test statistical analysis was performed; the error bars are ±SEM; * $p < 0.05$ for mCherry compared to mCherry-MK-STYX. Scale bar, 10 μm. The quantified data is the cumulative output of $n = 3$ independent biological experiments. Taking all fields of view into account, cell density was comparable for cells expressing mCherry and mCherry-MK-STYX. Representative images were chosen to illustrate the subcellular distribution of HDAC6.

To confirm at higher resolution that HDAC6 was indeed cytosolic and nuclear in cells expressing mCherry-MK-STYX, cells were analyzed by confocal microscopy (Figure 2). Z-stack imaging revealed that HDAC6 is cytosolic in control cells expressing mCherry (Figure 2A), whereas in cells expressing mCherry-MK-SYTX, HDAC6 localizes both in the cytosol and nucleus (Figure 2B). In addition, HDAC6 was also solely cytosolic in cells that did not express mCherry-MK-STYX within the same population of transfected cells (Figure 2B). These Z-stack confocal images validate that MK-STYX shifts the subcellular localization of HDAC6 from cytosolic to whole cell (cytosolic and nuclear).

2.2. MK-STYX Alters the Post-Translational Modifications of HDAC6

MK-STYX maintains the characteristic protein tyrosine phosphatase three-dimensional fold and has the ability to bind phosphorylated residues [1,3,35,36]. HDAC6 has multiple phosphorylation sites. Moreover, phosphorylation sites such as Ser22, Tyr570, and Ser458 regulate the deacetylase activity of HDAC6 [37]. To analyze the impact of MK-STYX on the phosphorylation status of HDAC6, we transfected HEK/293 cells with expression plasmids for GFP or GFP-MK-STYX in the absence or presence of serum. Intriguingly, the cells overexpressing MK-STYX showed less HDAC6 phosphorylation at Ser22, compared to control cells expressing GFP (Figure 3A), suggesting that MK-STYX may inhibit a kinase that phosphorylates HDAC6 or promote a phosphatase that dephosphorylates HDAC6 at Ser22. In addition, there was a decrease in tyrosine phosphorylation of proteins in cells overexpressing MK-STYX (Figure 3B). Because acetylation of HDAC6 is also important for HDAC6 function [25,38], we also examined whether MK-STYX alters acetylation. Total protein acetylation was decreased in stressful (absence of serum) conditions in control cells and cells overexpressing MK-STYX. In addition, total protein acetylation was decreased in the presence of serum in cells overexpressing MK-STYX (Figure 3C).

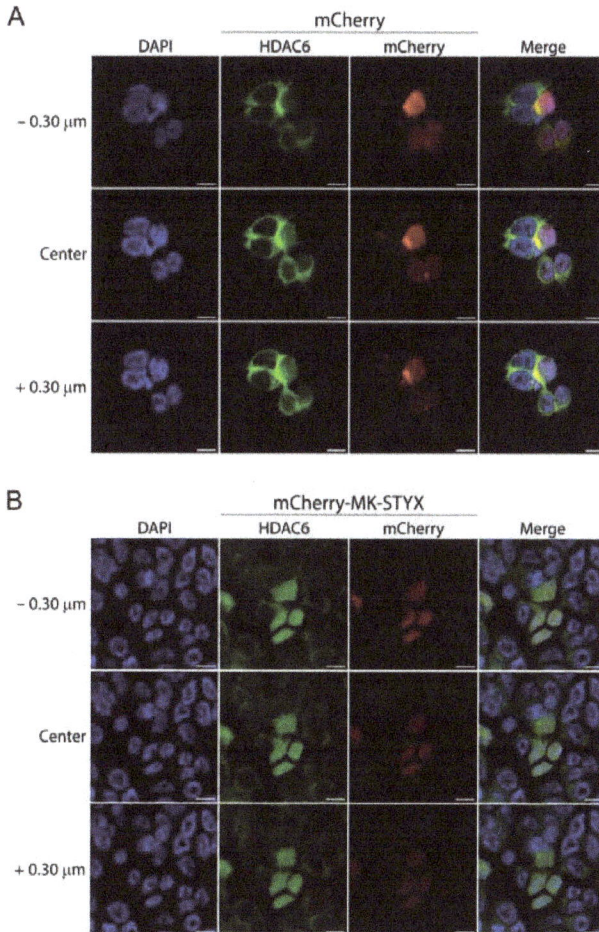

Figure 2. Confocal microscopy validates that the subcellular distribution of HDAC6 is altered in the presence of MK-STYX. (**A**) Representative examples of the subcellular distribution of endogenous HDAC6, detected with an anti-HDAC6 antibody (anti-HDAC6) and anti-rabbit conjugated to FITC in HEK/293 cells, cultured in the presence of serum and transfected with mCherry (control vector) or (**B**) mCherry-MK-STYX. Z-stacks at 0.15 μm were performed by confocal microcopy, starting below the center of the cells. Representative images are shown above (−0.30 μm) and below (+0.30 μm) the center image. Scale bar, 10 μm. Taking all fields of view into account, cell density was comparable for cells expressing mCherry and mCherry-MK-STYX. Representative images were chosen to illustrate the subcellular distribution of HDAC6.

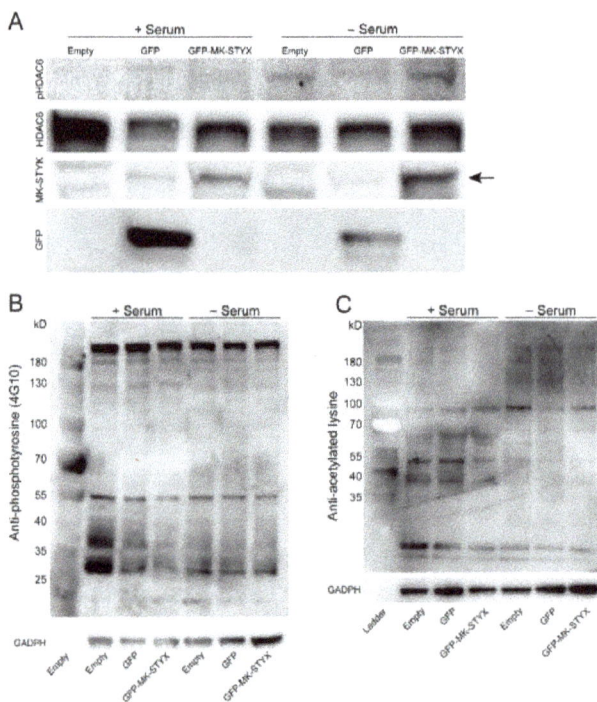

Figure 3. MK-STYX decreases the phosphorylation of HDAC6. (**A**) Twenty-four hours post-transfection cells were serum starved (− serum) or not (+ serum), lysed 24 h thereafter, and immunoblotted. Anti-phospho-HDAC6 antibody showed that MK-STYX decreased HDAC6 phosphorylation in non-serum starved cells relative to the cells overexpressing green fluorescence protein (GFP). However, MK-STYX increased HDAC6 phosphorylation in cells depleted of serum relative to GFP-expressing cells. These blots were stripped and probed for HDAC6 as a loading control. The blot was also stripped and probed with anti-GFP antibody to detect GFP (~27 kDa) or anti-STYXL1 (MK-STYX). Anti-STYXL1 antibody showed overexpressed GFP-MK-STYX (~67 kDa; indicated by the black arrow) relative to GFP (control vector) and non-transfected cells to confirm transfection. Three biologically independent replicate experiments were performed. (**B**) Samples were also used to detect effects of MK-STYX on tyrosine phosphorylation of proteins in the presence and absence of serum with the anti-phosphotyrosine antibody (clone 4G10). Cells expressing GFP-MK-STYX showed decreased phosphorylation relative to the controls in the presence of serum. The blot was stripped and probed with anti-GADPH (glyceraldehyde-3-diphosphatedehydrogenase; 37 kDa) antibody for a loading control. (**C**) Samples were also analyzed for acetylation with anti-lysine acetylation antibody to determine any effects of MK-STYX on acetylated proteins. In the presence or absence of serum, acetylation was decreased in cells expressing MK-STYX relative to control cells. The blot was stripped and probed with anti-GAPDH for a loading control. Three biologically independent replicates were performed.

2.3. MK-STYX Increases Detyrosination of Tubulin

HDAC6 utilizes the microtubule network to transport ubiquitinated proteins; therefore, it is associated with the microtubule network [39–41]. To determine whether MK-STYX affects this microtubule network, we transfected HEK/293 cells with mCherry or mCherry-MK-STYX constructs and observed the organization of microtubules in the presence or absence of nocodazole, which depolymerizes microtubules. MK-STYX did not alter the microtubule network in the absence or presence of nocodazole; microtubules in MK-STYX-overexpressing cells were similar to control cells (Figure 4A). Nocodazole disrupted microtubules (as expected) in both control and MK-STYX-expressing cells (Figure 4B).

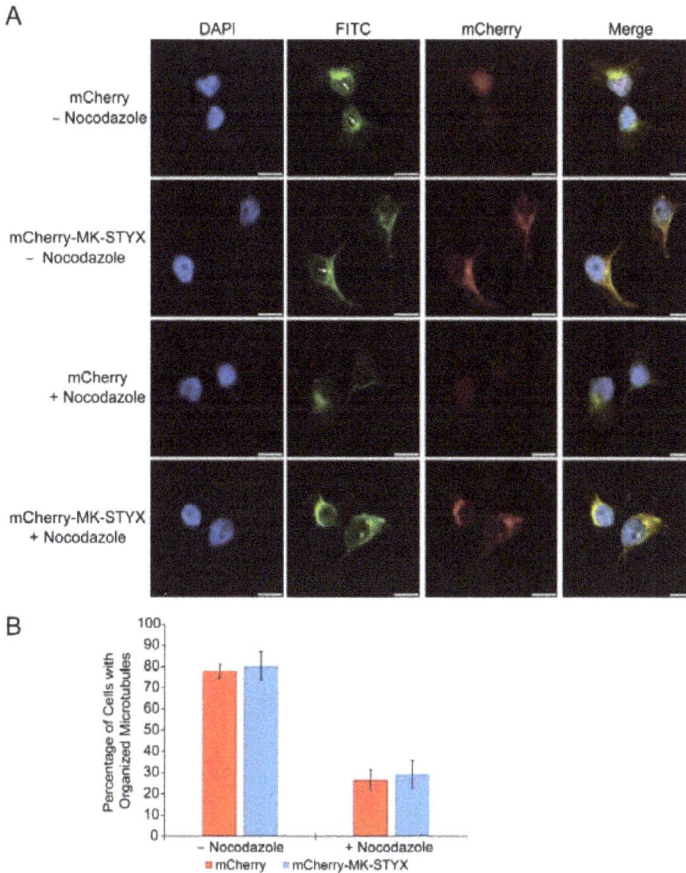

Figure 4. Microtubule organization is stable in the presence of MK-STYX. (**A**) Representative examples of microtubule organization, detected with an anti-tubulin antibody conjugated with FITC, in cells treated or not treated with nocodazole and in cells expressing mCherry or mCherry-MK-STYX, as indicated. (**B**) The number of cells with organized microtubules was scored in the presence or absence of nocodazole. In the absence of nocodazole, microtubules were well-organized in control cells (mCherry) and cells overexpressing MK-STYX, i.e., the microtubules nucleated from the centrosome (white arrow). However, microtubule organization was disrupted (tubulin was diffuse in the cytoplasm) in the presence of nocodazole in both control cells, as expected, and those expressing mCherry-MK-STYX. 100 cells were scored per replicate; three biologically independent replicates were performed. Scale bar, 10 μm. The quantified data is the cumulative output of $n = 3$ independent biological experiments. Cell density was comparable for mCherry and mCherry-MK-STYX. Representative images were chosen to illustrate the subcellular distribution of tubulin.

Because HDAC6 interaction with the microtubule network is dependent on the acetylation and detyrosination of microtubules [23], we also tested whether MK-STYX alters these post-translational modifications (Figure 5). The acetylation of tubulin was significantly decreased in control cells expressing mCherry and treated with nocodazole (paired *t*-test; $p < 0.05$) (Figure 5A); however, tubulin acetylation was unchanged in cells overexpressing mCherry-MK-STYX (Figure 5A), suggesting that

MK-STYX sustains the acetylation of tubulin. In addition, MK-STYX significantly increased the detyrosination of tubulin in the absence of nocodazole (paired *t*-test; $p < 0.05$) (Figure 5B), whereas the detyrosination of tubulin was decreased in control and MK-STYX-overexpressing cells under nocodazole treatment (Figure 5B).

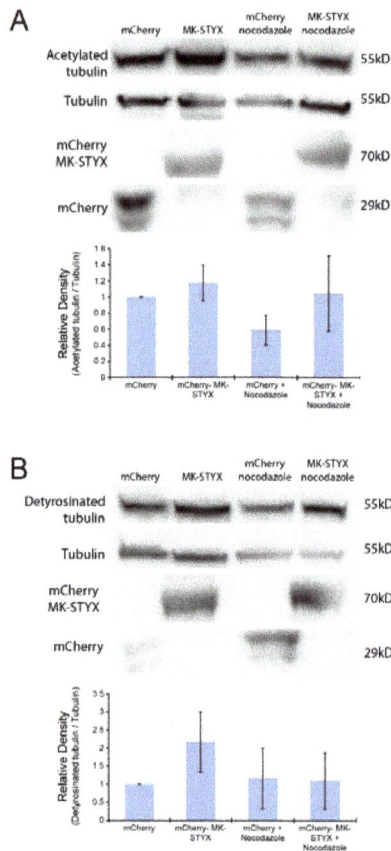

Figure 5. MK-STYX increases acetylated and detyrosinated tubulin. Cells were transfected with expression plasmids for mCherry or mCherry-MK-STYX. Twenty-four hours post-transfection, cells treated with nocodazole or not were lysed and analyzed by immunoblotting. We examined cells expressing mCherry constructs by fluorescence microscopy to confirm transfection. (**A**) Blots were probed with anti-acetylated tubulin and showed that acetylated tubulin (55 kDa) is significantly decreased (paired *t*-test; $p < 0.05$) in the presence of nocodazole in control cells (mCherry-expressing) relative to cells expressing mCherry-MK-STYX. Cells overexpressing MK-STYX sustain acetylated tubulin in the presence of nocodazole. The blots were stripped and probed for tubulin as the loading control; they were also stripped and probed with anti-mCherry to confirm expression of mCherry (27 kDa) and mCherry-MK-STYX (67 kDa). (**B**) Lysates were also analyzed for detyrosinated tubulin (55 kDa) by detection with anti-detyrosinated tubulin. A significant increase (paired *t*-test; $p < 0.05$) in detyrosinated tubulin was observed in mCherry-MK-STYX-expressing cells in the absence of nocodazole, whereas detyrosination was decreased in cells expressing mCherry-MK-STYX in the presence of nocodazole. Blots were stripped and probed with anti-tubulin antibody for a loading control, then stripped a second time and probed with anti-mCherry antibody to confirm transfection. The error bars are ±SEM; three biologically independent replicate experiments were performed.

3. Discussion

HDAC6 catalyzes the deacetylation of various proteins such as tubulin and heat shock proteins. As such, it is an essential regulator of acetylation balance, which is critical for maintaining homeostasis [20]. A disruption of this balance may result in the development of human diseases such as cancer, Parkinson's, and chronic obstructive pulmonary disease [20–23,42,43], making it an appealing target for drug therapy and further investigation. This report investigated the effects of the pseudophosphatase MK-STYX on HDAC6. Because both HDAC6 and MK-STYX interact with the stress granule nucleator G3BP-1 and they have antagonistic roles in stress granule formation, promotion and inhibition [4,28], it was important to determine whether MK-STYX affects HDAC6 subcellular location and phosphorylation state, as well as the post-translational modification of its substrate, microtubules. We showed that MK-STYX caused a proportion of this cytosolic protein to localize in the nucleus. Furthermore, in stress conditions the number of HDAC6 aggregates significantly decreased in the presence of MK-STYX. Immunoblots showed that MK-STYX also affected the post-translational modification of HDAC6. Phosphorylation of HDAC6 at Ser22 and total protein tyrosine phosphorylation and acetylation decreased in the presence of MK-STYX. Lastly, post-translational modification of microtubules, which are important for the interaction of motor proteins and HDAC6 [23,44], was also affected by MK-STYX. The acetylation of tubulin increased in cells overexpressing MK-STYX that were treated with nocodazole, which depolymerizes microtubules, compared to control cells treated with nocodazole. In the absence of nocodazole, however, the detyrosination of tubulin was significantly increased in the presence of MK-STYX compared to control cells. Taken together, these data illustrate that MK-STYX influences HDAC6 dynamics at multiple levels: Its subcellular localization, its post-translational modification (phosphorylation), and the post-translational modification of its tubulin substrate. Considering that these dynamics of HDAC6 are crucial for its function of regulating cellular homeostasis, these interactions between MK-STYX and HDAC6 provide insight into the role that MK-STYX plays as a signaling molecule in the stress response pathway [4]. To our knowledge, these are the first studies to identify that MK-STYX serves a role in HDAC6 dynamics.

HDAC6 is a unique member of the class II deacetylases that contains two functional catalytic domains [20,45]. Because of its nuclear export signal (NES) and SE14 domain, HDAC6 was originally thought to be exclusively cytoplasmic; however, numerous reports have revealed that HDAC6 is also present in the nucleus [23,38]. Within the nucleus, HDAC serves as a transcription factor for several biological functions such as ubiquitination, autophagy, and cell motility [23,46–49]. Furthermore, mouse HDAC6 has been shown to undergo nucleocytoplasmic trafficking [23,49]. The subcellular localization of HDAC6 appears to be cell line dependent; in Jurkat cells, it is more nuclear than cytoplasmic, compared to HEK/293 cells, where it is mostly cytoplasmic [23]. Our present study suggests that MK-STYX causes a conformational change to expose a nuclear localization signal (NLS) of HDAC6 that moves it to the nucleus. Prior reports have shown that acetylation of HDAC6 regulates its subcellular localization; acetylation of lysine 40 is important for the retention of HDAC in the cytosol [21,23,38]. Further investigation of how the dynamics (subcellular localization and post-translational modification) of HDAC6 are altered by MK-STYX may provide important insights into the nuclear and cytosolic functions of HDAC6.

HDAC serves as a mediator of various signaling pathways, including signaling by heat shock factor 1, STAT (signal transducers and activator transcription factors), CREB (cyclic AMP responsive element binding protein), Akt, NFκB, and p53 [48,50]. In particular, HDAC6 regulates ubiquitination and autophagy [22,39]. HDAC6 inhibitors and knockout of HDAC6 have been shown to prevent autophagy [39] and HDAC6 localizes to ubiquitinated organelles such as the mitochondria [51]. Furthermore, HDAC6 is known to bind to monoubiquitinated and polyubiquitinated proteins with high affinity [50]. In the present study, the disappearance of HDAC6 aggregates under stress conditions suggests that MK-STYX has a role in the proteasome pathway. HDAC6 forms aggregates such as aggresomes [52] and stress granules [28,53] under stress conditions [53]. Moreover, HDAC6 is a modulator of lytic granules [54]. Although MK-STYX does not localize to stress granules, it does localize to aggresomes (unpublished data), indicating that MK-STYX may be a modulator of various types of cytosolic granules.

HDAC6 is known to modulate various cytosolic granules through microtubules and motor proteins [52,53]. Microtubules are important structural proteins that provide mechanical function for cells. They are composed of α β-tubulin isoforms, which display a chemical diversity of post-translational modifications such as acetylation, detyrosination, glycylation, phosphorylation, and glutamylation [44]. These modifications of tubulin regulate its properties and recognition by various effectors [44]. HDAC6 mediates the acetylation of the ubiquitin ligase TRIM50 [55] and microtubules [38]. Furthermore, acetylation of HDAC6 prevents it from deacetylating tubulin and promoting HDAC6 import into the nucleus [38]. Prior studies have shown that nocodazole decreases tubulin acetylation [30,31], however, it is important to note that modifications of tubulin are very dependent on the cell line used; post-translational modifications may increase, decrease, or remain unaltered in various cells [30]. In the present study, we showed that acetylation of microtubules significantly decreased in control cells treated with nocodazole. Intriguingly, α-tubulin acetylation was sustained in the presence of MK-STYX, suggesting that MK-STYX inhibits the function of HDAC6. However, the nuclear localization of HDAC6 in the presence of MK-STYX that we observed contradicts the notion of an inactive HDAC6. Prior reports have concluded that when the B subunit of HDAC6 is acetylated, it is unable to deacetylate tubulin and is solely cytosolic [23,38]. Therefore, the acetylation of HDAC6 in the presence of MK-STYX may be at other sites that do not hinder the NLS, but only hinder deacetylase activity. Alternatively, the acetylation sites could be the same and MK-STYX could alter the localization of HDAC6 through other mechanisms. Here, we showed that MK-STYX also increases the detyrosination of α-tubulin, which is important for processes such as wound healing, tubulin stability, and chromosomal direction [44,56].

MK-STYX impacts the dynamics of HDAC6, through altering the subcellular localization and post-translational modification of HDAC6, and the post-translational modification of microtubules, a substrate of HDAC6. These data establish a strong link for the role of MK-STYX in HDAC6 signaling. Further, results support MK-STYX as a regulator in the stress response pathway, while exposing a possible role in the ubiquitin proteasome system. HDAC6 has been previously linked to other phosphatase family members; for example, HDAC6 and protein phosphatase 1 (PP1) form a complex [49,57,58] and HDAC inhibitors disrupt the complex [49]. As another example, Shp2 (non-receptor PTP; encoded by the *Ptpn11* gene) disrupts microtubule regulation by cooperating with HDAC6 to reduce the acetylation and stability of microtubules [59]. Shp2 downregulates RhoA-Dia signaling, resulting in HDAC6-mediated reduction of acetylated microtubules and ERK hyperactivation [59]. HDAC6 has also been linked to the tumor suppressor PTEN (tensin homology phosphatase); combined treatment with celecoxib (cyclooxygenase-2 inhibitor) and an HDAC6 inhibitor activates the PTEN/AKT pathway [15]. The exact mechanism of how MK-STYX regulates HDAC6 signaling requires further study; however, our data highlight the important role that MK-STYX has in this pathway. Moreover, our data corroborate and provide more depth to our previous reports demonstrating the role of MK-STYX in neurite formation [4–6]. Acetylation, tyrosination, and detyrosination are important post-translational modifications of microtubules, which have roles in growth cone formation, axon formation, and platelet marginal bands [44]. Our previous reports show that MK-STYX decreases RhoA activation [7] and increases growth cones [6]. Intriguingly, HDAC6 has a role in RhoA-Dia signaling [59]. Here, we demonstrate that MK-STYX modulates the dynamics of HDAC6, providing more insight and depth to our understanding of MK-STYX as a key signaling molecule in diverse cellular pathways. MK-STYX continues to be an important and interesting atypical member of the PTP family to investigate.

4. Materials and Methods

4.1. Antibodies

The following antibodies were used: Anti-phospho-HDAC6 (pSer22) antibody (Sigma-Aldrich, Burlington, MA, USA; SAB4504190;); anti-HDAC6 (Cell Signaling, Danvers, MA, USA; 7558); anti-GFP antibody (Thermo Fisher, Grand Island, NY, USA c; MA5-15256); anti-phosphotyrosine, clone 4G10 (Millipore; 05-321); anti-acetylated-lysine antibody (Cell Signaling; 9441); anti-STYXL1 antibody (Sigma-Aldrich; S9823); anti-β-tubulin polyclonal antibody (Thermo Fisher; PA1-21153); monoclonal acetylated microtubule antibody (Sigma, Burlington, MA, USA; T7451); anti-tubulin, detyrosinated antibody (Sigma; AB3201); anti-monoclonal anti-mCherry antibody (SAB2702291); monoclonal anti-FLAG M2-FTTC, Clone M2 antibody (Sigma-Aldrich; F4049); monoclonal anti-β-tubulin-FITC (Sigma-Aldrich; F2043); anti-GAPDH (Cell Signaling; 5174).

4.2. Cell Culture and Transient Transfection

HEK/293 (ATCC) cells were maintained at 37 °C, 5% CO_2 in Dulbecco's Modified Eagle medium (DMEM, Invitrogen, Grand Island, NY, USA) supplemented with 10% fetal bovine serum (FBS). Transfections were performed using Lipofectamine 2000 Reagent (Invitrogen); cells were transfected with expression plasmids pMT2, mCherry, pMT2-FLAG-MK-STYX-FLAG, or mCherry-MK-STYX. Cells were either not stressed or stressed by depletion of serum, or treated with nocodazole for the subsequent experiments, and analyzed by fluorescence microscopy or immunoblotting. When serum starvation experiments were required, cells were maintained in DMEM supplemented with 0.1 % FBS for 8–12 h.

4.3. Transient Transfection and Cell Imaging

For immunofluorescence assays, HEK/293 cells were grown to 80–90% confluence and 2×10^5 cells were plated onto lysine treated coverslips in 6-well dishes (Nunc, Grand Island, NY, USA). Twelve to eighteen hours post-plating, cells at 40–60% confluence were transfected with 2 μg of mCherry or mCherry-MK-STYX expression plasmid DNA and 4 μL of Lipofectamine 2000 Reagent (Invitrogen) per well, according to the manufacturer's protocol. The medium was replaced 5 h after transfection. Twenty-four hours post-transfection, cells were serum-starved with DMEM supplemented with 0.1% FBS serum for 8–12, washed with PBS and fixed with 3.7% formaldehyde. The coverslips were mounted to a slide using GelMount containing 4′,6-diamidino-2′-phenylinodole dihydrochloride (DAPI, Sigma) (0.5 mg/mL).

For experiments examining the effect of MK-STYX on HDAC6 subcellular localization, cells were stained with anti-HDAC6 antibody (1:250) dilution (Cell Signaling) for 1 h, then probed with anti-rabbit conjugated Cy3 antibody (1:250) (Cell Signaling). To visualize whether MK-STYX disrupted the microtubules, cells were transfected with mCherry or mCherry-MK-STYX. Anti-β-tubulin-FITC (1:250) was used as the marker to determine the organization of microtubules in the presence or absence of nocodazole. Post-staining, the coverslips were mounted to a slide using GelMount containing DAPI.

Cells were scored for localization of HDAC6 (cytoplasmic or whole cell) and microtubule organization or disruption. Samples were scored blind with regard to treatment and were scored independently by at least two different individuals. At least three biologically independent replicate transfections were performed and at least 100 cells were scored per replicate. Counting and image collection were performed on a Nikon ECLIPSE Ti inverted fluorescence microscope. A Nikon A1Rsi confocal microscope Ti-E-PFS (Nikon Inc., Melville, NY, USA) with a 60× oil objective was utilized to obtain Z-stacks of cells. The 488-nm line of krypton-argon laser with a band-pass of 525/50 nm emission filter was used for GFP detection, the 561-nm line with a band-pass emission filter was used for mCherry and the 405-nm line with a band-pass of 450/500 emission filter was used to detect DAPI. Z-stacks of cells were taken at 0.15 μm. NIS-Elements Basic Research software (version 3.10,

Nikon, Brighton, MI, USA) was used for image acquisition and primary image processing and Adobe Photoshop™ and Illustrator™ were used for secondary image processing.

4.4. Nocodazole Treatment

Twenty-four hours post-transfection, HEK/293 cells were treated with 6 μg/mL of nocodazole for 30 min, then immediately washed with phosphate buffer saline (PBS) and immediately fixed or lysed for immunoblotting analysis. Fixation studies were analyzed for the organization of microtubules. Immunoblot studies were analyzed for whether the post-translational modifications (acetylation and detyrosination) of microtubules were changed.

4.5. Immunoblotting

HEK/293 cells were transfected with pMT2, pMT2-FLAG-MK-STYX-FLAG, GFP, mCherry, GFP-MK-STYX, or mCherry-MK-STYX expression plasmids, then lysed and analyzed by Western blotting. Cells were harvested in lysis buffer (50 mM HEPES, pH 7.2, 150 mM NaCl, 10% glycerol, 10 mM NaF, 1 mM Na_3VO_4, 1% Nonidet P-40 alternative (Calbiochem, Burlington, MA, USA), and protease inhibitor cocktail tablets (Roche, Branchburg, NJ, USA). Lysates were sonicated, centrifuged at 14,000× g for 10 min, and the supernatant protein concentration was determined by NanoDrop quantification. Lysates were resolved by 4%–20% Bis-Tris gels and transferred to PVDF by the eBlot L1 (Genscript, Piscataway, NJ, USA). Chemiluminescence was detected using a BioRad ChemiDoc MP imaging system for immunoblot analysis with anti-phospho-HDAC6 (pSer22); anti-HDAC6; anti-phosphotyrosine, clone 4G10; anti-acetylated-lysine; anti-STYXL1; anti-GAPDH; anti-β-tubulin polyclonal; monoclonal acetylated microtubule; anti-β-tubulin, detyrosinated; anti-GFP; and monoclonal anti-mCherry antibodies, followed by chemiluminescent detection. When warranted, blots were stripped (200 mM glycine, 3.5 mM SDS, 1% Tween 20) and re-probed.

4.6. Statistical Analysis

Paired *t*-tests were used to determine the statistical significance of differences between the subcellular localization of HDAC6 in the absence or presence of MK-STYX under non-stressed and stressed conditions with a significance level of $p < 0.05$ (Figure 1). Paired *t*-tests were also used to determine whether MK-STYX disrupted microtubule organization (Figure 3) and to compare the post-translational modification of control cells (expressing mCherry; Takara Bio, Mountain View, CA, USA) with nocodazole treatment and/or cells overexpressing MK-STYX (mCherry-MK-STYX). To compare all samples to each other, analysis of variance (ANOVA) was performed.

5. Conclusions

The shared function of MK-STYX and HDAC6 as regulators in the stress granule life cycle led to investigating whether MK-STYX has a role in HDAC6 dynamics. We have revealed that MK-STYX alters the subcellular distribution and phosphorylation of HDAC6 and sustained the acetylation and increased the detyrosination of tubulin, an HDAC6 substrate. These data enhance our understanding of the pseudophosphatase MK-STYX as a signaling regulator, pose new avenues to pursue in exploring its role in various signaling pathways, and serve as the foundation for further investigation of its dynamics with HDAC6.

Author Contributions: S.D.H. is responsible for developing the questions and conceptualization of the study, as well as training and supervising the six undergraduates and high school students who participated in the study. She performed experiments for the immunoblots (Figures 3 and 5). She also analyzed data, wrote, and edited the manuscript. Y.C. performed the initial experiments and performed the statistical analysis to address whether MK-STYX affected HDAC6, which contributed to Figures 1 and 3. She also completed an Honors thesis, which served as a template (original draft preparation) for the manuscript. D.A.B. performed the final immunoblots for Figure 3, all the immunoblots for Figure 5, and performed the statistical analysis for all immunoblots for the manuscript. He also assisted with writing the methodology and statistical analysis for the immunoblots and editing the manuscript. A.M.M. completed all experiments for the microtubule organization studies depicted in Figure 4, for the HEK/293 cell line. In addition, he provided insight and discussion for the post-translational modification studies for microtubules (Figure 5). He also helped train Emily Pickering to complete the fixation and image acquisition and analysis for Figure 1. A.T.R. performed the initial studies and analysis of the effect of MK-STYX on microtubule organization. Her work set the foundation for the nocodazole studies performed in this study for Figure 4. K.M.R., A.M.Z., and E.S.P. repeated initial studies of Y.C.'s for Figure 1 to obtain publication quality images. K.M.R. and A.M.Z. performed the experiments and E.S.P. prepared and analyzed samples for fluorescence microscopy analysis. K.M.R. also assisted with paper preparation; she researched and obtained the required articles for the microtubule post-translational modification studies.

Funding: This work was funded by the National Science Foundation (NSF) MCB 1113167 awarded to S.D.H.; a NSF Research Assistantship for High School Students awarded to A.T.R., who completed this work as a Lafayette High School (Williamsburg, VA, 23188, USA) student; a Grace Blank Award from the Biology Department awarded to S.D.H.; Dintersmith Honors Fellowship awarded to Y.C.; Biology department Ferguson Fellowships awarded to Y.C., A.M.M, and A.M.Z.; Roy R. Charles Center Summer Research Grants awarded to K.M.R. and A.M.Z.; a Howard Hughes Medical Research Institute grant through the Undergraduate Science Education Program to the College of William and Mary to A.M.M.; and the Llanso-Sherman Scholarship from the Biology Department awarded to A.M.M.

Acknowledgments: We are grateful for Lizabeth A. Allison for her continued generosity of sharing reagents, providing technical support to our lab, which is very important in a challenging funding climate, and editing our work. We also give the utmost thanks to Vincent Roggero for his continued technical assistance for image formatting. We are also grateful to William R. Eckberg for continuing to serve as an excellent mentor and editor to our work.

Conflicts of Interest: The authors declare no conflict of interests.

Abbreviations

ANOVA	analysis of variance
DMB	dynein binding domain
FITC	fluorescein isothiocyanate
GAPDH	glyceraldehyde-3-diphosphatedehydrogenase
GFP	green fluorescent protein
G3BP-1	Ras-GAP (GTPase-activating protein) SH3 (Src homology 3) domain binding protein-1
HAT	histone acetyltransferase
HDAC6	histone deacetylase isoform 6
MK-STYX	mitogen-activated protein kinase phosphoserine/threonine/tyrosine-binding protein
NES	nuclear export signal
PTPM1	PTP localized to the mitochondrion 1
PTP	protein tyrosine phosphatase
SEM	standard error mean
SHP2	non-receptor PTP; encoded by the *Ptpn11* gene
ZnF-UBP	zinc-finger ubiquitin binding protein

References

1. Wishart, M.J.; Dixon, J.E. Gathering STYX: Phosphatase-like form predicts functions for unique protein-interaction domains. *Trends Biochem. Sci.* **1998**, *23*, 301–306. [CrossRef]
2. Tonks, N.K. Protein tyrosine phosphatases: From genes, to function, to disease. *Nat. Rev. Mol. Cell. Biol.* **2006**, *7*, 833–846. [CrossRef] [PubMed]
3. Hinton, S.D.; Myers, M.P.; Roggero, V.R.; Allison, L.A.; Tonks, N.K. The pseudophosphatase MK-STYX interacts with G3BP and decreases stress granule formation. *Biochem. J.* **2010**, *427*, 349–357. [CrossRef] [PubMed]
4. Hinton, S.D. The role of pseudophosphatases as signaling regulators. *Biochim. Biophys. Acta* **2019**, *1866*, 167–174. [CrossRef] [PubMed]
5. Dahal, A.; Hinton, S.D. Antagonistic roles for STYX pseudophosphatases in neurite outgrowth. *Biochem. Soc. Trans.* **2017**, *45*, 381–387. [CrossRef] [PubMed]
6. Banks, D.A.; Dahal, A.; McFarland, A.G.; Flowers, B.M.; Stephens, C.A.; Swack, B.; Gugssa, A.; Anderson, W.A.; Hinton, S.D. MK-STYX Alters the Morphology of Primary Neurons, and Outgrowths in MK-STYX Overexpressing PC-12 Cells Develop a Neuronal Phenotype. *Front. Mol. Biosci.* **2017**, *4*, 76. [CrossRef] [PubMed]

7. Flowers, B.M.; Rusnak, L.E.; Wong, K.E.; Banks, D.A.; Munyikwa, M.R.; McFarland, A.G.; Hinton, S.D. The pseudophosphatase MK-STYX induces neurite-like outgrowths in PC12 cells. *PLoS ONE* **2014**, *9*, e114535. [CrossRef]

8. Niemi, N.M.; Lanning, N.J.; Klomp, J.A.; Tait, S.W.; Xu, Y.; Dykema, K.J.; Murphy, L.O.; Gaither, L.A.; Xu, H.E.; Furge, K.A.; et al. MK-STYX, a catalytically inactive phosphatase regulating mitochondrially dependent apoptosis. *Mol. Cell. Biol.* **2011**, *31*, 1357–1368. [CrossRef]

9. Niemi, N.M.; Sacoman, J.L.; Westrate, L.M.; Gaither, L.A.; Lanning, N.J.; Martin, K.R.; MacKeigan, J.P. The pseudophosphatase MK-STYX physically and genetically interacts with the mitochondrial phosphatase PTPMT1. *PLoS ONE* **2014**, *9*, e93896. [CrossRef]

10. Barr, J.E.; Munyikwa, M.R.; Frazier, E.A.; Hinton, S.D. The pseudophosphatase MK-STYX inhibits stress granule assembly independently of Ser149 phosphorylation of G3BP-1. *FEBS J.* **2013**, *280*, 273–284. [CrossRef] [PubMed]

11. Tourriere, H.; Chebli, K.; Zekri, L.; Courselaud, B.; Blanchard, J.M.; Bertrand, E.; Tazi, J. The RasGAP-associated endoribonuclease G3BP assembles stress granules. *J. Cell Biol.* **2003**, *160*, 823–831. [CrossRef]

12. Anderson, P.; Kedersha, N. Visibly stressed: The role of eIF2, TIA-1, and stress granules in protein translation. *Cell Stress Chaperones* **2002**, *7*, 213–221. [CrossRef]

13. Youn, J.Y.; Dunham, W.H.; Hong, S.J.; Knight, J.D.R.; Bashkurov, M.; Chen, G.I.; Bagci, H.; Rathod, B.; MacLeod, G.; Eng, S.W.M.; et al. High-Density Proximity Mapping Reveals the Subcellular Organization of mRNA-Associated Granules and Bodies. *Mol. Cell* **2018**, *69*, 517–532. [CrossRef]

14. Boyault, C.; Sadoul, K.; Pabion, M.; Khochbin, S. HDAC6, at the crossroads between cytoskeleton and cell signaling by acetylation and ubiquitination. *Oncogene* **2007**, *26*, 5468–5476. [CrossRef] [PubMed]

15. Zhang, G.; Gan, Y.H. Synergistic antitumor effects of the combined treatment with an HDAC6 inhibitor and a COX-2 inhibitor through activation of PTEN. *Oncol. Rep.* **2017**, *38*, 2657–2666. [CrossRef] [PubMed]

16. Drazic, A.; Myklebust, L.M.; Ree, R.; Arnesen, T. The world of protein acetylation. *Biochim. Biophys. Acta* **2016**, *1864*, 1372–1401. [CrossRef]

17. Simoes-Pires, C.; Zwick, V.; Nurisso, A.; Schenker, E.; Carrupt, P.A.; Cuendet, M. HDAC6 as a target for neurodegenerative diseases: What makes it different from the other HDACs? *Mol. Neurodegener.* **2013**, *8*, 7. [CrossRef]

18. Zeb, A.; Park, C.; Rampogu, S.; Son, M.; Lee, G.; Lee, K.W. Structure-Based Drug Designing Recommends HDAC6 Inhibitors To Attenuate Microtubule-Associated Tau-Pathogenesis. *ACS Chem. Neurosci.* **2018**. [CrossRef]

19. De Ruijter, A.J.; van Gennip, A.H.; Caron, H.N.; Kemp, S.; van Kuilenburg, A.B. Histone deacetylases (HDACs): Characterization of the classical HDAC family. *Biochem. J.* **2003**, *370*, 737–749. [CrossRef]

20. Li, T.; Zhang, C.; Hassan, S.; Liu, X.; Song, F.; Chen, K.; Zhang, W.; Yang, J. Histone deacetylase 6 in cancer. *J. Hematol. Oncol.* **2018**, *11*, 111. [CrossRef]

21. Lam, H.C.; Cloonan, S.M.; Bhashyam, A.R.; Haspel, J.A.; Singh, A.; Sathirapongsasuti, J.F.; Cervo, M.; Yao, H.; Chung, A.L.; Mizumura, K.; et al. Histone deacetylase 6-mediated selective autophagy regulates COPD-associated cilia dysfunction. *J. Clin. Investig.* **2013**, *123*, 5212–5230. [CrossRef]

22. Li, Z.Y.; Zhang, C.; Zhang, Y.; Chen, L.; Chen, B.D.; Li, Q.Z.; Zhang, X.J.; Li, W.P. A novel HDAC6 inhibitor Tubastatin A: Controls HDAC6-p97/VCP-mediated ubiquitination-autophagy turnover and reverses Temozolomide-induced ER stress-tolerance in GBM cells. *Cancer Lett.* **2017**, *391*, 89–99. [CrossRef]

23. Li, L.; Yang, X.J. Tubulin acetylation: Responsible enzymes, biological functions and human diseases. *Cell. Mol. Life Sci.* **2015**, *72*, 4237–4255. [CrossRef]

24. Verdel, A.; Khochbin, S. Identification of a new family of higher eukaryotic histone deacetylases. Coordinate expression of differentiation-dependent chromatin modifiers. *J. Biol. Chem.* **1999**, *274*, 2440–2445. [CrossRef]

25. Verdin, E.; Dequiedt, F.; Kasler, H.G. Class II histone deacetylases: Versatile regulators. *Trends Genet. TIG* **2003**, *19*, 286–293. [CrossRef]

26. Grozinger, C.M.; Hassig, C.A.; Schreiber, S.L. Three proteins define a class of human histone deacetylases related to yeast Hda1p. *Proc. Natl. Acad. Sci. USA* **1999**, *96*, 4868–4873. [CrossRef]

27. Ryu, H.W.; Won, H.R.; Lee, D.H.; Kwon, S.H. HDAC6 regulates sensitivity to cell death in response to stress and post-stress recovery. *Cell Stress Chaperones* **2017**, *22*, 253–261. [CrossRef]

28. Kwon, S.; Zhang, Y.; Matthias, P. The deacetylase HDAC6 is a novel critical component of stress granules involved in the stress response. *Genes Dev.* **2007**, *21*, 3381–3394. [CrossRef]

29. Gallouzi, I.E.; Parker, F.; Chebli, K.; Maurier, F.; Labourier, E.; Barlat, I.; Capony, J.P.; Tocque, B.; Tazi, J. A novel phosphorylation-dependent RNase activity of GAP-SH3 binding protein: A potential link between signal transduction and RNA stability. *Mol. Cell. Biol.* **1998**, *18*, 3956–3965. [CrossRef]

30. Bulinski, J.C.; Richards, J.E.; Piperno, G. Posttranslational modifications of alpha tubulin: Detyrosination and acetylation differentiate populations of interphase microtubules in cultured cells. *J. Cell Biol.* **1988**, *106*, 1213–1220. [CrossRef]

31. Vaughan, E.E.; Geiger, R.C.; Miller, A.M.; Loh-Marley, P.L.; Suzuki, T.; Miyata, N.; Dean, D.A. Microtubule acetylation through HDAC6 inhibition results in increased transfection efficiency. *Mol. Ther. J. Am. Soc. Gene Ther.* **2008**, *16*, 1841–1847. [CrossRef]

32. Kreis, T.E. Microtubules containing detyrosinated tubulin are less dynamic. *EMBO J.* **1987**, *6*, 2597–2606. [CrossRef]

33. Maharaj, K.; Powers, J.J.; Achille, A.; Deng, S.; Fonseca, R.; Pabon-Saldana, M.; Quayle, S.N.; Jones, S.S.; Villagra, A.; Sotomayor, E.M.; et al. Silencing of HDAC6 as a therapeutic target in chronic lymphocytic leukemia. *Blood Adv.* **2018**, *2*, 3012–3024.

34. Glozak, M.A.; Sengupta, N.; Zhang, X.; Seto, E. Acetylation and deacetylation of non-histone proteins. *Gene* **2005**, *363*, 15–23. [CrossRef]

35. Tonks, N.K. Pseudophosphatases: Grab and hold on. *Cell* **2009**, *139*, 464–465. [CrossRef]

36. Wishart, M.J.; Denu, J.M.; Williams, J.A.; Dixon, J.E. A single mutation converts a novel phosphotyrosine binding domain into a dual-specificity phosphatase. *J. Biol. Chem.* **1995**, *270*, 26782–26785. [CrossRef]

37. Du, J.; Zhang, L.; Zhuang, S.; Qin, G.J.; Zhao, T.C. HDAC4 degradation mediates HDAC inhibition-induced protective effects against hypoxia/reoxygenation injury. *J. Cell. Physiol.* **2015**, *230*, 1321–1331. [CrossRef]

38. Liu, Y.; Peng, L.; Seto, E.; Huang, S.; Qiu, Y. Modulation of histone deacetylase 6 (HDAC6) nuclear import and tubulin deacetylase activity through acetylation. *J. Biol. Chem.* **2012**, *287*, 29168–29174. [CrossRef]

39. Kaliszczak, M.; van Hechanova, E.; Li, Y.; Alsadah, H.; Parzych, K.; Auner, H.W.; Aboagye, E.O. The HDAC6 inhibitor C1A modulates autophagy substrates in diverse cancer cells and induces cell death. *Br. J. Cancer* **2018**, *119*, 1278–1287. [CrossRef]

40. Kawaguchi, Y.; Kovacs, J.J.; McLaurin, A.; Vance, J.M.; Ito, A.; Yao, T.P. The deacetylase HDAC6 regulates aggresome formation and cell viability in response to misfolded protein stress. *Cell* **2003**, *115*, 727–738. [CrossRef]

41. Ouyang, H.; Ali, Y.O.; Ravichandran, M.; Dong, A.; Qiu, W.; MacKenzie, F.; Dhe-Paganon, S.; Arrowsmith, C.H.; Zhai, R.G. Protein aggregates are recruited to aggresome by histone deacetylase 6 via unanchored ubiquitin C termini. *J. Biol. Chem.* **2012**, *287*, 2317–2327. [CrossRef]

42. Li, G.; Jiang, H.; Chang, M.; Xie, H.; Hu, L. HDAC6 alpha-tubulin deacetylase: A potential therapeutic target in neurodegenerative diseases. *J. Neurol. Sci.* **2011**, *304*, 1–8. [CrossRef]

43. Lee, J.Y.; Nagano, Y.; Taylor, J.P.; Lim, K.L.; Yao, T.P. Disease-causing mutations in parkin impair mitochondrial ubiquitination, aggregation, and HDAC6-dependent mitophagy. *J. Cell. Biol.* **2010**, *189*, 671–679. [CrossRef]

44. Roll-Mecak, A. How cells exploit tubulin diversity to build functional cellular microtubule mosaics. *Curr. Opin. Cell Biol.* **2018**, *56*, 102–108. [CrossRef] [PubMed]

45. Bertos, N.R.; Gilquin, B.; Chan, G.K.; Yen, T.J.; Khochbin, S.; Yang, X.J. Role of the tetradecapeptide repeat domain of human histone deacetylase 6 in cytoplasmic retention. *J. Biol. Chem.* **2004**, *279*, 48246–48254. [CrossRef]

46. Yao, T.P. The role of ubiquitin in autophagy-dependent protein aggregate processing. *Genes Cancer* **2010**, *1*, 779–786. [CrossRef]

47. Chen, L.; Fischle, W.; Verdin, E.; Greene, W.C. Duration of nuclear NF-kappaB action regulated by reversible acetylation. *Science* **2001**, *293*, 1653–1657. [CrossRef]

48. Chen, C.S.; Weng, S.C.; Tseng, P.H.; Lin, H.P.; Chen, C.S. Histone acetylation-independent effect of histone deacetylase inhibitors on Akt through the reshuffling of protein phosphatase 1 complexes. *J. Biol. Chem.* **2005**, *280*, 38879–38887. [CrossRef]

49. Verdel, A.; Curtet, S.; Brocard, M.P.; Rousseaux, S.; Lemercier, C.; Yoshida, M.; Khochbin, S. Active maintenance of mHDA2/mHDAC6 histone-deacetylase in the cytoplasm. *Curr. Biol.* **2000**, *10*, 747–749. [CrossRef]

50. Pernet, L.; Faure, V.; Gilquin, B.; Dufour-Guerin, S.; Khochbin, S.; Vourc'h, C. HDAC6-ubiquitin interaction controls the duration of HSF1 activation after heat shock. *Mol. Biol. Cell* **2014**, *25*, 4187–4194. [CrossRef]

51. Lee, J.Y.; Koga, H.; Kawaguchi, Y.; Tang, W.; Wong, E.; Gao, Y.S.; Pandey, U.B.; Kaushik, S.; Tresse, E.; Lu, J.; et al. HDAC6 controls autophagosome maturation essential for ubiquitin-selective quality-control autophagy. *EMBO J.* **2010**, *29*, 969–980. [CrossRef] [PubMed]

52. Galindo-Moreno, M.; Giraldez, S.; Saez, C.; Japon, M.A.; Tortolero, M.; Romero, F. Both p62/SQSTM1-HDAC6-dependent autophagy and the aggresome pathway mediate CDK1 degradation in human breast cancer. *Sci. Rep.* **2017**, *7*, 10078. [CrossRef] [PubMed]

53. Zheng, K.; Jiang, Y.; He, Z.; Kitazato, K.; Wang, Y. Cellular defence or viral assist: The dilemma of HDAC6. *J. Gen. Virol.* **2017**, *98*, 322–337. [CrossRef] [PubMed]

54. Nunez-Andrade, N.; Iborra, S.; Trullo, A.; Moreno-Gonzalo, O.; Calvo, E.; Catalan, E.; Menasche, G.; Sancho, D.; Vazquez, J.; Yao, T.P.; et al. HDAC6 regulates the dynamics of lytic granules in cytotoxic T lymphocytes. *J. Cell Sci.* **2016**, *129*, 1305–1311. [CrossRef] [PubMed]

55. Fusco, C.; Micale, L.; Augello, B.; Mandriani, B.; Pellico, M.T.; De Nittis, P.; Calcagni, A.; Monti, M.; Cozzolino, F.; Pucci, P.; et al. HDAC6 mediates the acetylation of TRIM50. *Cell. Signal.* **2014**, *26*, 363–369. [CrossRef] [PubMed]

56. Huebner, H.; Knoerr, B.; Betzler, A.; Hartner, A.; Kehl, S.; Baier, F.; Wachter, D.L.; Strick, R.; Beckmann, M.W.; Fahlbusch, F.B.; et al. Detyrosinated tubulin is decreased in fetal vessels of preeclampsia placentas. *Placenta* **2018**, *62*, 58–65. [CrossRef] [PubMed]

57. Canettieri, G.; Morantte, I.; Guzman, E.; Asahara, H.; Herzig, S.; Anderson, S.D.; Yates, J.R., 3rd; Montminy, M. Attenuation of a phosphorylation-dependent activator by an HDAC-PP1 complex. *Nat. Struct. Biol.* **2003**, *10*, 175–181. [CrossRef]

58. Brush, M.H.; Guardiola, A.; Connor, J.H.; Yao, T.P.; Shenolikar, S. Deactylase inhibitors disrupt cellular complexes containing protein phosphatases and deacetylases. *J. Biol. Chem.* **2004**, *279*, 7685–7691. [CrossRef]

59. Tien, S.C.; Chang, Z.F. Oncogenic Shp2 disturbs microtubule regulation to cause HDAC6-dependent ERK hyperactivation. *Oncogene* **2014**, *33*, 2938–2946. [CrossRef] [PubMed]

International Journal of
Molecular Sciences

MDPI

Review

Dual-Specificity Phosphatases in Neuroblastoma Cell Growth and Differentiation

Caroline E. Nunes-Xavier [1,2,*], **Laura Zaldumbide** [3], **Olaia Aurtenetxe** [1],
Ricardo López-Almaraz [4], **José I. López** [1,3] and **Rafael Pulido** [1,5,*]

[1] Biomarkers in Cancer Unit, Biocruces-Bizkaia Health Research Institute, Barakaldo, Bizkaia 48903, Spain; olaia.aurtenetxesaez@osakidetza.eus (O.A.); joseignacio.lopezfernandezdevillaverde@osakidetza.eus (J.I.L.)
[2] Department of Tumor Biology, Institute for Cancer Research, Oslo University Hospital HF Radiumhospitalet, Oslo 0424, Norway
[3] Department of Pathology, Cruces University Hospital, University of the Basque Country (UPV/EHU), Barakaldo, Bizkaia 48903, Spain; laura.zaldumbideduenas@osakidetza.eus
[4] Pediatric Oncology and Hematology, Cruces University Hospital, Barakaldo, Bizkaia 48903, Spain; ricardo.lopezalmaraz@osakidetza.eus
[5] IKERBASQUE, Basque Foundation for Science, Bilbao 48011, Spain
[*] Correspondence: caroliten@gmail.com (C.E.N.-X.); rpulidomurillo@gmail.com (R.P.)

Received: 13 February 2019; Accepted: 1 March 2019; Published: 7 March 2019

Abstract: Dual-specificity phosphatases (DUSPs) are important regulators of neuronal cell growth and differentiation by targeting proteins essential to neuronal survival in signaling pathways, among which the MAP kinases (MAPKs) stand out. DUSPs include the MAPK phosphatases (MKPs), a family of enzymes that directly dephosphorylate MAPKs, as well as the small-size atypical DUSPs, a group of low molecular-weight enzymes which display more heterogeneous substrate specificity. Neuroblastoma (NB) is a malignancy intimately associated with the course of neuronal and neuroendocrine cell differentiation, and constitutes the source of more common extracranial solid pediatric tumors. Here, we review the current knowledge on the involvement of MKPs and small-size atypical DUSPs in NB cell growth and differentiation, and discuss the potential of DUSPs as predictive biomarkers and therapeutic targets in human NB.

Keywords: neuroblastoma; neuronal differentiation; dual-specificity phosphatases; MAP kinases; MAP kinase phosphatases; atypical dual-specificity phosphatases

---

1. Introduction

Neuroblastoma (NB) is the most common malignancy diagnosed in the first year of life, with an average age at diagnosis of 18 months, and constitutes the most frequent extracranial solid tumor in infants, accounting for about 7% of total pediatric tumors. Although novel treatments have improved the survival of NB patients (about 80% 5-year survival), the high-risk forms of NB are the major cause of pediatric cancer death, rendering about 15% of pediatric cancer mortality [1,2]. NB is a neuroendocrine embryonal malignancy that derives from developing sympathetic neuronal cells from the peripheral nervous system, primarily from the adrenal gland medulla (about 40%), but also from the paraspinal sympathetic ganglia from the thorax, abdomen, neck, and pelvis [3–5]. NB arising from adrenal medulla display the worst prognosis, whereas those from the thorax, neck, and pelvis display better prognoses [6]. The current major criteria for staging and classification of NB cases are age, tumor histology subtype and differentiation, tumor spreading (about 20% of NB tumors are disseminated at diagnosis, with bone, bone marrow and liver as the more frequent metastatic niches), ploidy status and segmental chromosome aberrations, and *MYCN* (encoding an oncogenic transcription factor; about 25% of cases) and *ALK* (encoding an oncogenic receptor tyrosine kinase (RTK); about 3% of cases) genes

amplification (*MYCN* and *ALK* are physically linked at 2p24-2p23). Children with *MYCN* amplification regardless of age, those older than 12 months with disseminated tumors, or those older than 18 months with unfavorable histology, are considered to be part of the high-risk NB group (about 40% of cases). Very low-, low-, and intermediate-risk NB patients show a 5-year survival of 90–95%, whereas high-risk NB patients show a 5-year survival of 40–50% [7–9]. The risk group determines the therapeutic treatment of NB patients, from observation or surgery alone for low-risk patients to multimodal therapy for high-risk patients. Multimodal therapy includes surgery, chemotherapy and radiotherapy, myeloablative therapy followed by bone marrow autologous transplantation, and immunotherapy and retinoic acid (RA)-based maintenance therapies. Current high-risk NB clinical trials are testing the efficacy of drugs targeting specific drivers of NB or major pro-oncogenic proteins. These include, among others, ALK and other RTKs, components of MYCN downstream pathways such as ornithine decarboxylase (ODC1), and components of the PI3K/AKT/mTOR and RAS-ERK1/2 MAPK signaling pathways [10–14].

Familial NB accounts for 1–2% of NB cases, with two major genes showing germline mutations in association with the disease: the *ALK* RTK gene, which is mainly expressed in the developing nervous system [15,16]; and the *PHOX2B*, encoding a transcription factor essential in neuronal differentiation and development of the autonomic nervous system, which is mutated in patients with congenital central hypoventilation syndrome (CCHS) [17]. Comprehensive analyses of the somatic mutational status of human cancers have revealed that the mutational burden of NB tumors is relatively low, compared with other cancers, a property which is shared by most pediatric tumors [18–21]. In addition, evidence for the existence of sporadic NB susceptibility genes has been recently obtained by several genome-wide association (GWA) studies on NB tumors, although the applicability of these findings has yet to be translated to the clinics [22–24]. In a GWA study, single nucleotide polymorphisms within the *DUSP12* gene, which encodes a large-size atypical DUSP proposed to target MAPKs, were associated with low-risk NB [25,26]. In this review, we summarize the current knowledge on the role of MAPK phosphatases (MKPs) and related small-size atypical DUSPS in NB, and their potential as NB biomarker and drug targets.

2. Neuroblastoma Cell Growth and Differentiation

NB can be considered a neural crest-related developmental tissue disease in which alterations in neuronal differentiation, driven by the unbalanced action of pro-proliferative and pro-differentiation factors on the maturation and migration of neural crest cells and neuroblasts, play a fundamental etiologic role. Highly differentiated NB tumors have a favorable clinical outcome, and spontaneous regression linked to neuroblast apoptosis is frequent even in metastatic cases [5,27–29]. *MYCN* amplification, the major hallmark of high-risk NB, associates with poorly differentiated NB tumors [30–32], and signaling through ALK favors proliferation and/or survival depending on the maturity of the neural cells [33,34]. In addition, activation of the tyrosine kinase neurotrophin receptors TrkA and TrkB leads to apoptotic/differentiation neuroblast responses or to survival/proliferative effects, respectively [35–37]. Several animal models suitable to the study of NB differentiation and transformation have been generated, mainly centered in *MYCN* amplification and ALK hyperactivation [38,39]. NB cell differentiation can also be triggered in vitro by culturing NB cells in the presence of differentiation factors, such as retinoids (retinoic acid, RA), phorbol esters (phorbol 12-myristate 13-acetate, PMA), and neurotrophins (nerve growth factor, NGF; brain-derived neurotrophic factor, BDNF) [40,41]. Human and rodent cell lines commonly used to study NB cell differentiation include SH-SY5Y and other derivatives from the SK-N-SH cell line, IMR-32, SMS-KCNR (all NB; human), Neuro2A (NB; mouse), PC12 (pheochromocytoma; rat), and P19 (embryonic teratocarcinoma; mouse) cell lines, among others. Extensive experimental work using these model systems has provided a picture in which the major signaling pathways involved in the molecular effects of MYCN and ALK in NB are the RAS/MAPK, PI3K/AKT, and JAK/STAT pathways [41–44] (Figure 1).

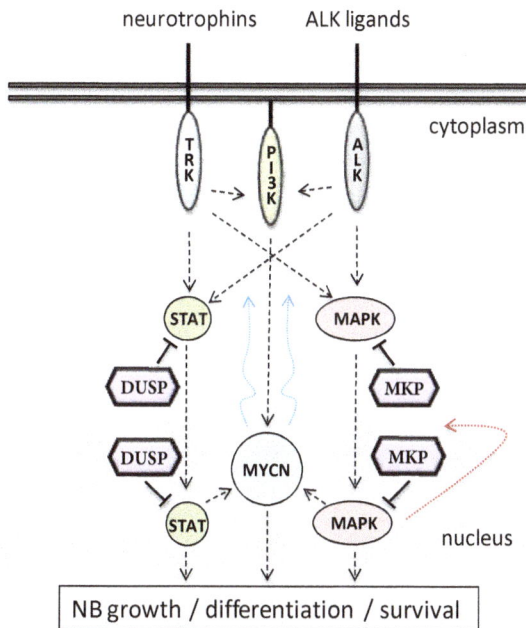

Figure 1. Schematic depiction of the major pathways involved in signaling through the ALK-MYCN axis in neuroblastoma (NB). ALK signals downstream mainly through the RAS-ERK1/2 MAP kinase (MAPK) pathway, as well as through the PI3K/AKT and JAK/STAT pathways, resulting in increased *MYCN* transcription and cell growth (straight dashed lines). Signaling through Trk neurotrophin receptors is also shown. MYCN transcriptional activity positively feeds the pathway by promoting *ALK* transcription (curved blue dotted lines), whereas transcriptional activity mediated by the MAPK nuclear effectors negatively feed-back the MAPK pathways by promoting the transcription of MAPK phosphatases (MKP) genes (curved red dotted line). Straight solid lines indicate direct dephosphorylation of protein substrates by MKPs or by other dual-specificity phosphatases (DUSPs). Dephosphorylation of MAPKs by MKPs is well documented, whereas evidence on the dephosphorylation of STATs or Trks by other DUSPs is limited. See text for more details.

Genes related to the RAS-ERK1/2 MAPK pathway display somatic alterations in about 5% of sporadic primary NB tumors, with a much higher percentage of alterations (especially in *ALK*, *NRAS* and *NF1* genes) in relapsed NB samples obtained after chemotherapy [45–47]. This makes the RAS-ERK1/2 axis a potential responsive pathway for therapeutic intervention in NB. MEK1/2 or RAF-MEK1/2 pharmacological inhibitors displayed good inhibitory growth and survival effects, and decreased pERK1/2 levels, on NB cell lines with mutations in the RAS-ERK1/2 axis and high ERK1/2 activation under basal conditions [48–51]. Interestingly, combined use of a MEK1/2 inhibitor and RA resulted in increased differentiation and inhibitor sensitization of NB cells [52]. On the other hand, ALK-addicted NB cells responded to MEK1/2 inhibitors with an increase in survival and proliferation, which was associated with ERK5 activation through the PI3K/AKT pathway [53,54]. This highlights the importance of precise molecular segregation of NB patients before testing MEK1/2-inhibition in combination with other therapies, and illustrates the key role of distinct ERK kinases in mediating the efficacy of MEK1/2 inhibitors in NB. The influence of alterations in JNK and p38 MAPK pathways in NB cell growth has been investigated less. Higher sensitivity to apoptotic stimuli has mainly been reported in NB cell lines upon pharmacological inhibition of JNKs or p38s, although an anti-apoptotic outcome after inhibition of these MAPKs has also been reported [55–59] (Table 1). A complex scenario emerges in NB in which the specific contribution of the effectors from the distinct MAPK pathways may differentially drive NB cell growth, differentiation, and apoptosis, with variable consequences in the response to chemotherapies. As it has been proposed for other cell types, it is possible that a locally- and timely-regulated MAPK activation is critical for the response of NB cells to apoptotic or stress conditions. The interference of MAPK function by the inhibition of MAPK activators or inactivators (including phosphatase inactivators) is an open therapeutic possibility that deserves further exploration in NB.

Table 1. MKPs and small-size atypical DUSPs in neuroblastoma cell growth and differentiation.

Gene/Protein	MAPK Substrates [1] Localization	Alterations in NB Cell Lines and NB Tumors
DUSP1/MKP1	JNK, p38 > ERK Nuclear	-SH-SY5Y cells: *DUSP1 KD:* ↓ H_2O_2-induced apoptosis, ↑ pERK1/2 [60]; ↓ DUSP1 mRNA upon AgNP treatment [61]; ↑ DUSP1 upon carbachol stimulation [62]. -SK-N-SH and SK-N-BE cells: ↓ DUSP1 protein in CD133+ cells resistant to chemotherapy [63]. -P19 cells: ↑ DUSP1 protein upon RA differentiation [64]. -GH4C1 neuroendocrine cells: ↑ DUSP1 upon TRH and EGF stimulation [65,66]. -SMS-KCNR cells: ↑ DUSP1 mRNA upon RA-induced differentiation [67]. -N1E-115 cells: ↑ DUSP1 upon hypoxia/re-oxygenation; *DUSP1 KD:* ↑ pJNK, ↑neuronal death; *DUSP1 OE:* ↓ pJNK, ↓ neuronal death [68].
DUSP4/MKP2	ERK, JNK > p38 Nuclear	-SH-SY5Y cells: ↓ DUSP4 mRNA upon AgNP treatment [61]. -Mouse ESCs: *DUSP4 KD:* ↓ neuronal differentiation, ↑ pERK1/2; ↑ DUSP4 in RA-induced differentiation [69]. -SH-SY5Y, KCNR cells: ↓ DUSP4 upon RA-induced differentiation [67]. -SK-N-AS cells: ↓ DUSP4 mRNA upon ALK inhibition, ↑ DUSP4 mRNA upon mutant ALK OE [70].
DUSP5	ERK Nuclear	-SiMa, Kelly, SH-SY5Y, CHP-134 cells: ↑ DUSP5P1/DUSP5 ratio compared to normal cells [71]. -SH-SY5Y, SMS-KCNR, IMR-32 cells: ↑ DUSP5 mRNA upon RA-induced neuronal differentiation; SH-SY5Y cells: ↑ DUSP5 mRNA upon EGF and PMA stimulation; *DUSP5 KD:* ↑ proliferation [67]. -SK-N-AS cells: ↓ DUSP5 mRNA upon ALK inhibition, ↑ DUSP5 mRNA upon mutant ALK OE [70]. -NB tumors: DUSP5 protein expression associated with poor prognosis and pERK1/2 expression [67].
DUSP6/MKP3	ERK Cytoplasmic	-PC12 cells: ↑ DUSP6 mRNA in FGF and NGF-mediated neuronal differentiation [72–74]. -P19 cells: ↑ DUSP6 protein upon RA differentiation, dependent on ERK1/2 activation [64]. -SH-SY5Y, BE(2)-C cells: ↑ DUSP6 mRNA upon RA-induced differentiation; *DUSP6 KD + RGS16 KD* in SH-SY5Y cells: ↓ RA-mediated proliferation arrest [75]. -IMR-32 cells: ↑DUSP6 mRNA upon RA-induced differentiation [67]. -SH-SY5Y cells: ↓ DUSP6 protein upon H_2O_2 stimulation [76]. -SK-N-AS cells: ↓ DUSP6 mRNA upon ALK inhibition, ↑ DUSP6 mRNA upon mutant ALK OE [70]. -BE(2)-C, LAN-1 cells: ↓ DUSP6 by a MYCN-SIRT1 complex [77].

Table 1. *Cont.*

Gene/Protein	MAPK Substrates [1] Localization	Alterations in NB Cell Lines and NB Tumors
DUSP7/MKPX	ERK Cytoplasmic	-NB tumors: Positive immunostaining, no clinical associations [67].
DUSP8	JNK, p38 Cytoplasmic/nuclear	-Mouse J1 ESCs: ↑ DUSP8 in RA-induced neuronal differentiation [78].
DUSP9/MKP4	ERK > p38 Cytoplasmic	-Mouse J1 ESCs:↓ DUSP9 in RA-induced neuronal differentiation [78]. -SH-SY5Y, SMS-KCNR cells: ↓ DUSP9 mRNA upon RA-induced differentiation [67]. -NB tumors: Mostly negative immunostaining, no clinical associations [67].
DUSP10/MKP5	JNK, p38 Cytoplasmic/nuclear	-Mouse J1 ESCs: ↑ DUSP10 in RA-induced neuronal differentiation [78].
DUSP16/MKP7	JNK, p38 Cytoplasmic/nuclear	-*DUSP16 -/- mice*: hydrocephalus and brain overgrowth [79]. -SMS-KCNR, IMR-32 cells: ↑ DUSP16 mRNA upon RA-induced differentiation [67].
DUSP3/VHR	(ERK, JNK) Cytoplasmic/nuclear	-SH-SY5Y cells: ↓ DUSP3 mRNA upon AgNP treatment [61].
DUSP13A/MDSP	Cytoplasmic	-SK-N-SH cells: physical association with pro-apoptotic ASK1, independent of phosphatase activity; *DUSP13A KD*: ↓ ASK1 kinase activity [80].
DUSP23/VHZ	(ERK, JNK) Cytoplasmic/nuclear	-Mouse J1 ESCs: ↑ DUSP23 in RA-induced neuronal differentiation; *DUSP23 KD*:↓ neuronal differentiation, ↑ pp38, ↓ pERK1/2, ↓ pJNK [78]. -NB tumors: ↑ Methylation in *MYCN*-amplified tumors, ↓ DUSP23 mRNA in patients with poor outcome [81].
DUSP26/MKP8	(p38) Nuclear	-Mouse J1 ESCs: ↑ DUSP26 in RA-induced neuronal differentiation [78]. -PC12 cells: ↑ DUSP26 mRNA in NGF-induced differentiation; *DUSP26 KD:* ↑ EGFR, ↑ NGF-induced differentiation; *DUSP26 OE:* ↓ EGF-induced cell growth, ↓ NGF-induced differentiation, ↑ cisplatin-induced apoptosis, ↓ pAKT, ↓ EGFR [82,83]. -Human NB cell lines: ↓ DUSP26 mRNA compared to normal adrenal gland [84]. -IMR-32 cells: *DUSP26 KD:* ↑ doxorubicin-induced apoptosis, ↓ proliferation, ↑ pp53, ↑ pp38; SH-SY5Y cells: *DUSP26 KD:* ↓proliferation; SK-N-SH cells: *DUSP26 OE:* ↓ doxorubicin-induced apoptosis, ↓ pp53 [85,86]. -NB tumors: ↑ DUSP26 protein in high-risk NB tumors [85].

[1] Note that substrate specificity towards MAPKs of small-size atypical DUSPs is debatable in some cases, and some effects on MAPK phosphorylation status have been reported to be likely mediated by scaffolding functions or by dephosphorylation of non-MAPK proteins. Abbreviations: AgNP—silver nanoparticles; ALK—anaplastic lymphoma kinase; EGF—epidermal growth factor; ESCs—embryonic stem cells; FGF—fibroblast growth factor; KD—knock-down; NB—neuroblastoma; NGF—nerve growth factor; OE—overexpression; PMA—phorbol 12-myristate 13-acetate; RA—retinoic acid; TRH—thyrotopin-releasing hormone; ↑—increase; ↓—decrease.

Finally, the involvement of the pro-survival PI3K/AKT and JAK/STAT pathways in NB has also been disclosed in a variety of studies, sometimes acting in coordination with MAPK activities; the inhibition of the effectors of these pathways as a NB therapeutic option is under scrutiny [44,87–90]. A functional ALK-MYCN axis operates in NB that positively controls cell growth through coordinated integration of these major signaling pathways [15,91,92] (Figure 1).

Int. J. Mol. Sci. **2019**, *20*, 1170

3. DUSPs in NB Cell Growth and Differentiation

DUSPs constitute a heterogeneous group of non-transmembrane enzymes within the class I Cys-based protein tyrosine phosphatase (PTP) family. They have in common the presence of a single catalytic PTP domain, which in the members of several DUSP subfamilies, has the dual capability to dephosphorylate both Ser/Thr- and Tyr-phosphorylated residues (pSer/pThr, pTyr) in proteins. Enzymes from other DUSP subgroups, however, have RNA, lipids, or other biomolecules as their major substrates [93–96]. Here, we will focus on the MAPK phosphatase PTP subfamily (MKPs; 11 genes in humans) and the MKP-related small-size atypical DUSPs (15 genes in humans) (Figure 2), two groups of DUSPs with high potential as anti-cancer drug targets and as regulators of NB cell growth and differentiation [97,98]. With the exception of the phosphatase inactive MKP STYXL1, MKPs are specialized in the selective dephosphorylation of the Thr and Tyr regulatory residues from the distinct MAPKs (ERKs, p38s, and JNKs), a group of Ser/Thr kinases that shuttle between the cytoplasm and the nucleus and have major physiologic roles as modulators of cell growth, differentiation, and apoptosis upon changes in extracellular cues. Accordingly, alterations in the outcome of the MAPK pathways have a notable impact on human disease [99–103]. MKPs are composed of a catalytic DUSP-PTP domain and a regulatory MAPK-binding domain, which is essential to control the specificity in the binding and dephosphorylation of the distinct MAPKs (Figure 2A). In addition to their role in MAPK dephosphorylation, which results in MAPK catalytic inactivation, MKPs also directly regulate the nuclear-cytoplasmic shuttling of MAPKs upon the binding of their MAPK-binding domains [104–106]. MKPs expression, subcellular localization, and function is highly regulated during physiologic and pathologic processes, and MKPs gene expression is induced in many cases upon activation of the specific MAPK pathways under their regulation. In consequence, expression of different MKPs has been associated with several forms of human cancer [107–109]. Small-size atypical DUSPs lack the regulatory MAPK-binding domain and constitute small enzymes, among which DUSP3 is the one more intensively studied [110] (Figure 2A). Most of small-size atypical DUSPs dephosphorylate pSer/pThr and pTyr from proteins (including MAPKs, STATs, and RTKs), whereas some of these enzymes have as physiologic substrates other biomolecules or, in the case of the small-size atypical DUSP STYX, are phosphatase inactive [111,112].

The functional role of MKPs and small-size atypical DUSPs as physiologic MAPK inactivators advocates for their high potential as important players in NB cell growth and differentiation (Figure 1). In Figure 2B (upper and middle panels), the mRNA expression profiles of MKPs and small-size atypical DUSPs in the adrenal gland (the more common source of NB cells) and SH-SY5Y NB cells (the more studied human NB cell line) are shown. In Figure 2B (bottom panel), the changes in the mRNA expression of MKPs and small-size atypical DUSPs from three human NB cell lines (SH-SY5Y, SMS-KCNR, and IMR-32) undergoing retinoic acid (RA)-induced differentiation are shown. Different RA-regulated expression patterns of these genes can be observed, suggesting a complex and cell-specific rearrangement of DUSPs gene expression during NB cell differentiation. Following is an account on the expression and function of MKPs and small-size atypical DUSPs in NB, and a summary of the information is provided in Table 1. MKPs have been grouped according to amino acid sequence conservation, subcellular localization and substrate specificity, as reported in References [98,113].

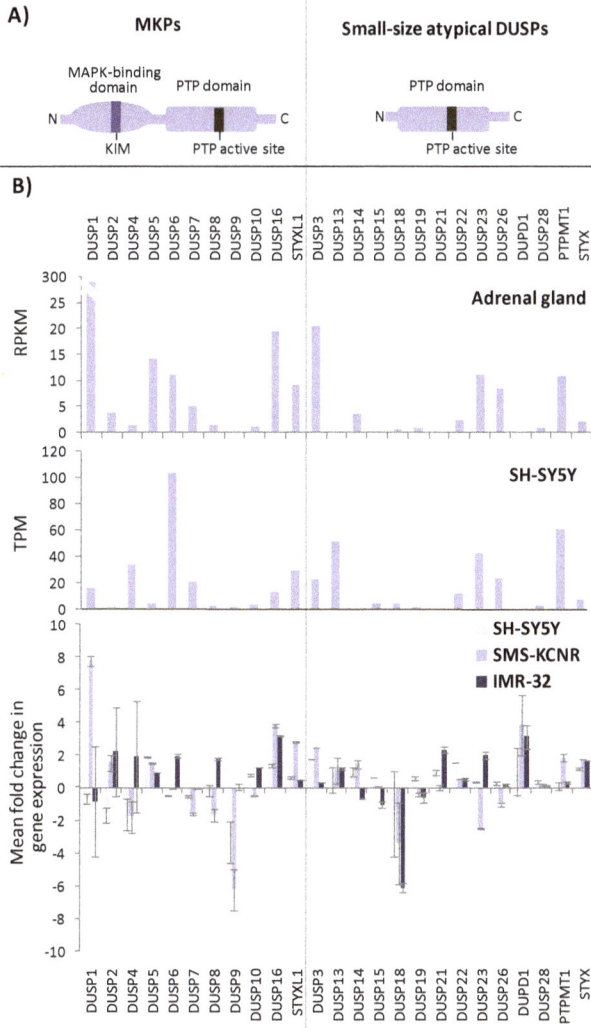

Figure 2. (**A**) Schematic depiction of the domain composition of MKPs and small-atypical DUSPs. KIM—kinase interaction motif; PTP—protein tyrosine phosphatase. (**B**) Upper panel: mRNA expression of MKPs and small-atypical DUSPs in adrenal gland. Data from GTEx (Genotype-Tissue Expression) data sets. RNA-seq data is reported as median RPKM (reads per kilobase per million reads mapped) (https://www.proteinatlas.org/). Middle panel: mRNA expression of MKPs and small-atypical DUSPs in SH-SY5Y human NB cells; data from the Human Protein Atlas. RNA-seq data is reported as median TPM (transcripts per kilobase of exon per million reads) (https://www. proteinatlas.org/). Bottom panel: mRNA expression analysis of MKPs and small-atypical DUSPs from SH-SY5Y, SMS-KCNR, and IMR-32 human NB cell lines treated with retinoic acid (RA). Cell lines were kept untreated or were treated for 10 days with RA, mRNA was extracted and RT-qPCR was performed using a set of MKP and small-atypical DUSP primers, as described in Reference [67]. Relative mRNA expression values are shown in Log$_2$ as fold change +SD of treated cells *versus* untreated cells, from at least two independent experiments. Mean fold change above 2 or below −2 was considered significant. MKP mRNA expression data from bottom panel has been previously published [67].

3.1. MKPs in NB Cell Growth and Differentiation

3.1.1. DUSP1, DUSP4, and DUSP5

DUSP1 (MKP1), DUSP4 (MKP2), and DUSP5 constitute a group of highly transcriptionally inducible MKPs with nuclear localization and distinct MAPK substrate specificity. DUSP1 mainly dephosphorylates JNKs and p38s, whereas DUSP4 and DUSP5 are more specific for ERK1/2 as a substrate [114–116]. DUSP1 is the founder of the MKP family and displays ubiquitous tissue distribution. DUSP1 has been widely involved in human disease, including immunological, inflammatory and cardiovascular diseases, cancer, and developmental nervous system diseases [117–122]. In human NB cell lines, DUSP1 has been shown to facilitate apoptotic processes, suggesting a negative role for this MKP in NB cell survival. H_2O_2 treatment of SH-SY5Y cells triggered apoptosis which was accompanied by early reduction and late induction of DUSP1 protein content, in inverse correlation with pERK1/2 levels, and siRNA suppression of DUSP1 expression attenuated H_2O_2-induced cell death [60]. It would be important to determine the contribution of DUSP1 mRNA transcription to increasing DUSP1 levels in SH-SY5Y cells during long-term H_2O_2 treatment. In this regard, treatment of SH-SY5Y cells with silver nanoparticles (AgNP) inducing neuronal differentiation increased reactive oxygen species (ROS) generation, which was accompanied by decreased mRNA levels in several DUSPs, including DUSP1, and increased pAKT and pERK1/2 levels [61,123]. In addition, carbachol treatment of SH-SY5Y cells resulted in rapid ERK1/2 activation and increased nuclear DUSP1 protein content [62], whereas long-term stimulation with the differentiating agent RA caused up-regulation of DUSP1 mRNA in SMS-KCNR NB cells, but not in SH-SY5Y cells [67]. Interestingly, CD133+ stem-cell subpopulations from the human NB cell lines SK-N-SH and SK-N-BE displayed high resistance to pro-apoptotic chemotherapeutic agents, in association with DUSP1 protein low expression levels and high phosphoERK1/2 (pERK1/2) and phosphop38 (pp38) content, suggesting that DUSP1 expression, at least in some NB cell populations, may facilitate chemotherapy efficacy [63]. Treatment of GH4C1 rat neuroendocrine cells with epidermal growth factor (EGF) or thyrotropin-releasing hormone (TRH) triggered DUSP1 mRNA translation [65], and P19 mouse embryonic stem cells (ESCs), induced to neuronal differentiation by RA treatment, also displayed increased nuclear DUSP1 protein content, coinciding with ERK1/2 inactivation [64]. On the other hand, an anti-apoptotic role has also been proposed for DUSP1 in NB cell lines in association with its action on JNKs. Hypoxia/re-oxygenation of N1E-115 mouse NB cells resulted in up-regulation of mRNA and protein DUSP1 levels, although JNK activation was not suppressed under these conditions. However, the siRNA knock-down of DUSP1 in N1E-115 cells induced to differentiate in the absence of serum resulted in increased JNK activation and apoptotic cell death, whereas DUSP1 over-expression caused the opposite effects [68]. These findings support the notion that the inhibitory action of nuclear DUSP1 on specific MAPKs at specific time-points could be a determinant of the survival of NB cells challenged with pro-apoptotic or pro-differentiation chemotherapies.

DUSP4 has been regarded as a potential tumor suppressor, as well as a promoter of chemotherapy resistance, in several human cancers [124–128]. DUSP4 expression augmented during RA-dependent neuronal differentiation of mouse ESCs, and shRNA silencing of DUSP4 expression in mouse ESCs decreased neuronal differentiation, in association with elevated pERK1/2 content and alterations in calcium homeostasis [69]. In contrast, DUSP4 mRNA was downregulated in SH-SY5Y cells subjected to neuronal differentiation by AgNP or RA [61,67]. In addition, human SK-N-AS NB cells subjected to pharmacologic ALK inhibition displayed mRNA downregulation of several ERK1/2 inactivators, including DUSP4, DUSP5, and DUSP6, whereas overexpression of hyperactive ALK caused upregulation of these MKPs [70]. This is suggestive of a negative ERK1/2 feed-back regulation, which could be operative during NB progression. Whether regulation of DUSP4 expression during NB cell growth is related to malignancy deserves further studies.

DUSP5, together with DUSP6, is an ERK1/2-specific MKP whose transcription is induced in an ERK1/2-dependent manner [129–132]. Some reports attribute tumor suppressor activities to DUSP5, and a dual role for this MKP in carcinogenesis, depending on the tissue and cellular context, has been proposed [109]. DUSP5 mRNA was upregulated on several human NB cell lines subjected to RA-induced differentiation. In SH-SY5Y cells, DUSP5 mRNA upregulation correlated with ERK1/2 activation conditions, illustrating that upregulation of ERK1/2-specific MKPs in NB cells is concomitant to the activation of ERK1/2 pathway, and DUSP5 siRNA knock-down resulted in increased cell proliferation. Importantly, high expression of DUSP5 protein was relatively frequent and also correlated with pERK1/2 expression in NB tumors, in association with poor patient prognosis, likely as a surrogate marker of ERK1/2 activation [67] (Figure 3). In another study, the expression in human cancer cells of DUSP5 and the DUSP5 pseudogene DUSP5P1 was compared, showing high ratios of DUSP5P1/DUSP5 expression in cancer cell lines, including NB cell lines, when compared to normal tissues [71]. In addition, DUSP5P1 expression has been found to be associated with poor prognosis in acute myeloid leukemia [133]. Further investigation is required to determine the role of DUSP5P1 in regulating DUSP5 function, and whether DUSP5P1 expression correlates with the prognosis of or tumor remission in NB.

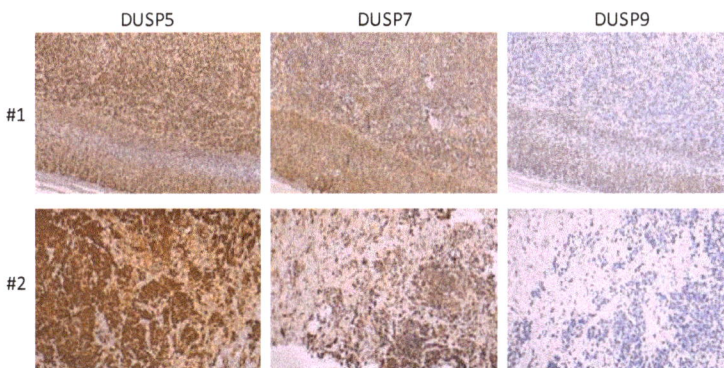

Figure 3. Immunostaining of selected MKPs from tissue sections from two NB tumors (#1: Stage I, unknown *MYCN* status, low-risk; #2: Stage IV, metastatic, *MYCN* amplification, high-risk). Representative staining patterns are shown. DUSP5 and DUSP7 mostly showed high-moderate immunoreactivity, whereas DUSP9 mostly displayed low immunoreactivity (magnification is ×100). Immunostaining was performed as described in [67].

3.1.2. DUSP6, DUSP7, and DUSP9

DUSP6 (MKP3), DUSP7 (MKPX), and DUSP9 (MKP4) are cytoplasmic MKPs with specificity towards ERK1/2 dephosphorylation. Similar to DUSP5, DUSP6 is inducible by growth and differentiating agents that activate the ERK1/2 pathway [132,134,135], and DUSP6 is also reported to play a tissue-specific dual tumor suppressive or pro-oncogenic role [105,136]. DUSP6 mRNA is induced with NGF- or fibroblast growth factor (FGF)-mediated neuronal differentiation in PC12 rat pheochromocytoma cells [72–74], as well as with RA-mediated neuronal differentiation of SH-SY5Y, BE(2)-C, and IMR-32 human NB cells [67,75]. On the other hand, SH-SY5Y cells challenged with H_2O_2 displayed reduced DUSP6 protein levels, which was associated with a higher pERK1/2 content and cell death [76]. This effect is likely to be mediated by caspase-3 cleavage of DUSP6, as described for other human cancer cell lines [104]. Besides, co-silencing of DUSP6 and RGS16 (a negative regulator of Ras G proteins also induced by RA) in SH-SY5Y cells increased pERK1/2 levels and decreased the proliferation arrest caused by RA in these cells [75]. Differentiation of P19 mouse cells with RA also increased DUSP6 protein levels dependent on ERK1/2 activation. Interestingly, DUSP1 protein accumulation in these conditions was independent on ERK1/2 activation [64]. As mentioned, DUSP6 mRNA induction can be triggered in NB cells by signaling through hyperactive ALK [70], and *DUSP6* transcription can be repressed by MYCN in BE(2)-C and LAN-1 NB cells [77]. Together, these findings illustrate the complexity in the negative and positive ERK1/2 regulatory feed-back loops that operate in NB cells by regulation of *DUSP6* gene transcription and DUSP6 protein stability. They prompt further investigations of the potential of DUSP6 as a modulator of NB malignancy.

Expression of both DUSP7 and DUSP9 protein has been detected in NB tumor samples, although, in the case of DUSP9, with a low frequency [67] (Figure 3). Human SH-SY5Y and SMS-KCNR cells, as well as mouse J1 ESCs, displayed downregulation of DUSP9 mRNA upon RA-induced differentiation, which in the case of mouse J1 cells was confirmed at the protein level [67,78].

3.1.3. DUSP8, DUSP10, and DUSP16

DUSP8, DUSP10 (MKP5), and DUSP16 (MKP7) are MKPs with substrate specificity towards JNKs and p38s MAPKs. These three MKPs are larger than the rest of the MKPs, with N- or C-terminal extensions in their amino acid sequences. DUSP10 and DUSP16 have been involved in autoimmunity and inflammation processes, and little is known about the physiological function of DUSP8. Interestingly, *DUSP16* gene loss caused perinatal lethality in mice, associated with hydrocephalus, brain overgrowth by expansion of neural progenitors, and increased midbrain pp38 content [79]. RA treatment has been reported to upregulate DUSP8 and DUSP10 in mouse J1 ESCs, and DUSP16 mRNA in human NB cell lines [67,78].

3.2. Small-Size Atypical DUSPs in NB Cell Growth and Differentiation

DUSP3 (VHR) is the prototype of small-size atypical DUSPs, and displays substrate specificity towards ERK1/2 and JNKs. However, some other proteins, including STAT5, FAK, growth factor receptors, and nuclear proteins regulating DNA damage repair, have been shown to be tyrosine-dephosphorylated by DUSP3 and have been proposed as DUSP3 physiological substrates [110,137]. As mentioned for other MKPs, DUSP3 mRNA was upregulated in SH-SY5Y cells differentiated in the presence of AgNP [61], as well as on SH-SY5Y or SMS-KCNR cells differentiated by RA (Figure 2B, bottom panel), but the potential role of DUSP3 in NB cell growth and differentiation remains unexplored.

DUSP13A and DUSP13B are two small-size atypical DUSPs encoded in the same gene and generated by the use of alternative open reading frames, whose physiologic substrate specificity is poorly known [138]. DUSP13A (MDSP) was found to be a potential regulator of apoptosis in human SK-N-SH NB cells in a phosphatase-independent manner, by virtue of its physical association with the p38- and JNK-activator protein apoptosis signal-regulated kinase 1 (ASK1). The knock-down of DUSP13A decreased the phosphorylation and activation of ASK1 [80].

DUSP23 (VHZ) has been reported to affect MAPKs activation both by phosphatase-dependent and -independent mechanisms [139,140], as well as the Ser- or Tyr-phosphorylation status of other proteins, including β-catenin and the transcription factor GCM1 [141,142]. DUSP23 mRNA and proteins were upregulated in RA-differentiated mouse ESCs, and siRNA knock-down of DUSP23 in these cells decreased neuronal differentiation, together with increased pp38 and decreased pERK1/2 and phosphoJNK (pJNK), likely by a combination of catalytic and scaffolding properties [78]. Noticeably, analysis of the methylation status of the *DUSP23* gene in NB tumors revealed higher methylation in *MYCN*-amplified tumors. In addition, DUSP23 mRNA was at lower levels in NB patients with poor outcomes, when compared to patients free of disease [81]. This suggests a tumor suppressive role for DUSP23 in NB. It would be important to analyze DUSP23 protein expression in these groups of NB patients.

DUSP26 (MKP8) was originally identified as a DUSP mainly expressed in embryonal cancers and displaying substrate specificity towards p38 in cells [143]. Subsequently, DUSP26 was also found in neuroendocrine tissues, and induced by NGF-stimulation in PC12 cells [82], as well as by RA in mouse ESCs [78]. DUSP26 overexpression in PC12 cells decreased EGFR and pAKT levels, which resulted in the suppression of NGF- and EGF-induced signaling. This effect was dependent on DUSP26 phosphatase activity, but was not associated with changes in MAPKs activation. Opposite consequences were observed with DUSP26 knock-down [82,83]. Additional studies in PC12 cells and in the zebrafish model sustain the possibility that DUSP26 could target for dephosphorylation specific receptor tyrosine kinases, including TrkA and FGFR1 [144]. In IMR-32 cells, DUSP26 has been proposed to increase N-cadherin-mediated cell-cell adhesion by dephosphorylation of the KIF3 motor complex component Kap3 [145]. A suppressive role of cell proliferation by DUSP26 targeting non-MAPK substrates has also been unveiled in epithelial cancer cells [84]. In contrast, an oncogenic role has been proposed for DUSP26 in human NB cells on the basis of DUSP26-mediated inhibition of p53 activity by specific Ser dephosphorylation and resistance to doxorubicin-induced apoptosis. Importantly, high-risk NB tumors displayed the highest levels of DUSP26 protein expression from a limited number of samples [85]. DUSP26 knock-down in SH-SY5Y cells inhibited cell growth in vitro and in xenograft mice, and this could be summarized by the pharmacological inhibition of DUSP26, which also increased phosphop53 (pp53) and pp38 levels [86]. Together, these results argue for both MAPK-dependent and -independent effects of DUSP26 in NB cell growth. It would be necessary to further define the oncogenic or tumor suppressive functions of DUSP26 in human NB before pointing to DUSP26 as a target for inhibition in these types of cancers.

4. Concluding Remarks

DUSPs have been proposed as cancer biomarkers in several human cancers, and recent advances in the understanding of the biology of DUSPs have made these enzymes potentially targetable molecules in cancer therapy: (1) DUSPs regulate cell growth, differentiation, and apoptosis through dephosphorylation of key effector proteins from major intracellular signaling pathways; (2) DUSPs present a well-defined catalytic mechanism towards well-defined substrates, especially those in the MAPK family; and (3) DUSPs activities can be efficiently repressed by small molecule inhibitors [96,98,108,109,146]. Signaling through the ALK-RAS-ERK1/2 MAPK pathway associates with NB tumor progression and relapse. Since the expression of some ERK1/2-specific MKPs, such as DUSP5 or DUSP6, is upregulated upon ERK1/2 activation, their expression in NB could be used as a surrogate marker of ALK and ERK1/2 activation status. In such scenario, inhibition of ERK1/2-MKPs is not desirable. Instead, increased activity of ERK1/2-MKPs would be therapeutically advantageous under the conditions of RAS-ERK1/2 pathway hyperactivation in NB, especially in high-risk NB committed to undifferentiation. On the other hand, the variable effects of JNK and p38 MAPKs on NB cell apoptosis endorse the possibility of a beneficial use of inhibitors of JNK- or p38-specific MKPs, such as DUSP1, DUSP8, DUSP10, or DUSP16, but only in NB cases in which the activity of JNKs or p38s favors tumor chemosensitivity. Finally, the possibility that specific small-atypical DUSPs directly target upstream components in the MAPK pathways, or receptor or effector proteins in other NB cell growth/survival pathways, such as Trks or STATs, deserves further exploration. A precise molecular definition of the pathways involved in NB cell growth dependence and drug resistance in the distinct groups of NB patients is desirable. This would help to bring into practice NB therapies based on MKPs or small-atypical DUSPs functional interference.

Author Contributions: conceptualization and writing—original draft preparation, C.E.N-X. and R.P.; writing—review and editing, C.E.N-X., L.Z., O.A., R.L.-A., J.I.L. and R.P.

Funding: This work was partially supported by the grants: BIO13/CI/001/BC from BIOEF (EITB maratoia), Basque Country, Spain; SAF2013-48812-R from the Ministerio de Educación y Ciencia (to R.P.), and SAF2016-79847-R from the Ministerio de Economía y Competitividad (Spain and Fondo Europeo de Desarrollo Regional) (to R.P. and J.I.L.); and 239813 from the Research Council of Norway (to C.E.N-X.).

Acknowledgments: We thank the Basque Country Biobank and the Basque Foundation for Science (Ikerbasque) for their help and support.

Conflicts of Interest: The authors declare no conflict of interest.

Abbreviations

ALK	Anaplastic lymphoma kinase
BDNGF	Brain-derived neurotrophic factor
CCHS	Congenital central hypoventilation syndrome
DUSP	Dual-specificity phosphatase
EGFR	Epidermal growth factor receptor
ESC	Embryonic stem cell
FGFR	Fibroblast growth factor receptor
GWA	Genome-wide association
IFN	Interferon
KIM	Kinase interaction motif
MAPK	Mitogen-activated protein kinase
MKP	MAPK phosphatase
NB	Neuroblastoma
NGF	Nerve growth factor
PMA	Phorbol 12-myristate 13-acetate
PTP	Protein tyrosine phosphatase
RA	Retinoic acid
RTK	Receptor tyrosine kinase
TRH	Thyrotropin-releasing hormone

References

1. Siegel, R.L.; Miller, K.D.; Jemal, A. Cancer statistics, 2018. *CA Cancer J. Clin.* **2018**, *68*, 7–30. [CrossRef] [PubMed]
2. Ward, E.; DeSantis, C.; Robbins, A.; Kohler, B.; Jemal, A. Childhood and adolescent cancer statistics, 2014. *CA Cancer J. Clin.* **2014**, *64*, 83–103. [CrossRef] [PubMed]
3. Maris, J.M.; Hogarty, M.D.; Bagatell, R.; Cohn, S.L. Neuroblastoma. *Lancet* **2007**, *369*, 2106–2120. [CrossRef]
4. Matthay, K.K.; Maris, J.M.; Schleiermacher, G.; Nakagawara, A.; Mackall, C.L.; Diller, L.; Weiss, W.A. Neuroblastoma. *Nat. Rev. Dis. Primers* **2016**, *2*, 16078. [CrossRef] [PubMed]
5. Tsubota, S.; Kadomatsu, K. Origin and initiation mechanisms of neuroblastoma. *Cell Tissue Res.* **2018**, *372*, 211–221. [CrossRef] [PubMed]
6. Vo, K.T.; Matthay, K.K.; Neuhaus, J.; London, W.B.; Hero, B.; Ambros, P.F.; Nakagawara, A.; Miniati, D.; Wheeler, K.; Pearson, A.D.; et al. Clinical, biologic, and prognostic differences on the basis of primary tumor site in neuroblastoma: A report from the international neuroblastoma risk group project. *J. Clin. Oncol.* **2014**, *32*, 3169–3176. [CrossRef] [PubMed]
7. Ahmed, A.A.; Zhang, L.; Reddivalla, N.; Hetherington, M. Neuroblastoma in children: Update on clinicopathologic and genetic prognostic factors. *Pediatr. Hematol. Oncol.* **2017**, *34*, 165–185. [CrossRef] [PubMed]
8. Park, J.R.; Bagatell, R.; London, W.B.; Maris, J.M.; Cohn, S.L.; Mattay, K.K.; Hogarty, M.; COG Neuroblastoma Committee. Children's Oncology Group's 2013 blueprint for research: Neuroblastoma. *Pediatr. Blood Cancer* **2013**, *60*, 985–993. [CrossRef] [PubMed]
9. Tolbert, V.P.; Matthay, K.K. Neuroblastoma: Clinical and biological approach to risk stratification and treatment. *Cell Tissue Res.* **2018**, *372*, 195–209. [CrossRef] [PubMed]
10. Amoroso, L.; Haupt, R.; Garaventa, A.; Ponzoni, M. Investigational drugs in phase II clinical trials for the treatment of neuroblastoma. *Expert Opin. Investig. Drugs* **2017**, *26*, 1281–1293. [CrossRef] [PubMed]
11. Berlanga, P.; Canete, A.; Castel, V. Advances in emerging drugs for the treatment of neuroblastoma. *Expert Opin. Emerg. Drugs* **2017**, *22*, 63–75. [CrossRef] [PubMed]
12. Greengard, E.G. Molecularly Targeted Therapy for Neuroblastoma. *Children* **2018**, *5*, 142. [CrossRef] [PubMed]
13. Johnsen, J.I.; Dyberg, C.; Fransson, S.; Wickstrom, M. Molecular mechanisms and therapeutic targets in neuroblastoma. *Pharmacol. Res.* **2018**, *131*, 164–176. [CrossRef] [PubMed]
14. Zage, P.E. Novel Therapies for Relapsed and Refractory Neuroblastoma. *Children* **2018**, *5*, 148. [CrossRef] [PubMed]

15. Janoueix-Lerosey, I.; Lopez-Delisle, L.; Delattre, O.; Rohrer, H. The ALK receptor in sympathetic neuron development and neuroblastoma. *Cell Tissue Res.* **2018**, *372*, 325–337. [CrossRef] [PubMed]

16. Trigg, R.M.; Turner, S.D. ALK in Neuroblastoma: Biological and Therapeutic Implications. *Cancers* **2018**, *10*, 113. [CrossRef] [PubMed]

17. Maloney, M.A.; Kun, S.S.; Keens, T.G.; Perez, I.A. Congenital central hypoventilation syndrome: Diagnosis and management. *Expert Rev. Respir. Med.* **2018**, *12*, 283–292. [CrossRef] [PubMed]

18. Alexandrov, L.B.; Nik-Zainal, S.; Wedge, D.C.; Aparicio, S.A.; Behjati, S.; Biankin, A.V.; Bignell, G.R.; Bolli, N.; Borg, A.; Borresen-Dale, A.L.; et al. Signatures of mutational processes in human cancer. *Nature* **2013**, *500*, 415–421. [CrossRef] [PubMed]

19. Grobner, S.N.; Worst, B.C.; Weischenfeldt, J.; Buchhalter, I.; Kleinheinz, K.; Rudneva, V.A.; Johann, P.D.; Balasubramanian, G.P.; Segura-Wang, M.; Brabetz, S.; et al. The landscape of genomic alterations across childhood cancers. *Nature* **2018**, *555*, 321–327. [CrossRef] [PubMed]

20. Lawrence, M.S.; Stojanov, P.; Mermel, C.H.; Robinson, J.T.; Garraway, L.A.; Golub, T.R.; Meyerson, M.; Gabriel, S.B.; Lander, E.S.; Getz, G. Discovery and saturation analysis of cancer genes across 21 tumour types. *Nature* **2014**, *505*, 495–501. [CrossRef] [PubMed]

21. Ma, X.; Liu, Y.; Liu, Y.; Alexandrov, L.B.; Edmonson, M.N.; Gawad, C.; Zhou, X.; Li, Y.; Rusch, M.C.; Easton, J.; et al. Pan-cancer genome and transcriptome analyses of 1,699 paediatric leukaemias and solid tumours. *Nature* **2018**, *555*, 371–376. [CrossRef] [PubMed]

22. Barr, E.K.; Applebaum, M.A. Genetic Predisposition to Neuroblastoma. *Children* **2018**, *5*, 119. [CrossRef] [PubMed]

23. Ritenour, L.E.; Randall, M.P.; Bosse, K.R.; Diskin, S.J. Genetic susceptibility to neuroblastoma: Current knowledge and future directions. *Cell Tissue Res.* **2018**, *372*, 287–307. [CrossRef] [PubMed]

24. Tolbert, V.P.; Coggins, G.E.; Maris, J.M. Genetic susceptibility to neuroblastoma. *Curr. Opin. Genet. Dev.* **2017**, *42*, 81–90. [CrossRef] [PubMed]

25. Cho, S.S.L.; Han, J.; James, S.J.; Png, C.W.; Weerasooriya, M.; Alonso, S.; Zhang, Y. Dual-Specificity Phosphatase 12 Targets p38 MAP Kinase to Regulate Macrophage Response to Intracellular Bacterial Infection. *Front. Immunol.* **2017**, *8*, 1259. [CrossRef] [PubMed]

26. Nguyen le, B.; Diskin, S.J.; Capasso, M.; Wang, K.; Diamond, M.A.; Glessner, J.; Kim, C.; Attiyeh, E.F.; Mosse, Y.P.; Cole, K.; et al. Phenotype restricted genome-wide association study using a gene-centric approach identifies three low-risk neuroblastoma susceptibility Loci. *PLoS Genet.* **2011**, *7*, e1002026. [CrossRef] [PubMed]

27. Brodeur, G.M. Spontaneous regression of neuroblastoma. *Cell Tissue Res.* **2018**, *372*, 277–286. [CrossRef] [PubMed]

28. Mohlin, S.A.; Wigerup, C.; Pahlman, S. Neuroblastoma aggressiveness in relation to sympathetic neuronal differentiation stage. *Semin. Cancer Biol.* **2011**, *21*, 276–282. [CrossRef] [PubMed]

29. Tomolonis, J.A.; Agarwal, S.; Shohet, J.M. Neuroblastoma pathogenesis: Deregulation of embryonic neural crest development. *Cell Tissue Res.* **2018**, *372*, 245–262. [CrossRef] [PubMed]

30. Gherardi, S.; Valli, E.; Erriquez, D.; Perini, G. MYCN-mediated transcriptional repression in neuroblastoma: The other side of the coin. *Front. Oncol.* **2013**, *3*, 42. [CrossRef] [PubMed]

31. Huang, M.; Weiss, W.A. Neuroblastoma and MYCN. *Cold Spring Harb. Perspect. Med.* **2013**, *3*, a014415. [CrossRef] [PubMed]

32. Westermark, U.K.; Wilhelm, M.; Frenzel, A.; Henriksson, M.A. The MYCN oncogene and differentiation in neuroblastoma. *Semin. Cancer Biol.* **2011**, *21*, 256–266. [CrossRef] [PubMed]

33. Kramer, M.; Ribeiro, D.; Arsenian-Henriksson, M.; Deller, T.; Rohrer, H. Proliferation and Survival of Embryonic Sympathetic Neuroblasts by MYCN and Activated ALK Signaling. *J. Neurosci.* **2016**, *36*, 10425–10439. [CrossRef] [PubMed]

34. Reiff, T.; Huber, L.; Kramer, M.; Delattre, O.; Janoueix-Lerosey, I.; Rohrer, H. Midkine and Alk signaling in sympathetic neuron proliferation and neuroblastoma predisposition. *Development* **2011**, *138*, 4699–4708. [CrossRef] [PubMed]

35. Brodeur, G.M.; Minturn, J.E.; Ho, R.; Simpson, A.M.; Iyer, R.; Varela, C.R.; Light, J.E.; Kolla, V.; Evans, A.E. Trk receptor expression and inhibition in neuroblastomas. *Clin. Cancer Res.* **2009**, *15*, 3244–3250. [CrossRef] [PubMed]

36. Harel, L.; Costa, B.; Fainzilber, M. On the death Trk. *Dev. Neurobiol.* **2010**, *70*, 298–303. [CrossRef] [PubMed]

37. Thiele, C.J.; Li, Z.; McKee, A.E. On Trk—The TrkB signal transduction pathway is an increasingly important target in cancer biology. *Clin. Cancer Res.* **2009**, *15*, 5962–5967. [CrossRef] [PubMed]

38. Casey, M.J.; Stewart, R.A. Zebrafish as a model to study neuroblastoma development. *Cell Tissue Res.* **2018**, *372*, 223–232. [CrossRef] [PubMed]

39. Kiyonari, S.; Kadomatsu, K. Neuroblastoma models for insights into tumorigenesis and new therapies. *Expert Opin. Drug Discov.* **2015**, *10*, 53–62. [CrossRef] [PubMed]

40. Edsjo, A.; Holmquist, L.; Pahlman, S. Neuroblastoma as an experimental model for neuronal differentiation and hypoxia-induced tumor cell dedifferentiation. *Semin. Cancer Biol.* **2007**, *17*, 248–256. [CrossRef] [PubMed]

41. Palmer, R.H.; Vernersson, E.; Grabbe, C.; Hallberg, B. Anaplastic lymphoma kinase: Signalling in development and disease. *Biochem. J.* **2009**, *420*, 345–361. [CrossRef] [PubMed]

42. Megison, M.L.; Gillory, L.A.; Beierle, E.A. Cell survival signaling in neuroblastoma. *Anti-Cancer Agents Med. Chem.* **2013**, *13*, 563–575. [CrossRef]

43. Roskoski, R., Jr. Anaplastic lymphoma kinase (ALK): Structure, oncogenic activation, and pharmacological inhibition. *Pharmacol. Res.* **2013**, *68*, 68–94. [CrossRef] [PubMed]

44. Stafman, L.L.; Beierle, E.A. Cell Proliferation in Neuroblastoma. *Cancers* **2016**, *8*, 13. [CrossRef] [PubMed]

45. Eleveld, T.F.; Oldridge, D.A.; Bernard, V.; Koster, J.; Daage, L.C.; Diskin, S.J.; Schild, L.; Bentahar, N.B.; Bellini, A.; Chicard, M.; et al. Relapsed neuroblastomas show frequent RAS-MAPK pathway mutations. *Nat. Genet.* **2015**, *47*, 864–871. [CrossRef] [PubMed]

46. Padovan-Merhar, O.M.; Raman, P.; Ostrovnaya, I.; Kalletla, K.; Rubnitz, K.R.; Sanford, E.M.; Ali, S.M.; Miller, V.A.; Mosse, Y.P.; Granger, M.P.; et al. Enrichment of Targetable Mutations in the Relapsed Neuroblastoma Genome. *PLoS Genet.* **2016**, *12*, e1006501. [CrossRef] [PubMed]

47. Pugh, T.J.; Morozova, O.; Attiyeh, E.F.; Asgharzadeh, S.; Wei, J.S.; Auclair, D.; Carter, S.L.; Cibulskis, K.; Hanna, M.; Kiezun, A.; et al. The genetic landscape of high-risk neuroblastoma. *Nat. Genet.* **2013**, *45*, 279–284. [CrossRef] [PubMed]

48. Eppstein, A.C.; Sandoval, J.A.; Klein, P.J.; Woodruff, H.A.; Grosfeld, J.L.; Hickey, R.J.; Malkas, L.H.; Schmidt, C.M. Differential sensitivity of chemoresistant neuroblastoma subtypes to MAPK-targeted treatment correlates with ERK, p53 expression, and signaling response to U0126. *J. Pediatr. Surg.* **2006**, *41*, 252–259. [CrossRef] [PubMed]

49. Hart, L.S.; Rader, J.; Raman, P.; Batra, V.; Russell, M.R.; Tsang, M.; Gagliardi, M.; Chen, L.; Martinez, D.; Li, Y.; et al. Preclinical Therapeutic Synergy of MEK1/2 and CDK4/6 Inhibition in Neuroblastoma. *Clin. Cancer Res.* **2017**, *23*, 1785–1796. [CrossRef] [PubMed]

50. Tanaka, T.; Higashi, M.; Kimura, K.; Wakao, J.; Fumino, S.; Iehara, T.; Hosoi, H.; Sakai, T.; Tajiri, T. MEK inhibitors as a novel therapy for neuroblastoma: Their in vitro effects and predicting their efficacy. *J. Pediatr. Surg.* **2016**, *51*, 2074–2079. [CrossRef] [PubMed]

51. Woodfield, S.E.; Zhang, L.; Scorsone, K.A.; Liu, Y.; Zage, P.E. Binimetinib inhibits MEK and is effective against neuroblastoma tumor cells with low NF1 expression. *BMC Cancer* **2016**, *16*, 172. [CrossRef] [PubMed]

52. Singh, A.; Ruan, Y.; Tippett, T.; Narendran, A. Targeted inhibition of MEK1 by cobimetinib leads to differentiation and apoptosis in neuroblastoma cells. *J. Exp. Clin. Cancer Res. CR* **2015**, *34*, 104. [CrossRef] [PubMed]

53. Umapathy, G.; El Wakil, A.; Witek, B.; Chesler, L.; Danielson, L.; Deng, X.; Gray, N.S.; Johansson, M.; Kvarnbrink, S.; Ruuth, K.; et al. The kinase ALK stimulates the kinase ERK5 to promote the expression of the oncogene MYCN in neuroblastoma. *Sci. Signal.* **2014**, *7*, ra102. [CrossRef] [PubMed]

54. Umapathy, G.; Guan, J.; Gustafsson, D.E.; Javanmardi, N.; Cervantes-Madrid, D.; Djos, A.; Martinsson, T.; Palmer, R.H.; Hallberg, B. MEK inhibitor trametinib does not prevent the growth of anaplastic lymphoma kinase (ALK)-addicted neuroblastomas. *Sci. Signal.* **2017**, *10*, eaam755. [CrossRef] [PubMed]

55. Chen, K.; Lv, F.; Xu, G.; Zhang, M.; Wu, Y.; Wu, Z. Phosphoproteomics reveals ALK promote cell progress via RAS/ JNK pathway in neuroblastoma. *Oncotarget* **2016**, *7*, 75968–75980. [CrossRef] [PubMed]

56. Cheng, J.; Fan, Y.H.; Xu, X.; Zhang, H.; Dou, J.; Tang, Y.; Zhong, X.; Rojas, Y.; Yu, Y.; Zhao, Y.; et al. A small-molecule inhibitor of UBE2N induces neuroblastoma cell death via activation of p53 and JNK pathways. *Cell Death Dis.* **2014**, *5*, e1079. [CrossRef] [PubMed]

57. Dedoni, S.; Olianas, M.C.; Onali, P. Interferon-beta counter-regulates its own pro-apoptotic action by activating p38 MAPK signalling in human SH-SY5Y neuroblastoma cells. *Apoptosis* **2014**, *19*, 1509–1526. [CrossRef] [PubMed]

58. Marengo, B.; De Ciucis, C.G.; Ricciarelli, R.; Furfaro, A.L.; Colla, R.; Canepa, E.; Traverso, N.; Marinari, U.M.; Pronzato, M.A.; Domenicotti, C. p38MAPK inhibition: A new combined approach to reduce neuroblastoma resistance under etoposide treatment. *Cell Death Dis.* **2013**, *4*, e589. [CrossRef] [PubMed]

59. Zhuang, J.; Li, Y.; Chi, Y. Role of p38 MAPK activation and mitochondrial cytochrome-c release in allicin-induced apoptosis in SK-N-SH cells. *Anti-Cancer Drugs* **2016**, *27*, 312–317. [CrossRef] [PubMed]

60. Kim, G.S.; Choi, Y.K.; Song, S.S.; Kim, W.K.; Han, B.H. MKP-1 contributes to oxidative stress-induced apoptosis via inactivation of ERK1/2 in SH-SY5Y cells. *Biochem. Biophys. Res. Commun.* **2005**, *338*, 1732–1738. [CrossRef] [PubMed]

61. Dayem, A.A.; Kim, B.; Gurunathan, S.; Choi, H.Y.; Yang, G.; Saha, S.K.; Han, D.; Han, J.; Kim, K.; Kim, J.H.; et al. Biologically synthesized silver nanoparticles induce neuronal differentiation of SH-SY5Y cells via modulation of reactive oxygen species, phosphatases, and kinase signaling pathways. *Biotechnol. J.* **2014**, *9*, 934–943. [CrossRef] [PubMed]

62. Rossler, O.G.; Henss, I.; Thiel, G. Transcriptional response to muscarinic acetylcholine receptor stimulation: Regulation of Egr-1 biosynthesis by ERK, Elk-1, MKP-1, and calcineurin in carbachol-stimulated human neuroblastoma cells. *Arch. Biochem. Biophys.* **2008**, *470*, 93–102. [CrossRef] [PubMed]

63. Vangipuram, S.D.; Wang, Z.J.; Lyman, W.D. Resistance of stem-like cells from neuroblastoma cell lines to commonly used chemotherapeutic agents. *Pediatr. Blood Cancer* **2010**, *54*, 361–368. [CrossRef] [PubMed]

64. Reffas, S.; Schlegel, W. Compartment-specific regulation of extracellular signal-regulated kinase (ERK) and c-Jun N-terminal kinase (JNK) mitogen-activated protein kinases (MAPKs) by ERK-dependent and non-ERK-dependent inductions of MAPK phosphatase (MKP)-3 and MKP-1 in differentiating P19 cells. *Biochem. J.* **2000**, *352 Pt 3*, 701–708.

65. Ryser, S.; Tortola, S.; Schlegel, W. Map kinase phosphatase-1 gene expression and regulation in neuroendocrine cells. *J. Recept. Signal Transduct. Res.* **2002**, *22*, 17–29. [CrossRef] [PubMed]

66. Ryser, S.; Tortola, S.; van Haasteren, G.; Muda, M.; Li, S.; Schlegel, W. MAP kinase phosphatase-1 gene transcription in rat neuroendocrine cells is modulated by a calcium-sensitive block to elongation in the first exon. *J. Biol. Chem.* **2001**, *276*, 33319–33327. [CrossRef] [PubMed]

67. Aurtenetxe, O.; Zaldumbide, L.; Erramuzpe, A.; Lopez, R.; Lopez, J.I.; Cortes, J.M.; Pulido, R.; Nunes-Xavier, C.E. DUSP5 expression associates with poor prognosis in human neuroblastoma. *Exp. Mol. Pathol.* **2018**, *105*, 272–278. [CrossRef] [PubMed]

68. Koga, S.; Kojima, S.; Kishimoto, T.; Kuwabara, S.; Yamaguchi, A. Over-expression of map kinase phosphatase-1 (MKP-1) suppresses neuronal death through regulating JNK signaling in hypoxia/re-oxygenation. *Brain Res.* **2012**, *1436*, 137–146. [CrossRef] [PubMed]

69. Kim, S.Y.; Han, Y.M.; Oh, M.; Kim, W.K.; Oh, K.J.; Lee, S.C.; Bae, K.H.; Han, B.S. DUSP4 regulates neuronal differentiation and calcium homeostasis by modulating ERK1/2 phosphorylation. *Stem Cells Dev.* **2015**, *24*, 686–700. [CrossRef] [PubMed]

70. Lambertz, I.; Kumps, C.; Claeys, S.; Lindner, S.; Beckers, A.; Janssens, E.; Carter, D.R.; Cazes, A.; Cheung, B.B.; De Mariano, M.; et al. Upregulation of MAPK Negative Feedback Regulators and RET in Mutant ALK Neuroblastoma: Implications for Targeted Treatment. *Clin. Cancer Res.* **2015**, *21*, 3327–3339. [CrossRef] [PubMed]

71. Staege, M.S.; Muller, K.; Kewitz, S.; Volkmer, I.; Mauz-Korholz, C.; Bernig, T.; Korholz, D. Expression of dual-specificity phosphatase 5 pseudogene 1 (DUSP5P1) in tumor cells. *PLoS ONE* **2014**, *9*, e89577. [CrossRef] [PubMed]

72. Mourey, R.J.; Vega, Q.C.; Campbell, J.S.; Wenderoth, M.P.; Hauschka, S.D.; Krebs, E.G.; Dixon, J.E. A novel cytoplasmic dual specificity protein tyrosine phosphatase implicated in muscle and neuronal differentiation. *J. Biol. Chem.* **1996**, *271*, 3795–3802. [CrossRef] [PubMed]

73. Muda, M.; Boschert, U.; Dickinson, R.; Martinou, J.C.; Martinou, I.; Camps, M.; Schlegel, W.; Arkinstall, S. MKP-3, a novel cytosolic protein-tyrosine phosphatase that exemplifies a new class of mitogen-activated protein kinase phosphatase. *J. Biol. Chem.* **1996**, *271*, 4319–4326. [CrossRef] [PubMed]

74. Vician, L.; Basconcillo, R.; Herschman, H.R. Identification of genes preferentially induced by nerve growth factor versus epidermal growth factor in PC12 pheochromocytoma cells by means of representational difference analysis. *J. Neurosci. Res.* **1997**, *50*, 32–43. [CrossRef]

75. Liu, T.; Bohlken, A.; Kuljaca, S.; Lee, M.; Nguyen, T.; Smith, S.; Cheung, B.; Norris, M.D.; Haber, M.; Holloway, A.J.; et al. The retinoid anticancer signal: Mechanisms of target gene regulation. *Br. J. Cancer* **2005**, *93*, 310–318. [CrossRef] [PubMed]

76. Mendell, A.L.; MacLusky, N.J. The testosterone metabolite 3alpha-androstanediol inhibits oxidative stress-induced ERK phosphorylation and neurotoxicity in SH-SY5Y cells through an MKP3/DUSP6-dependent mechanism. *Neurosci. Lett.* **2018**, *696*, 60–66. [CrossRef] [PubMed]

77. Marshall, G.M.; Liu, P.Y.; Gherardi, S.; Scarlett, C.J.; Bedalov, A.; Xu, N.; Iraci, N.; Valli, E.; Ling, D.; Thomas, W.; et al. SIRT1 promotes N-Myc oncogenesis through a positive feedback loop involving the effects of MKP3 and ERK on N-Myc protein stability. *PLoS Genet.* **2011**, *7*, e1002135. [CrossRef] [PubMed]

78. Kim, S.Y.; Oh, M.; Lee, K.S.; Kim, W.K.; Oh, K.J.; Lee, S.C.; Bae, K.H.; Han, B.S. Profiling analysis of protein tyrosine phosphatases during neuronal differentiation. *Neurosci. Lett.* **2016**, *612*, 219–224. [CrossRef] [PubMed]

79. Zega, K.; Jovanovic, V.M.; Vitic, Z.; Niedzielska, M.; Knaapi, L.; Jukic, M.M.; Partanen, J.; Friedel, R.H.; Lang, R.; Brodski, C. Dusp16 Deficiency Causes Congenital Obstructive Hydrocephalus and Brain Overgrowth by Expansion of the Neural Progenitor Pool. *Front. Mol. Neurosci.* **2017**, *10*, 372. [CrossRef] [PubMed]

80. Park, J.E.; Park, B.C.; Kim, H.A.; Song, M.; Park, S.G.; Lee, D.H.; Kim, H.J.; Choi, H.K.; Kim, J.T.; Cho, S. Positive regulation of apoptosis signal-regulating kinase 1 by dual-specificity phosphatase 13A. *Cell. Mol. Life Sci. CMLS* **2010**, *67*, 2619–2629. [CrossRef] [PubMed]

81. Caren, H.; Djos, A.; Nethander, M.; Sjoberg, R.M.; Kogner, P.; Enstrom, C.; Nilsson, S.; Martinsson, T. Identification of epigenetically regulated genes that predict patient outcome in neuroblastoma. *BMC Cancer* **2011**, *11*, 66. [CrossRef] [PubMed]

82. Wang, J.Y.; Lin, C.H.; Yang, C.H.; Tan, T.H.; Chen, Y.R. Biochemical and biological characterization of a neuroendocrine-associated phosphatase. *J. Neurochem.* **2006**, *98*, 89–101. [CrossRef] [PubMed]

83. Wang, J.Y.; Yang, C.H.; Yeh, C.L.; Lin, C.H.; Chen, Y.R. NEAP causes down-regulation of EGFR, subsequently induces the suppression of NGF-induced differentiation in PC12 cells. *J. Neurochem.* **2008**, *107*, 1544–1555. [CrossRef] [PubMed]

84. Patterson, K.I.; Brummer, T.; Daly, R.J.; O'Brien, P.M. DUSP26 negatively affects the proliferation of epithelial cells, an effect not mediated by dephosphorylation of MAPKs. *Biochim. Biophys. Acta* **2010**, *1803*, 1003–1012. [CrossRef] [PubMed]

85. Shang, X.; Vasudevan, S.A.; Yu, Y.; Ge, N.; Ludwig, A.D.; Wesson, C.L.; Wang, K.; Burlingame, S.M.; Zhao, Y.J.; Rao, P.H.; et al. Dual-specificity phosphatase 26 is a novel p53 phosphatase and inhibits p53 tumor suppressor functions in human neuroblastoma. *Oncogene* **2010**, *29*, 4938–4946. [CrossRef] [PubMed]

86. Shi, Y.; Ma, I.T.; Patel, R.H.; Shang, X.; Chen, Z.; Zhao, Y.; Cheng, J.; Fan, Y.; Rojas, Y.; Barbieri, E.; et al. NSC-87877 inhibits DUSP26 function in neuroblastoma resulting in p53-mediated apoptosis. *Cell Death Dis.* **2015**, *6*, e1841. [CrossRef] [PubMed]

87. Dedoni, S.; Olianas, M.C.; Onali, P. Interferon-beta induces apoptosis in human SH-SY5Y neuroblastoma cells through activation of JAK-STAT signaling and down-regulation of PI3K/Akt pathway. *J. Neurochem.* **2010**, *115*, 1421–1433. [CrossRef] [PubMed]

88. King, D.; Yeomanson, D.; Bryant, H.E. PI3King the lock: Targeting the PI3K/Akt/mTOR pathway as a novel therapeutic strategy in neuroblastoma. *J. Pediatr. Hematol./Oncol.* **2015**, *37*, 245–251. [CrossRef] [PubMed]

89. Sattu, K.; Hochgrafe, F.; Wu, J.; Umapathy, G.; Schonherr, C.; Ruuth, K.; Chand, D.; Witek, B.; Fuchs, J.; Li, P.K.; et al. Phosphoproteomic analysis of anaplastic lymphoma kinase (ALK) downstream signaling pathways identifies signal transducer and activator of transcription 3 as a functional target of activated ALK in neuroblastoma cells. *FEBS J.* **2013**, *280*, 5269–5282. [CrossRef] [PubMed]

90. Yan, S.; Li, Z.; Thiele, C.J. Inhibition of STAT3 with orally active JAK inhibitor, AZD1480, decreases tumor growth in Neuroblastoma and Pediatric Sarcomas In vitro and In vivo. *Oncotarget* **2013**, *4*, 433–445. [CrossRef] [PubMed]

91. Hasan, M.K.; Nafady, A.; Takatori, A.; Kishida, S.; Ohira, M.; Suenaga, Y.; Hossain, S.; Akter, J.; Ogura, A.; Nakamura, Y.; et al. ALK is a MYCN target gene and regulates cell migration and invasion in neuroblastoma. *Sci. Rep.* **2013**, *3*, 3450. [CrossRef] [PubMed]

92. Schonherr, C.; Ruuth, K.; Kamaraj, S.; Wang, C.L.; Yang, H.L.; Combaret, V.; Djos, A.; Martinsson, T.; Christensen, J.G.; Palmer, R.H.; et al. Anaplastic Lymphoma Kinase (ALK) regulates initiation of transcription of MYCN in neuroblastoma cells. *Oncogene* **2012**, *31*, 5193–5200. [CrossRef] [PubMed]

93. Alonso, A.; Pulido, R. The extended human PTPome: A growing tyrosine phosphatase family. *FEBS J.* **2016**, *283*, 1404–1429. [CrossRef] [PubMed]

94. Pulido, R.; Hooft van Huijsduijnen, R. Protein tyrosine phosphatases: Dual-specificity phosphatases in health and disease. *FEBS J.* **2008**, *275*, 848–866. [CrossRef] [PubMed]

95. Pulido, R.; Stoker, A.W.; Hendriks, W.J. PTPs emerge as PIPs: Protein tyrosine phosphatases with lipid-phosphatase activities in human disease. *Hum. Mol. Genet.* **2013**, *22*, R66–R76. [CrossRef] [PubMed]

96. Rios, P.; Nunes-Xavier, C.E.; Tabernero, L.; Kohn, M.; Pulido, R. Dual-specificity phosphatases as molecular targets for inhibition in human disease. *Antioxid. Redox Signal.* **2014**, *20*, 2251–2273. [CrossRef] [PubMed]

97. Huang, C.Y.; Tan, T.H. DUSPs, to MAP kinases and beyond. *Cell Biosci.* **2012**, *2*, 24. [CrossRef] [PubMed]

98. Nunes-Xavier, C.; Roma-Mateo, C.; Rios, P.; Tarrega, C.; Cejudo-Marin, R.; Tabernero, L.; Pulido, R. Dual-specificity MAP kinase phosphatases as targets of cancer treatment. *Anti-Cancer Agents Med. Chem.* **2011**, *11*, 109–132. [CrossRef]

99. Caunt, C.J.; Keyse, S.M. Dual-specificity MAP kinase phosphatases (MKPs): Shaping the outcome of MAP kinase signalling. *FEBS J.* **2013**, *280*, 489–504. [CrossRef] [PubMed]

100. Kim, E.K.; Choi, E.J. Pathological roles of MAPK signaling pathways in human diseases. *Biochim. Biophys. Acta* **2010**, *1802*, 396–405. [CrossRef] [PubMed]

101. Owens, D.M.; Keyse, S.M. Differential regulation of MAP kinase signalling by dual-specificity protein phosphatases. *Oncogene* **2007**, *26*, 3203–3213. [CrossRef] [PubMed]

102. Plotnikov, A.; Zehorai, E.; Procaccia, S.; Seger, R. The MAPK cascades: Signaling components, nuclear roles and mechanisms of nuclear translocation. *Biochim. Biophys. Acta* **2011**, *1813*, 1619–1633. [CrossRef] [PubMed]

103. Raman, M.; Chen, W.; Cobb, M.H. Differential regulation and properties of MAPKs. *Oncogene* **2007**, *26*, 3100–3112. [CrossRef] [PubMed]

104. Cejudo-Marín, R.; Tárrega, C.; Nunes-Xavier, C.E.; Pulido, R. Caspase-3 cleavage of DUSP6/MKP3 at the interdomain region generates active MKP3 fragments that regulate ERK1/2 subcellular localization and function. *J. Mol. Biol.* **2012**, *420*, 128–138. [CrossRef] [PubMed]

105. Kidger, A.M.; Rushworth, L.K.; Stellzig, J.; Davidson, J.; Bryant, C.J.; Bayley, C.; Caddye, E.; Rogers, T.; Keyse, S.M.; Caunt, C.J. Dual-specificity phosphatase 5 controls the localized inhibition, propagation, and transforming potential of ERK signaling. *Proc. Natl. Acad. Sci. USA* **2017**, *114*, E317–E326. [CrossRef] [PubMed]

106. Tarrega, C.; Rios, P.; Cejudo-Marin, R.; Blanco-Aparicio, C.; van den Berk, L.; Schepens, J.; Hendriks, W.; Tabernero, L.; Pulido, R. ERK2 shows a restrictive and locally selective mechanism of recognition by its tyrosine phosphatase inactivators not shared by its activator MEK1. *J. Biol. Chem.* **2005**, *280*, 37885–37894. [CrossRef] [PubMed]

107. Kidger, A.M.; Keyse, S.M. The regulation of oncogenic Ras/ERK signalling by dual-specificity mitogen activated protein kinase phosphatases (MKPs). *Semin. Cell Dev. Biol.* **2016**, *50*, 125–132. [CrossRef] [PubMed]

108. Low, H.B.; Zhang, Y. Regulatory Roles of MAPK Phosphatases in Cancer. *Immune Netw.* **2016**, *16*, 85–98. [CrossRef] [PubMed]

109. Seternes, O.M.; Kidger, A.M.; Keyse, S.M. Dual-specificity MAP kinase phosphatases in health and disease. *Biochim. Biophys. Acta. Mol. Cell Res.* **2019**, *1866*, 124–143. [CrossRef] [PubMed]

110. Pavic, K.; Duan, G.; Kohn, M. VHR/DUSP3 phosphatase: Structure, function and regulation. *FEBS J.* **2015**, *282*, 1871–1890. [CrossRef] [PubMed]

111. Alonso, A.; Nunes-Xavier, C.E.; Bayon, Y.; Pulido, R. The Extended Family of Protein Tyrosine Phosphatases. *Methods Mol. Biol.* **2016**, *1447*, 1–23. [CrossRef] [PubMed]

112. Patterson, K.I.; Brummer, T.; O'Brien, P.M.; Daly, R.J. Dual-specificity phosphatases: Critical regulators with diverse cellular targets. *Biochem. J.* **2009**, *418*, 475–489. [CrossRef] [PubMed]

113. Dickinson, R.J.; Keyse, S.M. Diverse physiological functions for dual-specificity MAP kinase phosphatases. *J. Cell Sci.* **2006**, *119*, 4607–4615. [CrossRef] [PubMed]

114. Kutty, R.G.; Talipov, M.R.; Bongard, R.D.; Lipinski, R.A.J.; Sweeney, N.L.; Sem, D.S.; Rathore, R.; Ramchandran, R. Dual Specificity Phosphatase 5-Substrate Interaction: A Mechanistic Perspective. *Compr. Physiol.* **2017**, *7*, 1449–1461. [CrossRef] [PubMed]

115. Lawan, A.; Shi, H.; Gatzke, F.; Bennett, A.M. Diversity and specificity of the mitogen-activated protein kinase phosphatase-1 functions. *Cell. Mol. Life Sci. CMLS* **2013**, *70*, 223–237. [CrossRef] [PubMed]

116. Lawan, A.; Torrance, E.; Al-Harthi, S.; Shweash, M.; Alnasser, S.; Neamatallah, T.; Schroeder, J.; Plevin, R. MKP-2: Out of the DUSP-bin and back into the limelight. *Biochem. Soc. Trans.* **2012**, *40*, 235–239. [CrossRef] [PubMed]

117. Collins, L.M.; Downer, E.J.; Toulouse, A.; Nolan, Y.M. Mitogen-Activated Protein Kinase Phosphatase (MKP)-1 in Nervous System Development and Disease. *Mol. Neurobiol.* **2015**, *51*, 1158–1167. [CrossRef] [PubMed]

118. Li, C.Y.; Yang, L.C.; Guo, K.; Wang, Y.P.; Li, Y.G. Mitogen-activated protein kinase phosphatase-1: A critical phosphatase manipulating mitogen-activated protein kinase signaling in cardiovascular disease (review). *Int. J. Mol. Med.* **2015**, *35*, 1095–1102. [CrossRef] [PubMed]

119. Moosavi, S.M.; Prabhala, P.; Ammit, A.J. Role and regulation of MKP-1 in airway inflammation. *Respir. Res.* **2017**, *18*, 154. [CrossRef] [PubMed]

120. Ralph, J.A.; Morand, E.F. MAPK phosphatases as novel targets for rheumatoid arthritis. *Expert Opin. Ther. Targets* **2008**, *12*, 795–808. [CrossRef] [PubMed]

121. Shen, J.; Zhang, Y.; Yu, H.; Shen, B.; Liang, Y.; Jin, R.; Liu, X.; Shi, L.; Cai, X. Role of DUSP1/MKP1 in tumorigenesis, tumor progression and therapy. *Cancer Med.* **2016**, *5*, 2061–2068. [CrossRef] [PubMed]

122. Wancket, L.M.; Frazier, W.J.; Liu, Y. Mitogen-activated protein kinase phosphatase (MKP)-1 in immunology, physiology, and disease. *Life Sci.* **2012**, *90*, 237–248. [CrossRef] [PubMed]

123. Abdal Dayem, A.; Lee, S.B.; Choi, H.Y.; Cho, S.G. Silver Nanoparticles: Two-Faced Neuronal Differentiation-Inducing Material in Neuroblastoma (SH-SY5Y) Cells. *Int. J. Mol. Sci.* **2018**, *19*, 1470. [CrossRef] [PubMed]

124. Chen, M.; Zhang, J.; Berger, A.H.; Diolombi, M.S.; Ng, C.; Fung, J.; Bronson, R.T.; Castillo-Martin, M.; Thin, T.H.; Cordon-Cardo, C.; et al. Compound haploinsufficiency of Dok2 and Dusp4 promotes lung tumorigenesis. *J. Clin. Investig.* **2018**, *129*, 215–222. [CrossRef] [PubMed]

125. Hijiya, N.; Tsukamoto, Y.; Nakada, C.; Tung Nguyen, L.; Kai, T.; Matsuura, K.; Shibata, K.; Inomata, M.; Uchida, T.; Tokunaga, A.; et al. Genomic Loss of DUSP4 Contributes to the Progression of Intraepithelial Neoplasm of Pancreas to Invasive Carcinoma. *Cancer Res.* **2016**, *76*, 2612–2625. [CrossRef] [PubMed]

126. Ichimanda, M.; Hijiya, N.; Tsukamoto, Y.; Uchida, T.; Nakada, C.; Akagi, T.; Etoh, T.; Iha, H.; Inomata, M.; Takekawa, M.; et al. Downregulation of dual-specificity phosphatase 4 enhances cell proliferation and invasiveness in colorectal carcinomas. *Cancer Sci.* **2018**, *109*, 250–258. [CrossRef] [PubMed]

127. Kang, X.; Li, M.; Zhu, H.; Lu, X.; Miao, J.; Du, S.; Xia, X.; Guan, W. DUSP4 promotes doxorubicin resistance in gastric cancer through epithelial-mesenchymal transition. *Oncotarget* **2017**, *8*, 94028–94039. [CrossRef] [PubMed]

128. Menyhart, O.; Budczies, J.; Munkacsy, G.; Esteva, F.J.; Szabo, A.; Miquel, T.P.; Gyorffy, B. DUSP4 is associated with increased resistance against anti-HER2 therapy in breast cancer. *Oncotarget* **2017**, *8*, 77207–77218. [CrossRef] [PubMed]

129. Buffet, C.; Hecale-Perlemoine, K.; Bricaire, L.; Dumont, F.; Baudry, C.; Tissier, F.; Bertherat, J.; Cochand-Priollet, B.; Raffin-Sanson, M.L.; Cormier, F.; et al. DUSP5 and DUSP6, two ERK specific phosphatases, are markers of a higher MAPK signaling activation in BRAF mutated thyroid cancers. *PLoS ONE* **2017**, *12*, e0184861. [CrossRef] [PubMed]

130. Higa, T.; Takahashi, H.; Higa-Nakamine, S.; Suzuki, M.; Yamamoto, H. Up-regulation of DUSP5 and DUSP6 by gonadotropin-releasing hormone in cultured hypothalamic neurons, GT1-7 cells. *Biomed. Res.* **2018**, *39*, 149–158. [CrossRef] [PubMed]

131. Kucharska, A.; Rushworth, L.K.; Staples, C.; Morrice, N.A.; Keyse, S.M. Regulation of the inducible nuclear dual-specificity phosphatase DUSP5 by ERK MAPK. *Cell. Signal.* **2009**, *21*, 1794–1805. [CrossRef] [PubMed]

132. Nunes-Xavier, C.E.; Tárrega, C.; Cejudo-Marín, R.; Frijhoff, J.; Sandin, A.; Ostman, A.; Pulido, R. Differential up-regulation of MAP kinase phosphatases MKP3/DUSP6 and DUSP5 by Ets2 and c-Jun converge in the control of the growth arrest versus proliferation response of MCF-7 breast cancer cells to phorbol ester. *J. Biol. Chem.* **2010**, *285*, 26417–26430. [CrossRef] [PubMed]

133. Zhou, L.Y.; Yin, J.Y.; Tang, Q.; Zhai, L.L.; Zhang, T.J.; Wang, Y.X.; Yang, D.Q.; Qian, J.; Lin, J.; Deng, Z.Q. High expression of dual-specificity phosphatase 5 pseudogene 1 (DUSP5P1) is associated with poor prognosis in acute myeloid leukemia. *Int. J. Clin. Exp. Pathol.* **2015**, *8*, 16073–16080. [PubMed]

134. Ekerot, M.; Stavridis, M.P.; Delavaine, L.; Mitchell, M.P.; Staples, C.; Owens, D.M.; Keenan, I.D.; Dickinson, R.J.; Storey, K.G.; Keyse, S.M. Negative-feedback regulation of FGF signalling by DUSP6/MKP-3 is driven by ERK1/2 and mediated by Ets factor binding to a conserved site within the DUSP6/MKP-3 gene promoter. *Biochem. J.* **2008**, *412*, 287–298. [CrossRef] [PubMed]

135. Li, C.; Scott, D.A.; Hatch, E.; Tian, X.; Mansour, S.L. Dusp6 (Mkp3) is a negative feedback regulator of FGF-stimulated ERK signaling during mouse development. *Development* **2007**, *134*, 167–176. [CrossRef] [PubMed]

136. Ahmad, M.K.; Abdollah, N.A.; Shafie, N.H.; Yusof, N.M.; Razak, S.R.A. Dual-specificity phosphatase 6 (DUSP6): A review of its molecular characteristics and clinical relevance in cancer. *Cancer Biol. Med.* **2018**, *15*, 14–28. [CrossRef] [PubMed]

137. Monteiro, L.F.; Ferruzo, P.Y.M.; Russo, L.C.; Farias, J.O.; Forti, F.L. DUSP3/VHR: A Druggable Dual Phosphatase for Human Diseases. *Rev. Physiol. Biochem. Pharmacol.* **2018**. [CrossRef]

138. Chen, H.H.; Luche, R.; Wei, B.; Tonks, N.K. Characterization of two distinct dual specificity phosphatases encoded in alternative open reading frames of a single gene located on human chromosome 10q22.2. *J. Biol. Chem.* **2004**, *279*, 41404–41413. [CrossRef] [PubMed]

139. Takagaki, K.; Satoh, T.; Tanuma, N.; Masuda, K.; Takekawa, M.; Shima, H.; Kikuchi, K. Characterization of a novel low-molecular-mass dual-specificity phosphatase-3 (LDP-3) that enhances activation of JNK and p38. *Biochem. J.* **2004**, *383*, 447–455. [CrossRef] [PubMed]

140. Wu, Q.; Li, Y.; Gu, S.; Li, N.; Zheng, D.; Li, D.; Zheng, Z.; Ji, C.; Xie, Y.; Mao, Y. Molecular cloning and characterization of a novel dual-specificity phosphatase 23 gene from human fetal brain. *Int. J. Biochem. Cell Biol.* **2004**, *36*, 1542–1553. [CrossRef] [PubMed]

141. Gallegos, L.L.; Ng, M.R.; Sowa, M.E.; Selfors, L.M.; White, A.; Zervantonakis, I.K.; Singh, P.; Dhakal, S.; Harper, J.W.; Brugge, J.S. A protein interaction map for cell-cell adhesion regulators identifies DUSP23 as a novel phosphatase for beta-catenin. *Sci. Rep.* **2016**, *6*, 27114. [CrossRef] [PubMed]

142. Lin, F.Y.; Chang, C.W.; Cheong, M.L.; Chen, H.C.; Lee, D.Y.; Chang, G.D.; Chen, H. Dual-specificity phosphatase 23 mediates GCM1 dephosphorylation and activation. *Nucleic Acids Res.* **2011**, *39*, 848–861. [CrossRef] [PubMed]

143. Vasudevan, S.A.; Skoko, J.; Wang, K.; Burlingame, S.M.; Patel, P.N.; Lazo, J.S.; Nuchtern, J.G.; Yang, J. MKP-8, a novel MAPK phosphatase that inhibits p38 kinase. *Biochem. Biophys. Res. Commun.* **2005**, *330*, 511–518. [CrossRef] [PubMed]

144. Yang, C.H.; Yeh, Y.J.; Wang, J.Y.; Liu, Y.W.; Chen, Y.L.; Cheng, H.W.; Cheng, C.M.; Chuang, Y.J.; Yuh, C.H.; Chen, Y.R. NEAP/DUSP26 suppresses receptor tyrosine kinases and regulates neuronal development in zebrafish. *Sci. Rep.* **2017**, *7*, 5241. [CrossRef] [PubMed]

145. Tanuma, N.; Nomura, M.; Ikeda, M.; Kasugai, I.; Tsubaki, Y.; Takagaki, K.; Kawamura, T.; Yamashita, Y.; Sato, I.; Sato, M.; et al. Protein phosphatase Dusp26 associates with KIF3 motor and promotes N-cadherin-mediated cell-cell adhesion. *Oncogene* **2009**, *28*, 752–761. [CrossRef] [PubMed]

146. Lazo, J.S.; McQueeney, K.E.; Burnett, J.C.; Wipf, P.; Sharlow, E.R. Small molecule targeting of PTPs in cancer. *Int. J. Biochem. Cell Biol.* **2018**, *96*, 171–181. [CrossRef] [PubMed]

International Journal of
Molecular Sciences

MDPI

Review

Dual-Specificity Phosphatase Regulation in Neurons and Glial Cells

Raquel Pérez-Sen *, María José Queipo, Juan Carlos Gil-Redondo, Felipe Ortega, Rosa Gómez-Villafuertes⬦, María Teresa Miras-Portugal and Esmerilda G. Delicado *

Departamento de Bioquímica y Biología Molecular, Facultad de Veterinaria, Instituto Universitario de Investigación en Neuroquímica (IUIN), Instituto de Investigación Sanitaria del Hospital Clínico San Carlos (IdiSSC), Universidad Complutense Madrid, 28040 Madrid, Spain; mqueipo@ucm.es (M.J.Q.); jugil@ucm.es (J.C.G.-R.); fortegao@ucm.es (F.O.); marosa@ucm.es (R.G.-V.); mtmiras@ucm.es (M.T.M.-P.)
* Correspondence: rpsen@ucm.es (R.P.-S.); esmerild@ucm.es (E.G.D.); Tel.: +34-91-394-3892 (R.P.-S. & E.G.D.)

Received: 28 March 2019; Accepted: 19 April 2019; Published: 23 April 2019

Abstract: Dual-specificity protein phosphatases comprise a protein phosphatase subfamily with selectivity towards mitogen-activated protein (MAP) kinases, also named MKPs, or mitogen-activated protein kinase (MAPK) phosphatases. As powerful regulators of the intensity and duration of MAPK signaling, a relevant role is envisioned for dual-specificity protein phosphatases (DUSPs) in the regulation of biological processes in the nervous system, such as differentiation, synaptic plasticity, and survival. Important neural mediators include nerve growth factor (NGF) and brain-derived neurotrophic factor (BDNF) that contribute to *DUSP* transcriptional induction and post-translational mechanisms of DUSP protein stabilization to maintain neuronal survival and differentiation. Potent *DUSP* gene inducers also include cannabinoids, which preserve DUSP activity in inflammatory conditions. Additionally, nucleotides activating P2X7 and P2Y13 nucleotide receptors behave as novel players in the regulation of DUSP function. They increase cell survival in stressful conditions, regulating DUSP protein turnover and inducing *DUSP* gene expression. In general terms, in the context of neural cells exposed to damaging conditions, the recovery of DUSP activity is neuroprotective and counteracts pro-apoptotic over-activation of p38 and JNK. In addition, remarkable changes in DUSP function take place during the onset of neuropathologies. The restoration of proper DUSP levels and recovery of MAPK homeostasis underlie the therapeutic effect, indicating that DUSPs can be relevant targets for brain diseases.

Keywords: dual-specificity phosphatases; MAP kinases; nucleotide receptors; P2X7; P2Y13; BDNF; cannabinoids; granule neurons; astrocytes

1. Introduction

Dual-specificity phosphatases (DUSPs) comprise a family of protein tyrosine phosphatases (PTPs) with wide substrate selectivity and are known as powerful regulators of biological processes. They are Cys-based phosphatases and exhibit dual activity towards Ser/Thr and Tyr residues in their catalytic substrates. Among the DUSP family, MKPs or mitogen-activated protein kinase (MAPK) phosphatases form a small group of 10–12 phosphatases that selectively dephosphorylate MAP kinases. Although MAPK signaling can be modulated at different levels, MKPs exert tight control, deactivating MAPKs with a 10–100 stronger potency than the upstream kinases. Therefore, they are emerging as the most potent regulators of the duration and magnitude of MAPK signaling and function [1–3]. While DUSPs encompass a greater family of phosphatases, this term will be used in the present review to refer to the MAPK-specific group.

Many of the physiological implications of DUSPs, including their substrate selectivity, modes of regulation, and function, have been elucidated through the genetic expression or deletion of a specific DUSP subtype in cell lines and heterologous expression systems. However, the way in which they are regulated by extracellular messengers, trophic factors, or toxic insults still remains poorly understood. The best-known extracellular mediators regulating DUSPs are trophic factors. Indeed, DUSPs form part of the feedback regulation of MAPK signaling during proliferative responses induced by mitogens. Initially, mitogens activate MAP kinases, which in turn promote *DUSP* expression and/or their stabilization to terminate MAPK actions. In this context, the participation of DUSPs in the regulation of proliferative processes is well-documented and DUSPs have become a major focus of cancer research. In agreement, significant changes of *DUSP* expression levels in tumor cells support their role as important markers of the stage, progression, and prognosis of certain types of cancer. Some DUSPs evolve with a gain of function and behave as tumor suppressors. On the other hand, the loss of *DUSP* expression is usually associated with sustained proliferative responses and chemo-resistance in other cancer types, as seen with *DUSP6* hypermethylation in pancreatic cancer [4–6].

Data from the knock-out mice models have also been important to identify the important role that DUSPs have in the regulation of immune responses and metabolic homeostasis. With regard to this, genetic knock-out models of DUSP1 and *DUSP10* present a pro-inflammatory phenotype, characterized by an increase in the production and release of pro-inflammatory cytokines. In contrast to this, DUSP2 seems to function as a positive regulator of inflammatory responses [7–9].

From the perspective of the nervous system, however, little is known about the function of DUSP proteins and how extracellular messengers can regulate them. In fact, most studies concerning intracellular cascades activated in neural cells by neurotransmitters or neurotrophic factors have classically focused on signaling kinases and have always eluded the deactivation mechanisms. The present study covers the different modes of regulation of DUSP proteins described in neuronal and glial cell populations, especially for DUSP1 and DUSP6, which appear to be the most studied DUSPs in neural cells. Special attention will be paid to the role of growth factors and neurotrophins, such as the brain-derived neurotrophic factor (BDNF) and nerve growth factor (NGF), which provide relevant examples of the complex regulation of *DUSP* expression through different intracellular mechanisms in neural cells. In addition to neurotrophins, endogenous cannabinoids, well-known messengers in the nervous system and key regulators of synaptic activity, also have the ability to modulate DUSP function at different levels with relevant importance in pathological contexts. It is interesting to point out that extracellular nucleotides, ATP, and analogues become novel players of DUSP function in neural models. They activate nucleotide receptors and share the intracellular coupling to MAPK mechanisms of activation and deactivation with trophic factors [10,11]. Furthermore, DUSP regulation by nucleotides is essential for cell maintenance and survival [12–15]. The physiological meaning of DUSP regulation in neural systems is further analyzed concerning processes of neuroprotection against different cytotoxic insults and in pathological situations, neurodegenerative diseases, neurological disorders, and brain tumors. From this point of view, DUSPs represent new targets for extracellular mediators whose modulation seems to be essential for the maintenance and homeostasis of the nervous system.

2. Overview of Regulatory Mechanisms of DUSP Activity

The DUSP family has been classified into three groups based on sequence homology, subcellular localization, and substrate specificity. All of them share a common structure consisting of an N-terminal regulatory domain and a C-terminal catalytic domain. Some types also contain specific motifs responsible for specific subcellular location [1]. The first group includes four inducible nuclear phosphatases. DUSP1/MKP-1 was the first MKP to be cloned and characterized and has been extensively studied in terms of regulation and function in different cellular systems. DUSP1 is a nuclear and inducible phosphatase whose specificity can vary from extracellular signal-regulated protein kinase (ERK) to the stress kinases, p38 and c-Jun N-terminal kinase (JNK), depending on the cellular model and the cellular context. Other members of this subgroup are DUSP2/PAC-1, DUSP4/MKP-2, and DUSP5. The second group includes three cytoplasmic ERK-specific DUSPs: DUSP6/MKP-3, DUSP7/MKP-X, and DUSP9/MKP-4. DUSP6 is the most representative ERK-directed phosphatase that can be slowly induced and primarily acts in the cytoplasmic compartment. The third group includes DUSP8/hVH-5, DUSP10/MKP-5, and DUSP16/MKP-7, which specifically dephosphorylate stress kinases p38 and JNK and can be either nuclear or cytoplasmic. Different DUSP subtypes are co-expressed in the same cell, but they do not exhibit redundant activities and usually function towards different MAPK substrates and in different subcellular compartments.

The regulation of MAPK signaling by DUSPs is far more complex than initially expected, taking into account the different levels at which these phosphatases can be modulated. Modulation by transcriptional, post-transcriptional, post-translational, and epigenetic mechanisms has been described and is covered in excellent reviews [1,3,6,16–18].

The expression levels and regulation of DUSPs are cell-type- and context-specific and differ depending on their constitutive or inducible nature. The inducible protein phosphatases, such as DUSP1, behave as immediate early genes and are rapidly induced in response to growth factors, cytokines, and stressful stimuli. Nevertheless, even constitutive phosphatases, such as DUSP6, can be transcriptionally induced with delayed kinetics by several extracellular mediators like mitogens and growth factors, which activate MAPK signaling. Therefore, both MAPKs and DUSPs act in a coordinated fashion through negative feedback loops to regulate the proper duration and magnitude of MAPK signaling. The best-known model is provided by DUSP6, whose transcriptional expression is strongly dependent on ERK activity, through the activation of the E twenty-six (Ets) family of transcription factors [19,20]. However, cross-talk regulation is also possible and DUSPs can dephosphorylate different MAPK substrates to those responsible for their expression. In fibroblasts, the activation of p38 following UV radiation contributes to inducing *DUSP1* transcription, which in turn functions to dephosphorylate and attenuate JNK apoptotic signaling [21,22]. A wide variety of transcription factor downstream MAPKs can activate *DUSP* transcription, the most well-characterized of which are AP-1, SP-1, NFKB, and β-catenin [17]. Concerning *DUSP1* gene induction, glucocorticoid response elements (GRE) are present in the *DUSP1* promoter and recruit specific cofactors to facilitate *DUSP1* transcription. *DUSP1* transcriptional regulation lies behind the anti-inflammatory effect of glucocorticoids, such as dexamethasone [23,24]. Additionally, epigenetic modifications also take place through chromatin remodeling to facilitate *DUSP* transcriptional induction. The enhanced acetylation and phosphorylation of histone H3 during the arrival of stressful stimuli precede *DUSP1* transcriptional activation in mouse fibroblasts [25].

In addition, different modes of regulation can converge with transcriptional induction to finely modulate *DUSP* expression levels, such as mRNA stabilization. During apoptotic signaling in HeLa cells, MAPK activation plays a double role by favoring the transcription of *DUSP1* and, at the same time, increasing the translation of several genes that help to stabilize *DUSP1* mRNA and increase its half-life, such as the RNA binding protein HuR. These mechanisms of *DUSP1* up-regulation help to counteract JNK apoptotic signaling [26,27]. *DUSP6* mRNA stabilization also forms part of the mechanism by which *DUSP6* expression levels increase during hypoxia. This regulation depends on mitogen extracellular-activated kinase (MEK)-ERK and hypoxia-inducible factor (HIF)-I activity. Two novel factors have been identified as regulating DUSP6 stability (Tristetraprolin (TTP) and PUM2 in HEK293 cells [4].

Besides transcriptional induction, several other modes of regulation take place upon MAPK substrate binding through the kinase interacting motif (KIM) domain. Conformational changes near the catalytic site are behind the increase in DUSP activity [6,28–30]. In addition, like other PTPs, DUSP proteins are sensitive to oxidative stress. The cysteine residues located in the catalytic cleft are susceptible to reversible oxidation by reactive oxygen species (ROS), which renders phosphatases catalytically inactive [31–33]. The oxidized sulfonic acid intermediate forms disulphide bonds with nearby cysteines to prevent the irreversible formation of higher oxidation states. With respect to DUSP6, multiple cysteine residues in a distal domain function to bridge the active site and prevent it from excessive oxidation and irreversible inactivation [33,34].

Regarding post-translational mechanisms, these mainly involve processes of DUSP phosphorylation triggered by MAPKs. The final effect can vary, from enhancing DUSP turnover to stabilizing the DUSP protein and increasing its half-life [17]. Several residues have been identified that, upon phosphorylation, target DUSP proteins for proteasomal degradation. The best-known example is DUSP1, which is phosphorylated in Ser-323 and Ser-329 residues in the catalytic domain by ERK to allow the interaction with specific ubiquitin ligase, resulting in DUSP degradation [35,36]. However, in adjacent residues at the carboxyl terminal sequence, ERK-mediated phosphorylation of Ser-359 and Ser-364 exerts the opposite effect, augmenting DUSP1 protein stability. The same mechanism occurs in corresponding residues of DUSP4 [35,37]. Far from the catalytic domain, Ser-174 and Ser-197 in the amino terminus of DUSP6 are also susceptible to ERK-mediated phosphorylation that decreases its half-life [38,39]. Besides MAPKs, other upstream kinases gain importance in DUSP regulation, as happens with mTOR-dependent DUSP6 phosphorylation in Ser-159 that also leads to protein degradation [40]. Moreover, casein kinase-2α (CK2α) and protein kinase A (PKA) can differently affect both DUSP catalytic activity and MAPK binding and recognition [41,42]. Besides phosphorylation, other mechanisms of regulation have been described, such as DUSP1 acetylation. DUSP1 is acetylated in response to inflammatory stimuli and the activation of Toll-like receptor signaling. Modification at the Lys-57 residue in the KIM domain by p300 increases its association with p38 and its catalytic activity to stop inflammatory signaling [43]. Figure 1 summarizes the regulatory mechanisms of DUSP activity previously described.

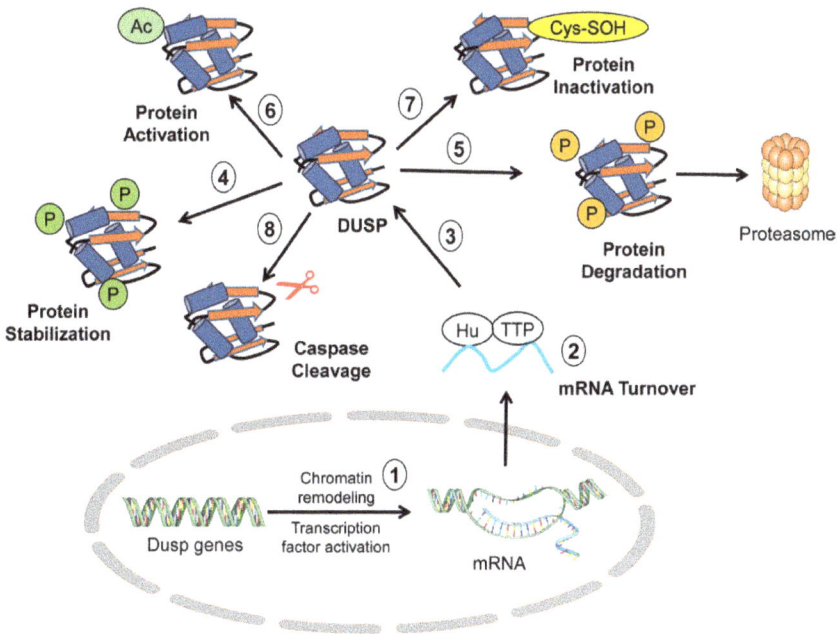

Figure 1. Regulation of dual-specificity protein phosphatases (DUSP) protein levels and activity. DUSPs can be regulated at different levels by their own substrates, mitogen-activated protein (MAP) kinases at (**1**) transcriptional, (**2**) post-transcriptional, and (**3**) post-translational levels by several mechanisms. The latter involve (**4,5**) phosphorylation, (**6**) acetylation, (**7**) cysteine oxidation, and (**8**) caspase–mediated cleavage. These modifications can positively or negatively affect DUSP activities, as well as act as a shuttle to subcellular compartmentalization.

Finally, a novel regulatory mechanism described for DUSP6 connects it with pro-apoptotic signaling. The inter-domain linker region of DUSP6 is located between the N-terminal MAPK-binding domain and the C-terminal catalytic domain and is the target of the apoptosis executioner caspase-3. In response to apoptotic stimuli, active caspase-3 cleaves DUSP6 at the level of the linker region and renders several truncated fragments with different catalytic activities that regulate ERK subcellular location and activity [44]. All these data together do not exclude the possibility that further new residues and mechanisms not yet identified could be relevant to regulating the kinetics and turnover of different DUSP types (Figure 1).

The complex regulatory network points out the importance of DUSPs that do not merely function to switch off intracellular cascades, but emerge as fine-tuners of MAPK signaling. Nevertheless, their role as MAPK regulators is not exclusive for DUSPs. It is necessary to take into account that constitutive Ser/Thr phosphatases also contribute to regulating the intensity of MAPK signaling, acting at multiple levels in upstream kinases. However, the lack of specific genetic tools and specific inhibitors makes it difficult to evaluate their real contribution. Besides, the control exerted by DUSPs at specific subcellular compartments and over specific MAPK substrates makes them better-quality regulators, allowing dynamic adaptations to different conditions and cellular contexts [2]. It is interesting to note that many of these regulatory mechanisms also take place in neural cells, which can have a special meaning in different physiological and pathological contexts.

3. DUSP in Neuronal Differentiation and Nervous System Development

Consistent with the role of MAPKs in proliferation and differentiation processes, MAPK deactivation is crucial to ensuring precise levels of neurite and axonal arborization. One of the most used in vitro models for neuronal differentiation studies has been rat PC12 pheochromocytoma cells. PC12 cells acquire a neuronal phenotype after treatment with NGF. In fact, NGF was the first neurotrophic factor reported to regulate *DUSP* expression in PC12 cells and embryonic sympathetic neurons. NGF induces *DUSP1*, *DUSP4*, and *DUSP6* gene expression during the first hours of incubation. *DUSP1* and *DUSP4* mRNAs appear during the first hour of incubation with neurotrophin, according to the inducible nature of these phosphatases, whereas *DUSP6* that appears later on, after three hours of incubation [45,46]. In previous studies, it has been demonstrated that NGF induces the sustained activation and nuclear translocation of ERK1/2 and cytosolic MAP kinase dephosphorylation and inactivation, observed after only 3 h of NGF treatment. The coincidence of the temporal window suggests that ERK could be the MAPK responsible for *DUSP6* induction. In addition, increases in the expression of DUSP1 and DUSP6 phosphatases are also detected during long-term incubation with NGF (one to four days), agreeing with the long-term actions of MAPKs in the context of NGF-induced PC12 cell differentiation [47]. Watezig and Herdergen proved that ERK1/2 and JNKs, but not p38, are crucially involved in the long-term differentiation of PC12 cells; JNKs alone are responsible for the phosphorylation of c-Jun and the expression of *DUSP1*. Supporting the role of DUSP in neuronal differentiation, *DUSP1* is also induced by NGF in embryonic dorsal root ganglion neurons [48]. In the P19 stem cell line differentiated with retinoic acid, *DUSP6* expression takes place earlier and more strongly, while the *DUSP1* transcript peaks are transient and slightly decrease during the course of differentiation. Only after the levels of DUSP6 and pERK decrease in the cytoplasmic fraction is DUSP1, together with pERK, detected and accumulated in the nucleus [49].

Continuing with neurotrophins, the elegant work of Jeanneteau and co-workers shows that the BDNF neurotrophin regulates the expression of *DUSP1* during the differentiation of mouse cortical neurons. BDNF provides different regulatory mechanisms of DUSP1, both at the level of gene transcription and protein stabilization. In primary cultures of mouse cortical neurons, BDNF acts as a strong inducer of the *DUSP1* mRNA. The transcriptional expression of the *DUSP1* gene is clearly dependent on ERK activation. Although *DUSP1* overexpression reveals broad substrate selectivity towards the three types of MAPKs, JNK appears to be the preferred DUSP1 substrate responsible for the effect of BDNF in axonal outgrowth and branching [50]. Using different conditional expression systems, it was possible to determine that BDNF also contributes to DUSP1 protein stabilization and sustains the increase in DUSP1 levels during BDNF stimulation. Additionally, the DUSP1 protein is lost upon the expression of some mutants of ERK-targeted residues, indicating that BDNF-induced DUSP1 phosphorylation at Ser-359 and Ser-364 might be responsible for protein stabilization and the increase in DUSP1 half-life [35,50]. The sustained DUSP1 activity is necessary to achieve adequate levels in phosphorylation of DUSP1 substrates that regulate cytoskeletal dynamics and result in permanent changes in axonal branching during differentiation. In this model, *DUSP1* primarily acts on JNK-phosphorylated targets that destabilize microtubules and allow cytoskeletal remodeling. The effect of BDNF in promoting axonal branching is dependent on the expression of *DUSP1* and is lost in cortical neurons obtained from *DUSP1*$^{-/-}$ knock-out mice [50]. Similar to cortical neurons, DUSP1 expression is also important in the development of the dopaminergic system during the critical period of striatal axogenesis. When overexpressed, DUSP1 increases the neuronal complexity of TH+ dopaminergic neurons, as noted by the increase in neurite length and neuronal branching. In this case, it is p38 that is responsible for *DUSP1* expression and mediates its effect on differentiation [51].

DUSP6 emerges as a critical regulator of fibroblast growth factor (FGF) downstream actions on cell proliferation and patterning during the development of zebrafish, chick, and mouse embryos [52]. In these systems, DUSP6 forms part of a signaling loop of ERK-activation in the FGF-activated pathway. First, DUSP6 localizes in specific sites or domains of FGF activity in developing embryos [53–55]. Second, FGF signaling is responsible for *DUSP6* induction, mainly dependent on the ERK-MAPK pathway, although the PI3K/Akt axis also seems to contribute to *DUSP6* expression in the mouse neural tube [53,54,56,57]. The impact of DUSP6 as a brake of FGF-mediated signaling is based on aberrant proliferation and alterations in neuronal cytoarchitecture observed in developing of chick limb and mouse neural plate under conditions of *DUSP6* overexpression [53,55]. The specificity of DUSP6 actions on ERK activity in embryos has been confirmed by silencing studies and by the use of a BCI DUSP1/6 specific inhibitor, which prolongs increases in pERK levels [54,58,59].

All these findings indicate that DUSPs also behave as critical players during neural cell development and stem cell differentiation. Particularly, *DUSP1* and *DUSP6* are induced in a spatio-temporal manner to limit the extent of proliferative MAPK signaling, which is necessary to refine neural cell populations, restrict over-proliferation, and ensure precise levels of synaptic connectivity. DUSP6 seems to be more associated with proliferative mediators, such as FGF. By contrast, a *DUSP1* expression peak occurs at later stages of embryo development, when neuronal networks are being refined in the prefrontal cortex, hippocampus, and striatum [50]. *DUSP1* expression correlates with synaptic activity and is required for proper dendritic growth and axonal arborization, as *DUSP1* gene deletion abrogated these processes. Furthermore, the overexpression of this protein phosphatase is associated with abnormal cytoarchitecture of newborn neurons [50,60].

4. DUSP in Neuroprotection; Role of Neurotrophins and Nucleotides

Neuroprotection covers the intracellular mechanisms triggered to recover cell viability in response to brain injury and neurodegenerative events. Common hallmarks of neuronal death include excitotoxicity and oxidative and genotoxic stress. In all these conditions, alterations in MAPK signaling are behind cell death elicited by different kinds of cytotoxic insults and apoptotic stimuli. The general rule points to sustained increases at the level of MAPK activation over time. This is particularly evident for stress kinases, p38 and JNK, but also for ERK proteins. Whether the increase in MAPK activation constitutes a defense mechanism or anticipates cell death is dependent on the kinetics and duration of MAPK activation and their action on cytosolic or nuclear downstream targets. For instance, prolonged MAPK activation over critical cytosolic substrates can induce cytoskeletal rearrangements that are toxic to the cell. On the other hand, the activation of transcriptional nuclear targets can be either a protective mechanism, when increasing the expression of survival genes, or detrimental, through the induction of the pro-apoptotic program. In terms of the balance between survival and apoptosis, fine-tuning of spatio-temporal dynamics of MAPK activation will determine the final outcome.

A great deal of evidence points towards a failure in MAPK deactivation mechanisms as major contributors of prolonged MAPK signaling. Among different protein phosphatases, dual-specificity phosphatases become dysregulated in damaging conditions and brain pathologies. Different mechanisms impair DUSP catalytic activity and can concur to elicit cell death in neural cells, such as increased turnover and down-regulation of DUSP proteins, transcriptional inhibition, and oxidative inactivation.

4.1. DUSP Regulation in Excitotoxicity and Oxidative Stress

Oxidative stress is common to several apoptotic insults, being majorly responsible for cell death in brain injury, neurotoxic conditions, ischemic insults, and neurodegenerative events. A direct link exists between oxidative stress and the ERK signaling alterations [2]. The prolonged ERK signaling during these conditions is mainly attributed to dysregulation of the PP2A Ser/Thr phosphatase. But also, ERK-directed dual-specificity phosphatases contribute to the ERK over-activation, as they become catalytically inactivated by the oxidation of key residues. Several studies revealed that excitotoxicity induced by high extracellular glutamate concentrations causes oxidative damage to DUSP phosphatases and cell death. Primary hippocampal and cortical neurons are particularly vulnerable to glutamate-induced oxidative toxicity [61–64]. Loss of cell survival correlates with sustained ERK activation and its persistent translocation to the nucleus [65]. Although the initial peak of ERK activation functions as a first line defense mechanism, the delayed increase in ERK activity is harmful [63]. The induction of cell death by several phosphatase inhibitors points to the failure of ERK-directed phosphatases in toxicity. Among them, the PP2A serine-threonine phosphatase seems to play a major role, but also the inactivation of tyrosine and dual-specificity phosphatases contributes to maintaining of the ERK activity in later stages of glutamate-induced excitotoxicity [65]. The mechanism underlying phosphatase inactivation might involve the reversible oxidation of key cysteine thiols in the DUSP catalytic domain during oxidant conditions, as DUSP6 catalytic activity recovers by the use of a DTT reducing agent [65]. In addition, degradation of the DUSP1 protein through the protein kinase C (PKC)δ-dependent pathway occurs during glutamate treatment and is responsible for augmented ERK signaling in hippocampal and cortical neurons. The use of proteasome inhibitors or PKCδ silencing restores DUSP1 protein levels and recovers cell survival [62]. Additionally, the overexpression of *DUSP1* and *DUSP6* by different genetic approaches re-establishes basal pERK levels and cell survival after glutamate-induced damage in hippocampal neurons [62,64]. This is in agreement with the process described in non-neural models, such as rat mesangial cells, in which *DUSP1* expression exerts a prominent cytoprotective role in oxidative conditions of exposure to hydrogen peroxide (H_2O_2) [66].

DUSP6 becomes down-regulated in oxidative conditions after H_2O_2 treatment and promotes persistent ERK activation in SH-SY5Y human female neuroblastoma cells. To the same extent as H_2O_2, treatment with a DUSP1/6 inhibitor, BCI, at critical concentrations, also induces cell death. This effect is accompanied by a robust increase in ERK phosphorylation and a concomitant decline of DUSP6 levels. Again, the recovery of *DUSP6* expression by 3α-androstanediol prevents toxicity induced by H_2O_2 challenge and BCI [67].

From the above results, it is clear that DUSP activity is required to maintain neuronal survival and homeostatic control of MAPK signaling (Figure 2). However, it is necessary to take into account that the final effect of increasing DUSP activity depends on the cell type and severity of the stimulus and whether the stress is in the acute or delayed phase. Strategies of *DUSP* overexpression can be detrimental if they prevent the pro-survival effect of ERK signaling, as described in SH-SY5Y human neuroblastoma cells challenged with H_2O_2. In this model, it is *DUSP1* knockdown that prevents apoptosis and exerts a cytoprotective role [68]. *DUSP6* overexpression can turn toxic to oligodendrocyte cultures submitted to glutamate excitotoxicity when abrogating ERK signaling that is required to elicit a full protective response during AMPA receptor activation. As a consequence of *DUSP6* overexpression, AMPA receptor signaling amplifies and calcium overload contributes to oligodendrocyte cell death. The fact that DUSP6 increases in rat optic nerves exposed to excitotoxicity and in rat optic nerves of patients with multiple sclerosis suggests that *DUSP6* overexpression might be a risk factor of vulnerability in early stages of this disease [69].

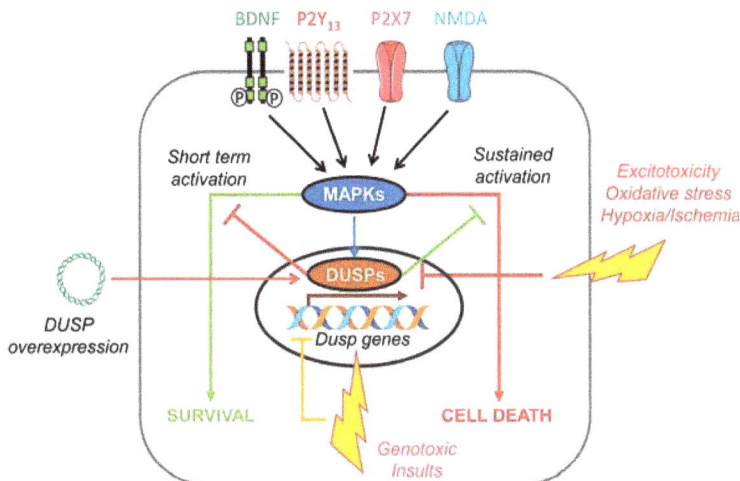

Figure 2. Scheme representing the neuroprotective role of some neurotrophins and nucleotide receptors through the regulation of MAPK and DUSP activity in neural cells. Tyrosine kinase receptors (BDNF, EGF, NGF and FGF), NMDA, and nucleotide ionotropic P2X7 and metabotropic P2Y$_{13}$ receptors regulate DUSP1, DUSP2, and DUSP6 levels through different mechanisms, involving transcriptional induction, turnover regulation, and protein stabilization. The reestablishment of DUSP activity is essential to avoid the over-activation of MAP kinases and cell death induced by different apoptotic insults in neurons and glial cells. Arrows indicate activation of MAPKs, DUSP expression and cell survival/death. T-bars indicate inhibition of MAPK activation and DUSP expression and activity.

4.2. DUSP Regulation in Genotoxic Stress

Altered MAPK signaling also promotes cell death in conditions of DNA damage characterized as genotoxic stress and, again, several mechanisms converge in both nuclear and cytoplasmic compartments to inactivate DUSP proteins. These include direct oxidative phosphatase inactivation, as oxidative damage is also inherent to genotoxic stress, and deficiencies in the *DUSP* transcription machinery. In fact, *DUSP*s are among the genes that become transcriptionally inhibited during exposition to DNA-damaging agents [70].

Exposure to the chemotherapeutic agent cisplatin in primary mice cortical neurons promotes genotoxic stress and is enough to induce long-lasting ERK activation that functions as a defense mechanism. In these conditions, the kinetics of ERK dephosphorylation lower and explain how phosphorylated ERK accumulates in response to the basal activity of NMDA and BDNF receptors. Further analysis demonstrates that a decline in *DUSP6* mRNA might be the consequence of transcriptional inhibition upon cisplatin treatment in cortical neurons. In fact, the use of transcription inhibitor actinomycin D not only diminishes *DUSP6* mRNA, but also prevents efficient ERK dephosphorylation. Among the extracellular mediators able to restore *DUSP6* expression, both NMDA and BDNF increase *DUSP6* transcription and contribute to neuroprotection. The neuroprotective effect of BDNF stimulation during cisplatin treatment depends on the ability to promote ERK pro-survival signaling (Figure 2) [71,72].

DUSP6 mRNA and protein levels also decrease during cisplatin treatment and correlate with ERK over-activation in cerebellar-cultured neurons. Although BDNF and growth factors represent strong signals in this neuronal model, extracellular nucleotides arise as significant regulators of DUSP proteins. The stimulation of several nucleotide receptors results in trophic behavior and the sharing of analogous mechanisms used to modulate signaling cascades with growth factors. In particular, the ERK-directed phosphatase DUSP6 appears to be a signaling target for the nucleotide receptor P2X7 (P2X7R) in primary cultures of rat cerebellar astrocytes and neurons. P2X7R is an ionotropic nucleotide receptor permeable to calcium and for which relevant functions in neural cells have been described [14,73–76]. In cerebellar neurons and astrocytes, ERK signaling activated by P2X7R is involved in the preservation of cell survival and differentiation [77–79]. With this experience, it was not surprising to find that P2X7R regulates DUSP6, one of the main targets of ERK. As expected, with the P2X7R agonist, 2′(3′)-*O*-(4-Benzoylbenzoyl)adenosine-5′-triphosphate (BzATP), is able to induce *DUSP6* gene transcription in an ERK-dependent way, which accounts for its protective effect in conditions of genotoxic stress induced by cisplatin and UV light (unpublished results) (Figure 2).

Interestingly, both astrocytes and granule neurons represent good examples of the fine regulation of DUSP6 activity in neural cells. *DUSP6* expression levels varied along with time of incubation with BzATP, exhibiting a biphasic profile. During short incubation periods, the DUSP6 protein rapidly and significantly decreases below basal levels, and then increases until recovery in neurons or beyond basal levels in astrocytes [80]. The first phase of DUSP6 protein loss is due to protein degradation, because it can be prevented by the use of a proteasome inhibitor. At least in astrocytes, it was demonstrated that ERK-mediated phosphorylation of DUSP6 at the Ser-197 residue could contribute to initial DUSP6 protein decline, because this residue is involved in targeting DUSP6 for proteasome degradation [38]. This biphasic mechanism was first described for DUSP6 by Jurek and co-workers [39]. In porcine aortic endothelial cells ectopically expressing the platelet-derived growth factor (PDGF) receptor, PDGFR activation promoted rapid DUSP6 degradation. Nevertheless, in this study, the residues involved in destabilizing the protein were Ser-300 and Ser-174 [39], different to what was found in astrocytes. Therefore, different residues could be susceptible to ERK-mediated phosphorylation for different stimuli and cell contexts.

The ERK-dependent transcription of the *DUSP6* gene is responsible for the recovery phase of DUSP6 protein levels in cerebellar neurons and astrocytes stimulated with the P2X7 agonist and in aortic cells stimulated with PDGF. The reestablishment of DUSP6 protein levels at later times helps to efficiently terminate ERK signaling [15,39]. The epidermal growth factor (EGF) reproduces the same pattern of biphasic regulation of DUSP6 in both cerebellar astrocytes and neurons, suggesting that this represents a common and universal mechanism shared by different mediators in neural and non-neural cells [15]. The physiological meaning of DUSP6 biphasic regulation allows ERK to regulate its own activity, by participating in positive feedforward and negative feedback mechanisms. Initially, DUSP6 down-regulation amplifies ERK-directed activity towards its cytosolic substrates to enlarge pro-survival signaling. This is followed by DUSP6 protein recovery to finish ERK activation and avoid inappropriate long-term activation of pro-apoptotic signaling (Figure 2). These data, together with the vulnerability of neurons and astrocytes to the BCI-specific DUSP1/6 inhibitor, indicate that the fine regulation and preservation of DUSP6 expression and activity is necessary to maintain cell survival, which can be accomplished by P2X7R stimulation.

As far as P2X7R provides a mechanism for reciprocal adaptation between ERK signaling and DUSP6 activity, examples of cross-regulation of MAPKs by DUSPs also occur with other nucleotide receptors [12,13]. In the same models of cerebellar neurons and astrocytes, the ADP-responding nucleotide metabotropic receptor P2Y$_{13}$ (P2Y$_{13}$R) is involved in the regulation of the nuclear and inducible phosphatase, DUSP2 (PAC-1). *DUSP2* was one of the over-represented genes related to protein phosphatase activity that were identified in microarray expression analysis of P2Y$_{13}$R-stimulated neurons. The transcriptional induction of *DUSP2* by P2Y$_{13}$R stimulation requires both PI3K- and ERK-dependent signaling [81]. The above results suggest that a certain level of specialization takes place in the specific DUSP regulated by different extracellular nucleotides and growth factors concurring in the same cellular type. It appears that DUSP6 is preferentially regulated by P2X7 and EGF receptors, while DUSP2 is the target of P2Y$_{13}$ receptors in cerebellar neurons and glial cells (Figure 2) [15].

As well as DUSP6, genotoxic stress challenge in cerebellar neurons also influences the activity of DUSP2. Both DUSP2 mRNA and protein levels decrease concomitantly with DNA damaging agents, UV light, and cisplatin, while ERK and p38 phosphorylation are increased. Interestingly, the preferred MAPK substrate for DUSP2 in cerebellar neurons seems to be p38, as *DUSP2* expression recovery mediated by P2Y$_{13}$R stimulation prevents the accumulation of the p38 phosphorylated form in the nucleus and increases cell survival [81]. Therefore, it is normalized DUSP2 expression that explains the neuroprotection elicited by P2Y$_{13}$R in granule neurons exposed to cytotoxic cisplatin.

Maintaining DUSP expression and activity is also protective in other neuronal types submitted to apoptotic stimuli. The withdrawal of trophic factors induces *DUSP1* expression in sympathetic neurons [82]. Similarly, in conditions of ER-induced stress, DUSP1 protein stabilization by phosphorylation in Ser-359 maintains survival in cerebellar neurons [83]. Even in non-neural cells exposed to different apoptotic stimuli, *DUSP1* induction is important to counteract apoptotic p38 and JNK signaling in HEK293 cells and in mouse embryo fibroblasts [21,22,84]. All these data together suggest that the induced expression of DUSPs is the mechanism shared by neural and non-neural cells to maintain survival.

4.3. DUSP in Hypoxia and Ischemia

In the neural context, DUSP1 forms part of the endogenous response to brain hypoxic and ischemic injury. Genome-wide gene expression analysis reveals the *DUSP1* gene as one of the survival genes upregulated in the rat hippocampal CA1 region in response to global cerebral ischemia [85]. Additionally, this phosphatase also appears to be involved in ischemic preconditioning in rat retina. *DUSP1* overexpression is essential to prevent the cell apoptotic program activated by JNK and p38 signaling in different cellular layers of the inner retina [86]. *DUSP1* expression also increases the survival of cortical neurons and neuroblastoma cells when submitted to hypoxia-deoxygenation, while *DUSP1* silencing abolishes this protective effect [87]. The neuroprotective role of *DUSP1* overexpression has also been reproduced in vivo in the middle cerebral artery occlusion (MCAo) rodent model of ischemia. Increased levels of the DUSP1 protein are detected in the peri-infarct region after stroke. Through counteracting the deleterious activation of JNK and p38 signaling, *DUSP1* expression prevents the apoptotic program in neurons and limits the inflammatory response in microglia. In agreement, DUSP1 pharmacological inhibition or genetic deletion increases infarct size, exacerbates neurological deficits, and worsens the final outcome, facilitating hemorrhagic transformation [88].

5. DUSP in Pain and Inflammation; Role of Cannabinoids

In certain pain paradigms that present pain hypersensitivity, activated MAPKs behave as transducers of signaling cascades involved in the initiation and maintenance of mechanical allodynia and inflammation. ERK and p38 proteins are mainly altered following peripheral nerve injury and mechanical-induced hypersensitivity in different spinal cell populations, especially in neurons and microglia. Furthermore, adaptations at the level of DUSPs also occur and DUSP proteins levels decrease in different pain conditions. As occurred in apoptotic conditions, the restoration of *DUSP* expression can have therapeutic value in the resolution of pain and inflammation. With regards to this, in a model of neuropathic pain, the overexpression of *DUSP1* phosphatase by genetic approaches in the spinal cord of rats is enough to prevent mechanical allodynia. This anti-nociceptive effect is followed by a decrease in pro-inflammatory mediators and lower levels of phosphorylated p38 in the spinal cord [89]. In agreement, the studies performed with genetic *DUSP6* knock-out mice also confirm that *DUSP6* expression is essential to achieve a spontaneous resolution of mechanical allodynia during post-operative pain [90]. The lack of the DUSP6 protein specifically alters the regulation of ERK in mice paw tissue after a surgical wound incision, indicating that also in the periphery, DUSP6 impairment contributes to maintaining post-operative allodynia [91].

Alternatively, *DUSP* expression recovery can be achieved by several pharmacological interventions. Among different mediators, endogenous cannabinoids behave as strong regulators of DUSP activity. Cannabinoids work as retrograde messengers that integrate the strength of synaptic inputs and activity to regulate synaptic transmission in the nervous system. When released in response to brain injury, they function to maintain survival and attenuate cell damage through the activation of CB1 and CB2 receptors. According to their neuroprotective role, cannabinoids have been revealed as potent inducers of *DUSP* expression, especially in inflammatory conditions.

Studies carried out in microglial cell models demonstrate that the stimulation of cannabinoid receptors, CB1 and CB2, limits inflammatory responses through the modulation of DUSP activity. In microglial cells treated with lipopolysaccharide (LPS), the CB2 agonist JWH015 increases the expression of both DUSP1 and DUSP6 protein phosphatases. This effect terminates ERK signaling and thereby prevents ERK-mediated migration and TNF-α production, two hallmarks of microglial reactivity [92]. Moreover, in BV-2 murine microglial cells submitted to an LPS inflammatory stimulus, *DUSP1* induction is mediated by the CB1/2 receptor activated by anandamide and WIN 55,212-2 agonists. The increase in DUSP1 protein levels switches off ERK signaling to limit the local immune response. Interestingly, *DUSP1* induction by cannabinoids is achieved through a novel mechanism that involves histone H3 phosphorylation, being independent of ERK signaling [93]. The regulation of *DUSP1* gene expression by histone remodeling also takes place following the arrival of different types of stressful stimuli, such as UV light, hydrogen peroxide, and heat shock in mouse embryo fibroblasts [25].

Endocannabinoids can also rescue retinal Müller glia from inflammatory damage induced by LPS. Anandamide and 2-arachidonoylglycerol acting through CB1 and CB2 receptors induce *DUSP1* expression, leading to the subsequent dephosphorylation of MAPKs, prevention of the NF-κB transcription complex, and synthesis of pro-inflammatory cytokines. Through the modulation of MAPK signaling, cannabinoids also increase the expression of tristetraprolin (TTP) in activated Müller cells, which binds and promotes the destabilization of AU-rich pro-inflammatory cytokine mRNAs, contributing to the suppression of cytokine protein levels [94].

More importantly, the effect of cannabinoids in *DUSP* expression is not restricted to cultured isolated cells, but also reproduced in vivo. Studies of peripheral nerve injury show that the intrathecal administration of CB2 agonist JWH015 also increases the expression levels of *DUSP1* and *DUSP6* phosphatases in both neuronal and microglial cells of the spinal cord in a rat nerve transection model of neuropathic pain. It is the restoration of phosphatase activity that re-establishes the physiological levels of pERK and p38 in spinal cord cells and relieves pain [95].

All these data together point out the crucial role of preserving *DUSP* expression in spinal cell populations to prevent alterations in MAPK activity during the establishment of pain. Therefore, DUSP phosphatases also behave as targets for the development of novel analgesic therapies and a new approach could be based on the activation of cannabinoid receptors in the spinal cord (Figure 3).

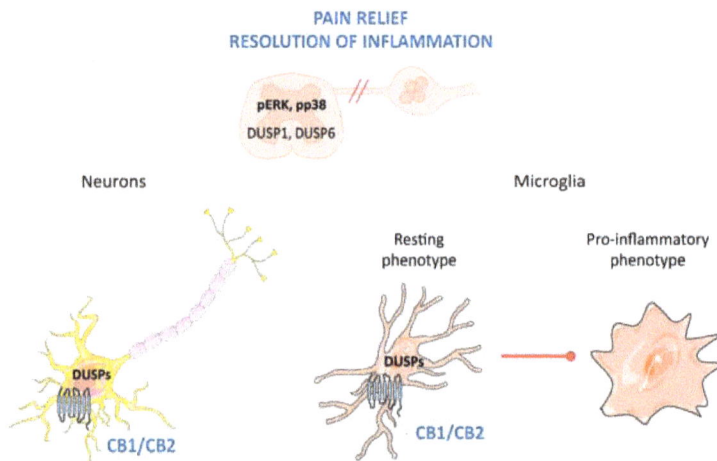

Figure 3. Role of cannabinoid receptors in pain relief and inflammation. CB1 and CB2 receptors increase *DUSP1* and *DUSP6* expression in neurons and microglial cells, which contributes to counteracting the over-activation of ERK and p38 in different pain models. In microglial cells, *DUSP1* expression induction takes place by a novel mechanism involving histone remodeling, which is independent of MAPK activation. DUSP activity prevents the microglia activation and the release of pro-inflammatory mediators.

6. DUSP in Brain Diseases

6.1. DUSP in Neurodegenerative Diseases

Concerning the important implications of DUSPs in neuroprotection, differentiation, and inflammation, it is not surprising to find that they become dysregulated in neurodegenerative diseases. Aberrant MAPK signaling represents a central feature of cell death pathways, such as oxidative stress, endoplasmic reticulum stress, mitochondrial dysfunction, and alterations in cellular proteostasis. During neurodegenerative events, the phosphorylation state of direct and indirect MAPKs substrates becomes altered and contributes to the assembly and stabilization of toxic protein aggregates characteristic of neurodegenerative diseases [96–100]. It is worth noting that among MAPKs, ERK signaling dysregulation can have a major impact on neurodegeneration, because it is essential for neurotransmission, synaptic plasticity, neuronal differentiation, and survival.

A misbalance between ERK activity and *DUSP* expression appears to be clear in Alzheimer's disease (AD). Downregulation of *DUSP6* is observed in post-mortem samples of AD patients, reaching a 50% decrease in frontal cortex lysates [101]. Taking into account that lower levels of *DUSP6* transcripts correlate with the over-expression of miR-125b observed in AD cortices, this study concludes that DUSP6 is a direct target of miR-125b and its dysregulation contributes to facilitating tau-mediated cytotoxicity. Indeed, when miR-125b is overexpressed in hippocampal neurons or directly injected into a mouse hippocampus, it reproduces several hallmarks of AD, such as tau hyperphosphorylation, sustained ERK activation, and DUSP6 decrease [101]. The potential neuroprotective role of *DUSP6* in AD is suggested by the results obtained in the C17.2 neural stem cell line, in which *DUSP6* overexpression protects against amyloid peptide fragment (Ab 31–35) toxicity and restores normal ERK signaling [102].

Concerning the relevant role of DUSP1 in the regulation of synaptic plasticity and neuronal morphology, impaired physiology of DUSP1 is also evident in Alzheimer's disease (AD). DUSP1 levels are also diminished in cortical tissues obtained from AD patients and correlate with tau pathology, cognitive decline, and high blood levels of glucocorticoids; the latter represents a risk factor for AD. In models of glucocorticoid resistance, *DUSP1* downregulation can be explained by the deactivation of BDNF signaling, taking into account that *DUSP1* is a transcriptional target of BDNF and glucocorticoids [103,104]. The delivery of minigene constructs overexpressing *DUSP1* in a mouse cortex and in primary cortical neurons has the ability to prevent tau hyperphosphorylation and other neurochemical deficits caused by deficient glucocorticoid and BDNF-dependent signaling [105].

Similarly, downregulation in *DUSP1* transcripts also occurs in the cortex and striatum of mice models of Huntington's disease (HD) and post-mortem samples of HD patients [106,107]. In addition, the loss of the DUSP1 protein and mRNA in mouse cortical and striatal neurons submitted to polyglutamine-expanded huntingtin toxicity correlates with a lack of dephosphorylating activity towards p38 and JNK. It is worth mentioning that neuroprotection is achieved when *DUSP1* expression recovers, both in cultured neurons and in rat striatum. The protective effect clearly involves ameliorating pro-apoptotic p38 and JNK signaling and preserving pro-survival ERK-dependent activity [107]. In line with this, the histone deacetylase inhibitor, sodium butyrate, is neuroprotective in the R6/2 transgenic model of HD and mediates the upregulation of *DUSP1* expression [108].

DUSP1 and DUSP6 phosphatases also behave as molecular markers of the progression and prognosis of Parkinson's disease (PD). In rat primary dopaminergic neurons challenged with 6-OHDA, a neurotoxin commonly used to model PD, 6-OHDA-induced cytotoxicity is dependent on a DUSP1 decrease and concomitant over-activation of p38 signaling. The protective effect is achieved by the overexpression of *DUSP1*. Moreover, overexpression also restores normal neuronal morphology and promotes neurite outgrowth and a normal degree of neuronal branching and complexity [51,109]. Transcriptome studies also reveal that *DUSP1* and *DUSP6* transcripts are among the most significantly upregulated genes in a mouse hemiparkinsonism model submitted to chronic L-Dopa treatment. *DUSP* gene expression induction can then function as a defense mechanism to overcome neural damage [110,111]. These results suggest that the search for *DUSP1* and *DUSP6* expression inducers can be a good therapeutic alternative in the treatment of neurodegenerative events.

6.2. DUSP in Neurological Disorders

In a different way to the specific loss of *DUSP* expression during apoptotic and neurodegenerative events, the opposite seems to occur in the context of neurological disorders, which seems to correlate with an increase in DUSP proteins. Moreover, the relationship between DUSPs and neurological disorders is supported by a strong genetic basis. The appearance of some *DUSP6* SNPs or genetic variants constitutes risk factors, among others, in some psychiatric diseases, such as bipolar disorder (BD). The *DUSP6* gene is located on chromosome region 12q22-q23 identified as bipolar disorder susceptibility loci. Wide genetic association studies reported a positive association between some missense mutations of *DUSP6* and BD, especially in females [112–114]. These missense mutations progress with a gain of function and weaken intracellular ERK signaling in response to lithium treatment. Therefore, changes in DUSP6 activity can have a strong impact on the therapeutic efficacy of the conventional BD treatment regimen [114]. In accordance with these data, hypoactivity in the ERK signaling cascade is reported in postmortem brains of BD patients [115,116].

With respect to DUSP1, this phosphatase appears to be more related to the pathophysiology of depressive behaviour and DUSP1 levels increase in the hippocampus of depressive patients and stressed rats [117]. The ERK-pathway is the most affected in depression and reflects on deficits in ERK-directed gene expression of key mediators of mood-relating function, such as BDNF and vascular endothelial growth factor (VEGF) [118]. Interestingly, the local overexpression of *DUSP1* in the hippocampus reproduces depressive symptoms in mice. In agreement, lack of the DUSP1 protein protects from depressive behaviour, indicating that inhibition of DUSP1 activity can be a promising therapeutic approach for the treatment of depression. In line with this, selective serotonin reuptake inhibitors (SSRIs) currently used as antidepressants are able to reduce DUSP1 levels [117,119]. The first attempt to assay DUSP1 inhibitors in the treatment of depression focused on Sanguinarine, a natural plant-derived alkaloid with the capacity to selectively inhibit DUSP1 [120]. The intracerebral infusion of Sanguinarine at the ventrolateral orbital cortex relieves the depressive behaviour of rats to a similar extent as that obtained with the systemic administration of the classical antidepressant fluoxetine. The specificity of Sanguinarine at the molecular level is demonstrated because it decreases DUSP1 cortical levels and promotes a concomitant increase in pERK [121].

6.3. DUSPs in Brain Tumors

Most of the dual phosphatases implicated in brain cancers have been identified in glioblastoma (GB), the most common, malignant, and lethal primary intracranial tumor in adults [122]. GB is characterized by its intrinsic aggressiveness and poor prognosis, showing a high heterogeneity at clinical, morphological, molecular, and cellular levels. Currently, GB is treated with surgical resection, radiotherapy, and daily chemotherapy with temozolomide, an oral alkylating agent that triggers tumor cells death by inducing DNA damage, but frequently, the tumor recurs. During the last decade, DUSPs have been proposed as relevant regulators of GB behavior, being involved in either tumor-suppressor or tumor-promoter mechanisms, depending on the phosphatase subtype and the specific context [123,124]. Analysis of a broad range of human GB tumors revealed that *DUSP1* and *DUSP6* exhibit the highest expression level compared to other *DUSPs* [125]. This is noteworthy considering the different but complementary subcellular distribution and MAPK specificity of *DUSP1* and *DUSP6*. Different groups have reported that DUSP1 is upregulated in response to low-oxygen conditions or chemotherapeutic exposure. Thus, hypoxia [125,126] and DNA-damaging agents, such as camptothecin [125] and temozolomide [127], induce *DUSP1* expression in GB cells. Moreover, both dexamethasone and rogisiglitazone reduce cell invasiveness through *DUSP1* induction in human malignant GB cells [128,129]. Recently, Arrizabalaga et al. (2017) studied DUSP1 levels in two sets of independent glioma human samples, revealing that a clinically high expression of *DUSP1* positively correlates with increased GB patient overall survival. This study also showed that the overexpression of DUSP1 in glioma cell lines significantly reduces cellular growth in vitro and in vivo, mainly by suppressing the proliferation and self-renewal ability of the glioma stem cells niche [127]. On the contrary, DUSP6 has been reported as a tumor-promoting factor in human GB [130]. Thus, *DUSP6* overexpression in primary and long-term cultures of human GB enhances tumor growth and increases resistance to cisplatin-mediated cell death in vitro and in vivo [131].

7. Concluding Remarks

The present work covers the data regarding the regulation of DUSP expression and activity by several extracellular messengers of physiological relevance in neural cells. DUSP regulatory mechanisms in cell populations of the nervous system match the modes of DUSP regulation found in peripheral tissues. Additionally, the identification of neurotrophins, cannabinoids, and nucleotides as new players in DUSP regulation place them in a position to intervene in processes of brain pathology.

It is important to remark that increasing *DUSP* expression or restoring DUSP levels is a valid approach to obtain neuroprotection in brain diseases. In agreement, the increased expression of phosphatases, mainly DUSP1 and DUSP6, exhibits a pro-survival effect against the arrival of stressful stimuli that compromise cell survival, such as H_2O_2, UV, cisplatin, hypoxic insults, and during excitotoxicity and ischemic injury. In studies performed in primary cultures of neurons and astrocytes, neuroprotection is achieved by restoring accurate *DUSP* expression levels and activity to counteract the pro-apoptotic activation of p38- and JNK-mediated signaling. These approaches are specifically important in vivo when neurodegeneration has been fully installed in brain tissue and the adaptive mechanisms are overwhelmed. It is not surprising to find striking changes in the expression levels and activity of these phosphatases in neurodegenerative diseases, as occurs with the downregulation of *DUSP1* and *DUSP6* in AD and HD. In these conditions, sustained changes in MAPKs accumulate over time and are responsible for several of the toxic events related to neurodegeneration, such as apoptosis, protein aggregation, and changes in cytoskeletal dynamics. However, it is necessary to take into account that the overexpression of *DUSPs* can also be detrimental if it prevents pro-survival ERK signaling, as demonstrated in some in vitro models, in which it is DUSP inhibition that prevents cell death. In addition, a balance between mitogenic and differentiating signals is required to ensure the proper processes of neuronal differentiation, arborization, and synaptic connectivity during development and DUSP activity needs to be accurately regulated in the precise stage of developing tissue. Other examples of the deleterious effect of increased DUSP activity are found in neurological disorders. DUSP6 and DUSP1 gains of function behave as markers of the progression and prognosis of bipolar disorder and depressive behavior, respectively, and they are responsible for deficient ERK signaling in these pathologies (Figure 4).

Concerning brain tumors, the great variability found in DUSP activity makes it difficult to apply a common therapeutic strategy based on modifying DUSP alterations. In glioblastomas, DUSPs can behave either as tumor suppressors or promoters, depending on the stage and progression of a specific cancer type. Much more needs to be addressed regarding brain tumors to establish a therapeutic approach based on inhibitors or inductors of DUSP. All these data together indicate that the actuation at the level of augmenting or inhibiting DUSP activity will depend on environmental changes and different pathological contexts (Figure 4).

From the perspective that the recovery of DUSP levels and activity is a valid approach to obtain neuroprotection during brain damage, it is important to better understand to what extent DUSP activity is modulated through specific neural mediators. In this role, neurotrophins and cannabinoids represent the most potent inducers of DUSP expression and stabilization in neural cells. The relevance of endogenous cannabinoids in the regulation of DUSP activity is based on their availability to use cannabinoid agonists in vivo for the resolution of pain and inflammation. Activation of BDNF signaling in the brain might ensure proper neuronal cytoarchitecture and cell survival through the regulation of DUSP1. At the same time, restoring BDNF signaling is imperative in depressive disorders. Therefore, different approaches to potentiate BDNF functions would be of great interest in brain pathology.

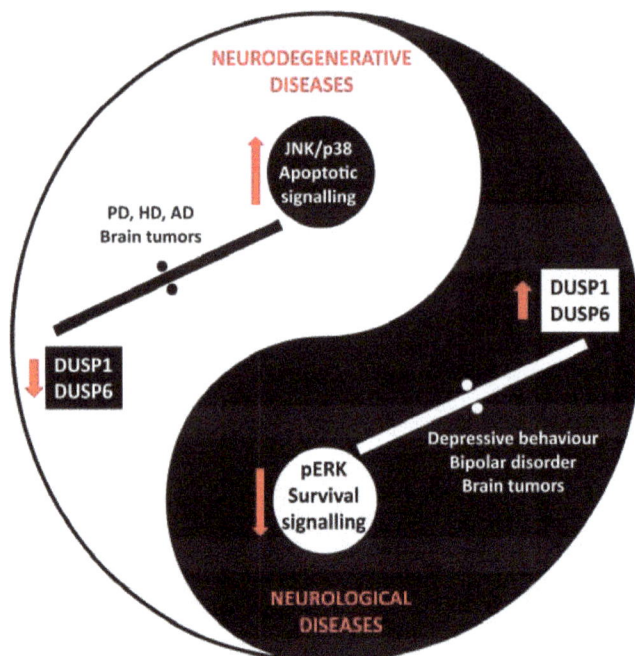

Figure 4. The yin and yang of DUSP activity in brain diseases. This figure summarizes the current available data about DUSP activity in brain diseases, but they only represent the tip of the iceberg. Whereas the increase in *DUSP* expression can be beneficial for neurodegenerative diseases, it is detrimental for neurological disorders. The restoring of the appropriate levels of DUSP activity will be determined by the injury type and the spatiotemporal context.

DUSPs also represent new elements in the nucleotide signaling network. In cerebellar cells, P2X7 and P2Y$_{13}$ nucleotide receptors specifically regulate DUSP6 and DUSP2, respectively. Additionally, nucleotides can integrate DUSP regulation elicited by other mediators. P2X7 and BDNF receptors interact at the level of glycogen synthase kinase 3- (GSK3)-coupled signaling in cerebellar neurons and amplify the survival responses elicited by BNDF [132]. It is tempting to speculate that such interactions can also take place at the level of DUSP regulation and could represent an alternative approach to potentiate and amplify BDNF-mediated signaling. The same could be true for nucleotide and cannabinoid receptors. Overall, DUSPs represent important convergent points of signaling networks that encompass the signaling cascades activated by different mediators in the nervous system.

Author Contributions: Conceptualization, R.P.-S. and E.G.D.; writing—original draft preparation, M.J.Q., J.C.G.-R., R.P.-S., and E.G.D.; writing—review and editing, R.G.-V., F.O., M.T.M.-P., R.P.-S., and E.G.D.

Funding: This work was funded by the Spanish Ministerio de Economia y Competitividad (MINECO, BFU 2014-53654-P) and the "Red de Excelencia Consolider-Ingenio Spanish Ion Channel Initiative" (BFU2015-70067REDC); by the Comunidad de Madrid (BRADE-CM S2013/ICE-2958); and by a Fundación Ramón Areces Grant (PR2018/ 16-02).

Acknowledgments: M.J.Q. is the recipient of an FPU fellowship, J.C.G.-R. is the recipient of an FPI fellowship, and F.O. is the recipient of a Ramón y Cajal contract (RYC-2013-13290).

Conflicts of Interest: The authors declare no conflict of interest. The funders had no role in the design of the study; in the collection, analyses, or interpretation of data; in the writing of the manuscript, or in the decision to publish the results.

Abbreviations

AD	Alzheimer's disease
BD	Bipolar disorder
BDNF	Brain-derived neurotrophic factor
BzATP	2′(3′)-*O*-(4-Benzoylbenzoyl)adenosine 5′-triphosphate
DUSP	Dual-specificity protein phosphatase
EGF	Epidermal growth factor
ERK	Extracellular signal regulated protein kinase
ETS	E twenty-six
GSK3	Glycogen synthase kinase 3
H_2O_2	Hydrogen peroxide
HD	Huntington's disease
JNKs	c-Jun N-terminal kinase
MAPK	Mitogen-activated protein kinase
MEK	Mitogen extracellular activated kinase
NGF	Nerve growth factor
PD	Parkinson's disease
PI3K	Phosphatidylinositol 3-kinase
PKA	Protein kinase A
PKC	Protein kinase C
PTP	Protein tyrosine phosphatase
TTP	Tristetraprolin

References

1. Dickinson, R.J.; Keyse, S.M. Diverse physiological functions for dual-specificity MAP kinase phosphatases. *J. Cell Sci.* **2006**, *119*, 4607–4615. [CrossRef]
2. Junttila, M.R.; Li, S.P.; Westermarck, J. Phosphatase-mediated crosstalk between MAPK signaling pathways in the regulation of cell survival. *FASEB J.* **2008**, *22*, 954–965. [CrossRef]
3. Patterson, K.I.; Brummer, T.; O'Brien, P.M.; Daly, R.J. Dual-specificity phosphatases: Critical regulators with diverse cellular targets. *Biochem. J.* **2009**, *418*, 475–489. [CrossRef]
4. Bermudez, O.; Pages, G.; Gimond, C. The dual-specificity MAP kinase phosphatases: Critical roles in development and cancer. *Am. J. Physiol. Cell Physiol.* **2010**, *299*, C189–C202. [CrossRef]
5. Nunes-Xavier, C.; Roma-Mateo, C.; Rios, P.; Tarrega, C.; Cejudo-Marin, R.; Tabernero, L.; Pulido, R. Dual-specificity MAP kinase phosphatases as targets of cancer treatment. *Anticancer Agents Med. Chem.* **2011**, *11*, 109–132. [CrossRef]
6. Owens, D.M.; Keyse, S.M. Differential regulation of MAP kinase signalling by dual-specificity protein phosphatases. *Oncogene* **2007**, *26*, 3203–3213. [CrossRef]
7. Lang, R.; Hammer, M.; Mages, J. DUSP meet immunology: Dual specificity MAPK phosphatases in control of the inflammatory response. *J. Immunol.* **2006**, *177*, 7497–7504. [CrossRef]
8. Jeffrey, K.L.; Camps, M.; Rommel, C.; Mackay, C.R. Targeting dual-specificity phosphatases: Manipulating MAP kinase signalling and immune responses. *Nat. Rev. Drug Discov.* **2007**, *6*, 391–403. [CrossRef]
9. Wu, Z.; Jiao, P.; Huang, X.; Feng, B.; Feng, Y.; Yang, S.; Hwang, P.; Du, J.; Nie, Y.; Xiao, G.; et al. MAPK phosphatase-3 promotes hepatic gluconeogenesis through dephosphorylation of forkhead box O1 in mice. *J. Clin. Investig.* **2010**, *120*, 3901–3911. [CrossRef]
10. Coddou, C.; Yan, Z.; Obsil, T.; Huidobro-Toro, J.P.; Stojilkovic, S.S. Activation and regulation of purinergic P2X receptor channels. *Pharmacol. Rev.* **2011**, *63*, 641–683. [CrossRef]
11. von Kugelgen, I.; Hoffmann, K. Pharmacology and structure of P2Y receptors. *Neuropharmacology* **2016**, *104*, 50–61. [CrossRef]
12. Perez-Sen, R.; Queipo, M.J.; Morente, V.; Ortega, F.; Delicado, E.G.; Miras-Portugal, M.T. Neuroprotection Mediated by P2Y13 Nucleotide Receptors in Neurons. *Comput. Struct. Biotechnol. J.* **2015**, *13*, 160–168. [CrossRef] [PubMed]

13. Perez-Sen, R.; Gomez-Villafuertes, R.; Ortega, F.; Gualix, J.; Delicado, E.G.; Miras-Portugal, M.T. An Update on P2Y13 Receptor Signalling and Function. *Adv. Exp. Med. Biol.* **2017**, *1051*, 139–168. [CrossRef]

14. Miras-Portugal, M.T.; Gomez-Villafuertes, R.; Gualix, J.; Diaz-Hernandez, J.I.; Artalejo, A.R.; Ortega, F.; Delicado, E.G.; Perez-Sen, R. Nucleotides in neuroregeneration and neuroprotection. *Neuropharmacology* **2016**, *104*, 243–254. [CrossRef]

15. Miras-Portugal, M.T.; Queipo, M.J.; Gil-Redondo, J.C.; Ortega, F.; Gomez-Villafuertes, R.; Gualix, J.; Delicado, E.G.; Perez-Sen, R. P2 receptor interaction and signalling cascades in neuroprotection. *Brain Res. Bull.* **2018**. [CrossRef]

16. Kondoh, K.; Nishida, E. Regulation of MAP kinases by MAP kinase phosphatases. *Biochim. Biophys. Acta* **2007**, *1773*, 1227–1237. [CrossRef]

17. Caunt, C.J.; Keyse, S.M. Dual-specificity MAP kinase phosphatases (MKPs): Shaping the outcome of MAP kinase signalling. *FEBS J.* **2013**, *280*, 489–504. [CrossRef]

18. Huang, C.Y.; Tan, T.H. DUSPs, to MAP kinases and beyond. *Cell Biosci.* **2012**, *2*, 24. [CrossRef]

19. Ekerot, M.; Stavridis, M.P.; Delavaine, L.; Mitchell, M.P.; Staples, C.; Owens, D.M.; Keenan, I.D.; Dickinson, R.J.; Storey, K.G.; Keyse, S.M. Negative-feedback regulation of FGF signalling by DUSP6/MKP-3 is driven by ERK1/2 and mediated by Ets factor binding to a conserved site within the DUSP6/MKP-3 gene promoter. *Biochem. J.* **2008**, *412*, 287–298. [CrossRef]

20. Furukawa, T.; Tanji, E.; Xu, S.; Horii, A. Feedback regulation of DUSP6 transcription responding to MAPK1 via ETS2 in human cells. *Biochem. Biophys. Res. Commun.* **2008**, *377*, 317–320. [CrossRef]

21. Staples, C.J.; Owens, D.M.; Maier, J.V.; Cato, A.C.; Keyse, S.M. Cross-talk between the p38alpha and JNK MAPK pathways mediated by MAP kinase phosphatase-1 determines cellular sensitivity to UV radiation. *J. Biol. Chem.* **2010**, *285*, 25928–25940. [CrossRef]

22. Franklin, C.C.; Srikanth, S.; Kraft, A.S. Conditional expression of mitogen-activated protein kinase phosphatase-1, MKP-1, is cytoprotective against UV-induced apoptosis. *Proc. Natl. Acad. Sci. USA* **1998**, *95*, 3014–3019. [CrossRef] [PubMed]

23. Kassel, O.; Sancono, A.; Kratzschmar, J.; Kreft, B.; Stassen, M.; Cato, A.C. Glucocorticoids inhibit MAP kinase via increased expression and decreased degradation of MKP-1. *EMBO J.* **2001**, *20*, 7108–7116. [CrossRef] [PubMed]

24. Shipp, L.E.; Lee, J.V.; Yu, C.Y.; Pufall, M.; Zhang, P.; Scott, D.K.; Wang, J.C. Transcriptional regulation of human dual specificity protein phosphatase 1 (DUSP1) gene by glucocorticoids. *PLoS ONE* **2010**, *5*, e13754. [CrossRef]

25. Li, J.; Gorospe, M.; Hutter, D.; Barnes, J.; Keyse, S.M.; Liu, Y. Transcriptional induction of MKP-1 in response to stress is associated with histone H3 phosphorylation-acetylation. *Mol. Cell. Biol.* **2001**, *21*, 8213–8224. [CrossRef]

26. Kuwano, Y.; Kim, H.H.; Abdelmohsen, K.; Pullmann, R., Jr.; Martindale, J.L.; Yang, X.; Gorospe, M. MKP-1 mRNA stabilization and translational control by RNA-binding proteins HuR and NF90. *Mol. Cell. Biol.* **2008**, *28*, 4562–4575. [CrossRef]

27. Kuwano, Y.; Gorospe, M. Protecting the stress response, guarding the MKP-1 mRNA. *Cell Cycle* **2008**, *7*, 2640–2642. [CrossRef]

28. Camps, M.; Nichols, A.; Gillieron, C.; Antonsson, B.; Muda, M.; Chabert, C.; Boschert, U.; Arkinstall, S. Catalytic activation of the phosphatase MKP-3 by ERK2 mitogen-activated protein kinase. *Science* **1998**, *280*, 1262–1265. [CrossRef]

29. Chen, P.; Hutter, D.; Yang, X.; Gorospe, M.; Davis, R.J.; Liu, Y. Discordance between the binding affinity of mitogen-activated protein kinase subfamily members for MAP kinase phosphatase-2 and their ability to activate the phosphatase catalytically. *J. Biol. Chem.* **2001**, *276*, 29440–29449. [CrossRef]

30. Zhang, Q.; Muller, M.; Chen, C.H.; Zeng, L.; Farooq, A.; Zhou, M.M. New insights into the catalytic activation of the MAPK phosphatase PAC-1 induced by its substrate MAPK ERK2 binding. *J. Mol. Biol.* **2005**, *354*, 777–788. [CrossRef]

31. Liu, R.M.; Choi, J.; Wu, J.H.; Gaston Pravia, K.A.; Lewis, K.M.; Brand, J.D.; Mochel, N.S.; Krzywanski, D.M.; Lambeth, J.D.; Hagood, J.S.; et al. Oxidative modification of nuclear mitogen-activated protein kinase phosphatase 1 is involved in transforming growth factor beta1-induced expression of plasminogen activator inhibitor 1 in fibroblasts. *J. Biol. Chem.* **2010**, *285*, 16239–16247. [CrossRef]

32. Tephly, L.A.; Carter, A.B. Differential expression and oxidation of MKP-1 modulates TNF-alpha gene expression. *Am. J. Respir. Cell Mol. Biol.* **2007**, *37*, 366–374. [CrossRef]

33. Tonks, N.K. Protein tyrosine phosphatases: From genes, to function, to disease. *Nat. Rev. Mol. Cell. Biol.* **2006**, *7*, 833–846. [CrossRef]

34. Seth, D.; Rudolph, J. Redox regulation of MAP kinase phosphatase 3. *Biochemistry* **2006**, *45*, 8476–8487. [CrossRef]

35. Brondello, J.M.; Pouyssegur, J.; McKenzie, F.R. Reduced MAP kinase phosphatase-1 degradation after p42/p44MAPK-dependent phosphorylation. *Science* **1999**, *286*, 2514–2517. [CrossRef]

36. Lin, Y.W.; Yang, J.L. Cooperation of ERK and SCFSkp2 for MKP-1 destruction provides a positive feedback regulation of proliferating signaling. *J. Biol. Chem.* **2006**, *281*, 915–926. [CrossRef]

37. Cagnol, S.; Rivard, N. Oncogenic KRAS and BRAF activation of the MEK/ERK signaling pathway promotes expression of dual-specificity phosphatase 4 (DUSP4/MKP2) resulting in nuclear ERK1/2 inhibition. *Oncogene* **2013**, *32*, 564–576. [CrossRef]

38. Marchetti, S.; Gimond, C.; Chambard, J.C.; Touboul, T.; Roux, D.; Pouyssegur, J.; Pages, G. Extracellular signal-regulated kinases phosphorylate mitogen-activated protein kinase phosphatase 3/DUSP6 at serines 159 and 197, two sites critical for its proteasomal degradation. *Mol. Cell. Biol.* **2005**, *25*, 854–864. [CrossRef]

39. Jurek, A.; Amagasaki, K.; Gembarska, A.; Heldin, C.H.; Lennartsson, J. Negative and positive regulation of MAPK phosphatase 3 controls platelet-derived growth factor-induced Erk activation. *J. Biol. Chem.* **2009**, *284*, 4626–4634. [CrossRef]

40. Bermudez, O.; Marchetti, S.; Pages, G.; Gimond, C. Post-translational regulation of the ERK phosphatase DUSP6/MKP3 by the mTOR pathway. *Oncogene* **2008**, *27*, 3685–3691. [CrossRef]

41. Castelli, M.; Camps, M.; Gillieron, C.; Leroy, D.; Arkinstall, S.; Rommel, C.; Nichols, A. MAP kinase phosphatase 3 (MKP3) interacts with and is phosphorylated by protein kinase CK2alpha. *J. Biol. Chem.* **2004**, *279*, 44731–44739. [CrossRef] [PubMed]

42. Dickinson, R.J.; Delavaine, L.; Cejudo-Marin, R.; Stewart, G.; Staples, C.J.; Didmon, M.P.; Trinidad, A.G.; Alonso, A.; Pulido, R.; Keyse, S.M. Phosphorylation of the kinase interaction motif in mitogen-activated protein (MAP) kinase phosphatase-4 mediates cross-talk between protein kinase A and MAP kinase signaling pathways. *J. Biol. Chem.* **2011**, *286*, 38018–38026. [CrossRef] [PubMed]

43. Cao, W.; Bao, C.; Padalko, E.; Lowenstein, C.J. Acetylation of mitogen-activated protein kinase phosphatase-1 inhibits Toll-like receptor signaling. *J. Exp. Med.* **2008**, *205*, 1491–1503. [CrossRef] [PubMed]

44. Cejudo-Marin, R.; Tarrega, C.; Nunes-Xavier, C.E.; Pulido, R. Caspase-3 cleavage of DUSP6/MKP3 at the interdomain region generates active MKP3 fragments that regulate ERK1/2 subcellular localization and function. *J. Mol. Biol.* **2012**, *420*, 128–138. [CrossRef]

45. Muda, M.; Boschert, U.; Dickinson, R.; Martinou, J.C.; Martinou, I.; Camps, M.; Schlegel, W.; Arkinstall, S. MKP-3, a novel cytosolic protein-tyrosine phosphatase that exemplifies a new class of mitogen-activated protein kinase phosphatase. *J. Biol. Chem.* **1996**, *271*, 4319–4326. [CrossRef] [PubMed]

46. Misra-Press, A.; Rim, C.S.; Yao, H.; Roberson, M.S.; Stork, P.J. A novel mitogen-activated protein kinase phosphatase. Structure, expression, and regulation. *J. Biol. Chem.* **1995**, *270*, 14587–14596. [CrossRef] [PubMed]

47. Waetzig, V.; Herdegen, T. The concerted signaling of ERK1/2 and JNKs is essential for PC12 cell neuritogenesis and converges at the level of target proteins. *Mol. Cell. Neurosci.* **2003**, *24*, 238–249. [CrossRef]

48. Peinado-Ramon, P.; Wallen, A.; Hallbook, F. MAP kinase phosphatase-1 mRNA is expressed in embryonic sympathetic neurons and is upregulated after NGF stimulation. *Brain Res. Mol. Brain Res.* **1998**, *56*, 256–267. [CrossRef]

49. Reffas, S.; Schlegel, W. Compartment-specific regulation of extracellular signal-regulated kinase (ERK) and c-Jun N-terminal kinase (JNK) mitogen-activated protein kinases (MAPKs) by ERK-dependent and non-ERK-dependent inductions of MAPK phosphatase (MKP)-3 and MKP-1 in differentiating P19 cells. *Biochem. J.* **2000**, *352*, 701–708.

50. Jeanneteau, F.; Deinhardt, K.; Miyoshi, G.; Bennett, A.M.; Chao, M.V. The MAP kinase phosphatase MKP-1 regulates BDNF-induced axon branching. *Nat. Neurosci.* **2010**, *13*, 1373–1379. [CrossRef]

51. Collins, L.M.; O'Keeffe, G.W.; Long-Smith, C.M.; Wyatt, S.L.; Sullivan, A.M.; Toulouse, A.; Nolan, Y.M. Mitogen-activated protein kinase phosphatase (MKP)-1 as a neuroprotective agent: Promotion of the morphological development of midbrain dopaminergic neurons. *Neuromol. Med.* **2013**, *15*, 435–446. [CrossRef]

52. Dickinson, R.J.; Eblaghie, M.C.; Keyse, S.M.; Morriss-Kay, G.M. Expression of the ERK-specific MAP kinase phosphatase PYST1/MKP3 in mouse embryos during morphogenesis and early organogenesis. *Mech. Dev.* **2002**, *113*, 193–196. [CrossRef]

53. Eblaghie, M.C.; Lunn, J.S.; Dickinson, R.J.; Munsterberg, A.E.; Sanz-Ezquerro, J.J.; Farrell, E.R.; Mathers, J.; Keyse, S.M.; Storey, K.; Tickle, C. Negative feedback regulation of FGF signaling levels by Pyst1/MKP3 in chick embryos. *Curr. Biol.* **2003**, *13*, 1009–1018. [CrossRef]

54. Echevarria, D.; Martinez, S.; Marques, S.; Lucas-Teixeira, V.; Belo, J.A. Mkp3 is a negative feedback modulator of Fgf8 signaling in the mammalian isthmic organizer. *Dev. Biol.* **2005**, *277*, 114–128. [CrossRef]

55. Smith, T.G.; Sweetman, D.; Patterson, M.; Keyse, S.M.; Munsterberg, A. Feedback interactions between MKP3 and ERK MAP kinase control scleraxis expression and the specification of rib progenitors in the developing chick somite. *Development* **2005**, *132*, 1305–1314. [CrossRef]

56. Kawakami, Y.; Rodriguez-Leon, J.; Koth, C.M.; Buscher, D.; Itoh, T.; Raya, A.; Ng, J.K.; Esteban, C.R.; Takahashi, S.; Henrique, D.; et al. MKP3 mediates the cellular response to FGF8 signalling in the vertebrate limb. *Nat. Cell. Biol.* **2003**, *5*, 513–519. [CrossRef]

57. Smith, T.G.; Karlsson, M.; Lunn, J.S.; Eblaghie, M.C.; Keenan, I.D.; Farrell, E.R.; Tickle, C.; Storey, K.G.; Keyse, S.M. Negative feedback predominates over cross-regulation to control ERK MAPK activity in response to FGF signalling in embryos. *FEBS Lett.* **2006**, *580*, 4242–4245. [CrossRef]

58. Tsang, M.; Maegawa, S.; Kiang, A.; Habas, R.; Weinberg, E.; Dawid, I.B. A role for MKP3 in axial patterning of the zebrafish embryo. *Development* **2004**, *131*, 2769–2779. [CrossRef]

59. Molina, G.; Vogt, A.; Bakan, A.; Dai, W.; Queiroz de Oliveira, P.; Znosko, W.; Smithgall, T.E.; Bahar, I.; Lazo, J.S.; Day, B.W.; et al. Zebrafish chemical screening reveals an inhibitor of Dusp6 that expands cardiac cell lineages. *Nat. Chem. Biol.* **2009**, *5*, 680–687. [CrossRef]

60. Jeanneteau, F.; Deinhardt, K. Fine-tuning MAPK signaling in the brain: The role of MKP-1. *Commun. Integr. Biol.* **2011**, *4*, 281–283. [CrossRef]

61. Stanciu, M.; Wang, Y.; Kentor, R.; Burke, N.; Watkins, S.; Kress, G.; Reynolds, I.; Klann, E.; Angiolieri, M.R.; Johnson, J.W.; et al. Persistent activation of ERK contributes to glutamate-induced oxidative toxicity in a neuronal cell line and primary cortical neuron cultures. *J. Biol. Chem.* **2000**, *275*, 12200–12206. [CrossRef]

62. Choi, B.H.; Hur, E.M.; Lee, J.H.; Jun, D.J.; Kim, K.T. Protein kinase Cdelta-mediated proteasomal degradation of MAP kinase phosphatase-1 contributes to glutamate-induced neuronal cell death. *J. Cell. Sci.* **2006**, *119*, 1329–1340. [CrossRef]

63. Luo, Y.; DeFranco, D.B. Opposing roles for ERK1/2 in neuronal oxidative toxicity: Distinct mechanisms of ERK1/2 action at early versus late phases of oxidative stress. *J. Biol. Chem.* **2006**, *281*, 16436–16442. [CrossRef]

64. Huang, X.; Liao, W.; Huang, Y.; Jiang, M.; Chen, J.; Wang, M.; Lin, H.; Guan, S.; Liu, J. Neuroprotective effect of dual specificity phosphatase 6 against glutamate-induced cytotoxicity in mouse hippocampal neurons. *Biomed. Pharmacother.* **2017**, *91*, 385–392. [CrossRef]

65. Levinthal, D.J.; Defranco, D.B. Reversible oxidation of ERK-directed protein phosphatases drives oxidative toxicity in neurons. *J. Biol. Chem.* **2005**, *280*, 5875–5883. [CrossRef]

66. Xu, Q.; Konta, T.; Nakayama, K.; Furusu, A.; Moreno-Manzano, V.; Lucio-Cazana, J.; Ishikawa, Y.; Fine, L.G.; Yao, J.; Kitamura, M. Cellular defense against H2O2-induced apoptosis via MAP kinase-MKP-1 pathway. *Free Radic. Biol. Med.* **2004**, *36*, 985–993. [CrossRef]

67. Mendell, A.L.; MacLusky, N.J. The testosterone metabolite 3alpha-androstanediol inhibits oxidative stress-induced ERK phosphorylation and neurotoxicity in SH-SY5Y cells through an MKP3/DUSP6-dependent mechanism. *Neurosci. Lett.* **2019**, *696*, 60–66. [CrossRef]

68. Kim, G.S.; Choi, Y.K.; Song, S.S.; Kim, W.K.; Han, B.H. MKP-1 contributes to oxidative stress-induced apoptosis via inactivation of ERK1/2 in SH-SY5Y cells. *Biochem. Biophys. Res. Commun.* **2005**, *338*, 1732–1738. [CrossRef]

69. Domercq, M.; Alberdi, E.; Sanchez-Gomez, M.V.; Ariz, U.; Perez-Samartin, A.; Matute, C. Dual-specific phosphatase-6 (Dusp6) and ERK mediate AMPA receptor-induced oligodendrocyte death. *J. Biol. Chem.* **2011**, *286*, 11825–11836. [CrossRef]

70. Hetman, M.; Vashishta, A.; Rempala, G. Neurotoxic mechanisms of DNA damage: Focus on transcriptional inhibition. *J. Neurochem.* **2010**, *114*, 1537–1549. [CrossRef]

71. Gozdz, A.; Habas, A.; Jaworski, J.; Zielinska, M.; Albrecht, J.; Chlystun, M.; Jalili, A.; Hetman, M. Role of N-methyl-D-aspartate receptors in the neuroprotective activation of extracellular signal-regulated kinase 1/2 by cisplatin. *J. Biol. Chem.* **2003**, *278*, 43663–43671. [CrossRef]

72. Gozdz, A.; Vashishta, A.; Kalita, K.; Szatmari, E.; Zheng, J.J.; Tamiya, S.; Delamere, N.A.; Hetman, M. Cisplatin-mediated activation of extracellular signal-regulated kinases 1/2 (ERK1/2) by inhibition of ERK1/2 phosphatases. *J. Neurochem.* **2008**, *106*, 2056–2067. [CrossRef]

73. Sperlagh, B.; Illes, P. P2X7 receptor: An emerging target in central nervous system diseases. *Trends Pharmacol. Sci.* **2014**, *35*, 537–547. [CrossRef]

74. Carrasquero, L.M.; Delicado, E.G.; Bustillo, D.; Gutierrez-Martin, Y.; Artalejo, A.R.; Miras-Portugal, M.T. P2X7 and P2Y13 purinergic receptors mediate intracellular calcium responses to BzATP in rat cerebellar astrocytes. *J. Neurochem.* **2009**, *110*, 879–889. [CrossRef]

75. Diaz-Hernandez, M.; del Puerto, A.; Diaz-Hernandez, J.I.; Diez-Zaera, M.; Lucas, J.J.; Garrido, J.J.; Miras-Portugal, M.T. Inhibition of the ATP-gated P2X7 receptor promotes axonal growth and branching in cultured hippocampal neurons. *J. Cell. Sci.* **2008**, *121*, 3717–3728. [CrossRef]

76. Neary, J.T.; Kang, Y. Signaling from P2 nucleotide receptors to protein kinase cascades induced by CNS injury: Implications for reactive gliosis and neurodegeneration. *Mol. Neurobiol.* **2005**, *31*, 95–103. [CrossRef]

77. Ortega, F.; Perez-Sen, R.; Delicado, E.G.; Miras-Portugal, M.T. P2X7 nucleotide receptor is coupled to GSK-3 inhibition and neuroprotection in cerebellar granule neurons. *Neurotox. Res.* **2009**, *15*, 193–204. [CrossRef]

78. Ortega, F.; Perez-Sen, R.; Delicado, E.G.; Teresa Miras-Portugal, M. ERK1/2 activation is involved in the neuroprotective action of P2Y13 and P2X7 receptors against glutamate excitotoxicity in cerebellar granule neurons. *Neuropharmacology* **2011**, *61*, 1210–1221. [CrossRef]

79. Carrasquero, L.M.; Delicado, E.G.; Sanchez-Ruiloba, L.; Iglesias, T.; Miras-Portugal, M.T. Mechanisms of protein kinase D activation in response to P2Y(2) and P2X7 receptors in primary astrocytes. *Glia* **2010**, *58*, 984–995. [CrossRef]

80. Queipo, M.J.; Gil-Redondo, J.C.; Morente, V.; Ortega, F.; Miras-Portugal, M.T.; Delicado, E.G.; Perez-Sen, R. P2X7 Nucleotide and EGF Receptors Exert Dual Modulation of the Dual-Specificity Phosphatase 6 (MKP-3) in Granule Neurons and Astrocytes, Contributing to Negative Feedback on ERK Signaling. *Front. Mol. Neurosci.* **2017**, *10*, 448. [CrossRef]

81. Morente, V.; Perez-Sen, R.; Ortega, F.; Huerta-Cepas, J.; Delicado, E.G.; Miras-Portugal, M.T. Neuroprotection elicited by P2Y13 receptors against genotoxic stress by inducing DUSP2 expression and MAPK signaling recovery. *Biochim. Biophys. Acta* **2014**, *1843*, 1886–1898. [CrossRef]

82. Kristiansen, M.; Hughes, R.; Patel, P.; Jacques, T.S.; Clark, A.R.; Ham, J. Mkp1 is a c-Jun target gene that antagonizes JNK-dependent apoptosis in sympathetic neurons. *J. Neurosci.* **2010**, *30*, 10820–10832. [CrossRef]

83. Li, B.; Yi, P.; Zhang, B.; Xu, C.; Liu, Q.; Pi, Z.; Xu, X.; Chevet, E.; Liu, J. Differences in endoplasmic reticulum stress signalling kinetics determine cell survival outcome through activation of MKP-1. *Cell Signal.* **2011**, *23*, 35–45. [CrossRef] [PubMed]

84. Cadalbert, L.; Sloss, C.M.; Cameron, P.; Plevin, R. Conditional expression of MAP kinase phosphatase-2 protects against genotoxic stress-induced apoptosis by binding and selective dephosphorylation of nuclear activated c-jun N-terminal kinase. *Cell Signal.* **2005**, *17*, 1254–1264. [CrossRef]

85. Kawahara, N.; Wang, Y.; Mukasa, A.; Furuya, K.; Shimizu, T.; Hamakubo, T.; Aburatani, H.; Kodama, T.; Kirino, T. Genome-wide gene expression analysis for induced ischemic tolerance and delayed neuronal death following transient global ischemia in rats. *J. Cereb. Blood Flow Metab.* **2004**, *24*, 212–223. [CrossRef]

86. Dreixler, J.C.; Bratton, A.; Du, E.; Shaikh, A.R.; Savoie, B.; Alexander, M.; Marcet, M.M.; Roth, S. Mitogen-activated protein kinase phosphatase-1 (MKP-1) in retinal ischemic preconditioning. *Exp. Eye Res.* **2011**, *93*, 340–349. [CrossRef] [PubMed]

87. Koga, S.; Kojima, S.; Kishimoto, T.; Kuwabara, S.; Yamaguchi, A. Over-expression of map kinase phosphatase-1 (MKP-1) suppresses neuronal death through regulating JNK signaling in hypoxia/re-oxygenation. *Brain Res.* **2012**, *1436*, 137–146. [CrossRef]

88. Liu, L.; Doran, S.; Xu, Y.; Manwani, B.; Ritzel, R.; Benashski, S.; McCullough, L.; Li, J. Inhibition of mitogen-activated protein kinase phosphatase-1 (MKP-1) increases experimental stroke injury. *Exp. Neurol.* **2014**, *261*, 404–411. [CrossRef]

89. Ndong, C.; Landry, R.P.; DeLeo, J.A.; Romero-Sandoval, E.A. Mitogen activated protein kinase phosphatase-1 prevents the development of tactile sensitivity in a rodent model of neuropathic pain. *Mol. Pain* **2012**, *8*, 34. [CrossRef]

90. Saha, M.; Skopelja, S.; Martinez, E.; Alvarez, D.L.; Liponis, B.S.; Romero-Sandoval, E.A. Spinal mitogen-activated protein kinase phosphatase-3 (MKP-3) is necessary for the normal resolution of mechanical allodynia in a mouse model of acute postoperative pain. *J. Neurosci.* **2013**, *33*, 17182–17187. [CrossRef] [PubMed]

91. Skopelja-Gardner, S.; Saha, M.; Alvarado-Vazquez, P.A.; Liponis, B.S.; Martinez, E.; Romero-Sandoval, E.A. Mitogen-activated protein kinase phosphatase-3 (MKP-3) in the surgical wound is necessary for the resolution of postoperative pain in mice. *J. Pain Res.* **2017**, *10*, 763–774. [CrossRef] [PubMed]

92. Romero-Sandoval, E.A.; Horvath, R.; Landry, R.P.; DeLeo, J.A. Cannabinoid receptor type 2 activation induces a microglial anti-inflammatory phenotype and reduces migration via MKP induction and ERK dephosphorylation. *Mol. Pain* **2009**, *5*, 25. [CrossRef] [PubMed]

93. Eljaschewitsch, E.; Witting, A.; Mawrin, C.; Lee, T.; Schmidt, P.M.; Wolf, S.; Hoertnagl, H.; Raine, C.S.; Schneider-Stock, R.; Nitsch, R.; et al. The endocannabinoid anandamide protects neurons during CNS inflammation by induction of MKP-1 in microglial cells. *Neuron* **2006**, *49*, 67–79. [CrossRef] [PubMed]

94. Krishnan, G.; Chatterjee, N. Endocannabinoids alleviate proinflammatory conditions by modulating innate immune response in muller glia during inflammation. *Glia* **2012**, *60*, 1629–1645. [CrossRef]

95. Landry, R.P.; Martinez, E.; DeLeo, J.A.; Romero-Sandoval, E.A. Spinal cannabinoid receptor type 2 agonist reduces mechanical allodynia and induces mitogen-activated protein kinase phosphatases in a rat model of neuropathic pain. *J. Pain* **2012**, *13*, 836–848. [CrossRef] [PubMed]

96. Bhore, N.; Wang, B.J.; Chen, Y.W.; Liao, Y.F. Critical Roles of Dual-Specificity Phosphatases in Neuronal Proteostasis and Neurological Diseases. *Int. J. Mol. Sci.* **2017**, *18*, 1963. [CrossRef] [PubMed]

97. Colucci-D'Amato, L.; Perrone-Capano, C.; di Porzio, U. Chronic activation of ERK and neurodegenerative diseases. *Bioessays* **2003**, *25*, 1085–1095. [CrossRef] [PubMed]

98. Zhu, X.; Lee, H.G.; Raina, A.K.; Perry, G.; Smith, M.A. The role of mitogen-activated protein kinase pathways in Alzheimer's disease. *Neurosignals* **2002**, *11*, 270–281. [CrossRef] [PubMed]

99. Reijonen, S.; Kukkonen, J.P.; Hyrskyluoto, A.; Kivinen, J.; Kairisalo, M.; Takei, N.; Lindholm, D.; Korhonen, L. Downregulation of NF-kappaB signaling by mutant huntingtin proteins induces oxidative stress and cell death. *Cell. Mol. Life Sci.* **2010**, *67*, 1929–1941. [CrossRef]

100. Gianfriddo, M.; Melani, A.; Turchi, D.; Giovannini, M.G.; Pedata, F. Adenosine and glutamate extracellular concentrations and mitogen-activated protein kinases in the striatum of Huntington transgenic mice. Selective antagonism of adenosine A2A receptors reduces transmitter outflow. *Neurobiol. Dis.* **2004**, *17*, 77–88. [CrossRef]

101. Banzhaf-Strathmann, J.; Benito, E.; May, S.; Arzberger, T.; Tahirovic, S.; Kretzschmar, H.; Fischer, A.; Edbauer, D. MicroRNA-125b induces tau hyperphosphorylation and cognitive deficits in Alzheimer's disease. *EMBO J.* **2014**, *33*, 1667–1680. [CrossRef] [PubMed]

102. Liao, W.; Zheng, Y.; Fang, W.; Liao, S.; Xiong, Y.; Li, Y.; Xiao, S.; Zhang, X.; Liu, J. Dual Specificity Phosphatase 6 Protects Neural Stem Cells from beta-Amyloid-Induced Cytotoxicity through ERK1/2 Inactivation. *Biomolecules* **2018**, *8*, 181. [CrossRef]

103. Jeanneteau, F.; Chao, M.V. Are BDNF and glucocorticoid activities calibrated? *Neuroscience* **2013**, *239*, 173–195. [CrossRef] [PubMed]

104. Menke, A.; Arloth, J.; Putz, B.; Weber, P.; Klengel, T.; Mehta, D.; Gonik, M.; Rex-Haffner, M.; Rubel, J.; Uhr, M.; et al. Dexamethasone stimulated gene expression in peripheral blood is a sensitive marker for glucocorticoid receptor resistance in depressed patients. *Neuropsychopharmacology* **2012**, *37*, 1455–1464. [CrossRef]

105. Arango-Lievano, M.; Peguet, C.; Catteau, M.; Parmentier, M.L.; Wu, S.; Chao, M.V.; Ginsberg, S.D.; Jeanneteau, F. Deletion of Neurotrophin Signaling through the Glucocorticoid Receptor Pathway Causes Tau Neuropathology. *Sci. Rep.* **2016**, *6*, 37231. [CrossRef]

106. Luthi-Carter, R.; Hanson, S.A.; Strand, A.D.; Bergstrom, D.A.; Chun, W.; Peters, N.L.; Woods, A.M.; Chan, E.Y.; Kooperberg, C.; Krainc, D.; et al. Dysregulation of gene expression in the R6/2 model of polyglutamine disease: Parallel changes in muscle and brain. *Hum. Mol. Genet.* **2002**, *11*, 1911–1926. [CrossRef]

107. Taylor, D.M.; Moser, R.; Regulier, E.; Breuillaud, L.; Dixon, M.; Beesen, A.A.; Elliston, L.; Silva Santos Mde, F.; Kim, J.; Jones, L.; et al. MAP kinase phosphatase 1 (MKP-1/DUSP1) is neuroprotective in Huntington's disease via additive effects of JNK and p38 inhibition. *J. Neurosci.* **2013**, *33*, 2313–2325. [CrossRef]

108. Ferrante, R.J.; Kubilus, J.K.; Lee, J.; Ryu, H.; Beesen, A.; Zucker, B.; Smith, K.; Kowall, N.W.; Ratan, R.R.; Luthi-Carter, R.; et al. Histone deacetylase inhibition by sodium butyrate chemotherapy ameliorates the neurodegenerative phenotype in Huntington's disease mice. *J. Neurosci.* **2003**, *23*, 9418–9427. [CrossRef]

109. Collins, L.M.; Downer, E.J.; Toulouse, A.; Nolan, Y.M. Mitogen-Activated Protein Kinase Phosphatase (MKP)-1 in Nervous System Development and Disease. *Mol. Neurobiol.* **2015**, *51*, 1158–1167. [CrossRef]

110. Heiman, M.; Heilbut, A.; Francardo, V.; Kulicke, R.; Fenster, R.J.; Kolaczyk, E.D.; Mesirov, J.P.; Surmeier, D.J.; Cenci, M.A.; Greengard, P. Molecular adaptations of striatal spiny projection neurons during levodopa-induced dyskinesia. *Proc. Natl. Acad. Sci. USA* **2014**, *111*, 4578–4583. [CrossRef]

111. Brehm, N.; Bez, F.; Carlsson, T.; Kern, B.; Gispert, S.; Auburger, G.; Cenci, M.A. A Genetic Mouse Model of Parkinson's Disease Shows Involuntary Movements and Increased Postsynaptic Sensitivity to Apomorphine. *Mol. Neurobiol.* **2015**, *52*, 1152–1164. [CrossRef] [PubMed]

112. Lee, K.Y.; Ahn, Y.M.; Joo, E.J.; Chang, J.S.; Kim, Y.S. The association of DUSP6 gene with schizophrenia and bipolar disorder: Its possible role in the development of bipolar disorder. *Mol. Psychiatry* **2006**, *11*, 425–426. [CrossRef] [PubMed]

113. Toyota, T.; Watanabe, A.; Shibuya, H.; Nankai, M.; Hattori, E.; Yamada, K.; Kurumaji, A.; Karkera, J.D.; Detera-Wadleigh, S.D.; Yoshikawa, T. Association study on the DUSP6 gene, an affective disorder candidate gene on 12q23, performed by using fluorescence resonance energy transfer-based melting curve analysis on the LightCycler. *Mol. Psychiatry* **2000**, *5*, 489–494. [CrossRef]

114. Kim, S.H.; Shin, S.Y.; Lee, K.Y.; Joo, E.J.; Song, J.Y.; Ahn, Y.M.; Lee, Y.H.; Kim, Y.S. The genetic association of DUSP6 with bipolar disorder and its effect on ERK activity. *Prog. Neuropsychopharmacol. Biol. Psychiatry* **2012**, *37*, 41–49. [CrossRef]

115. Kalkman, H.O. Potential opposite roles of the extracellular signal-regulated kinase (ERK) pathway in autism spectrum and bipolar disorders. *Neurosci. Biobehav. Rev.* **2012**, *36*, 2206–2213. [CrossRef] [PubMed]

116. McCarthy, M.J.; Wei, H.; Landgraf, D.; Le Roux, M.J.; Welsh, D.K. Disinhibition of the extracellular-signal-regulated kinase restores the amplification of circadian rhythms by lithium in cells from bipolar disorder patients. *Eur. Neuropsychopharmacol.* **2016**, *26*, 1310–1319. [CrossRef] [PubMed]

117. Duric, V.; Banasr, M.; Licznerski, P.; Schmidt, H.D.; Stockmeier, C.A.; Simen, A.A.; Newton, S.S.; Duman, R.S. A negative regulator of MAP kinase causes depressive behavior. *Nat. Med.* **2010**, *16*, 1328–1332. [CrossRef]

118. Wang, J.Q.; Mao, L. The ERK Pathway: Molecular Mechanisms and Treatment of Depression. *Mol. Neurobiol.* **2019**. [CrossRef]

119. Akbarian, S.; Davis, R.J. Keep the 'phospho' on MAPK, be happy. *Nat. Med.* **2010**, *16*, 1187–1188. [CrossRef] [PubMed]

120. Vogt, A.; Tamewitz, A.; Skoko, J.; Sikorski, R.P.; Giuliano, K.A.; Lazo, J.S. The benzo[c]phenanthridine alkaloid, sanguinarine, is a selective, cell-active inhibitor of mitogen-activated protein kinase phosphatase-1. *J. Biol. Chem.* **2005**, *280*, 19078–19086. [CrossRef]

121. Chen, Y.; Wang, H.; Zhang, R.; Wang, H.; Peng, Z.; Sun, R.; Tan, Q. Microinjection of sanguinarine into the ventrolateral orbital cortex inhibits Mkp-1 and exerts an antidepressant-like effect in rats. *Neurosci. Lett.* **2012**, *506*, 327–331. [CrossRef] [PubMed]

122. Ostrom, Q.T.; Bauchet, L.; Davis, F.G.; Deltour, I.; Fisher, J.L.; Langer, C.E.; Pekmezci, M.; Schwartzbaum, J.A.; Turner, M.C.; Walsh, K.M.; et al. The epidemiology of glioma in adults: A "state of the science" review. *Neuro-Oncology* **2014**, *16*, 896–913. [CrossRef]

123. Prabhakar, S.; Asuthkar, S.; Lee, W.; Chigurupati, S.; Zakharian, E.; Tsung, A.J.; Velpula, K.K. Targeting DUSPs in glioblastomas—wielding a double-edged sword? *Cell. Biol. Int.* **2014**, *38*, 145–153. [CrossRef]

124. Dedobbeleer, M.; Willems, E.; Freeman, S.; Lombard, A.; Goffart, N.; Rogister, B. Phosphatases and solid tumors: Focus on glioblastoma initiation, progression and recurrences. *Biochem. J.* **2017**, *474*, 2903–2924. [CrossRef] [PubMed]

125. Mills, B.N.; Albert, G.P.; Halterman, M.W. Expression Profiling of the MAP Kinase Phosphatase Family Reveals a Role for DUSP1 in the Glioblastoma Stem Cell Niche. *Cancer Microenviron.* **2017**, *10*, 57–68. [CrossRef]

126. Laderoute, K.R.; Mendonca, H.L.; Calaoagan, J.M.; Knapp, A.M.; Giaccia, A.J.; Stork, P.J. Mitogen-activated protein kinase phosphatase-1 (MKP-1) expression is induced by low oxygen conditions found in solid tumor microenvironments. A candidate MKP for the inactivation of hypoxia-inducible stress-activated protein kinase/c-Jun N-terminal protein kinase activity. *J. Biol. Chem.* **1999**, *274*, 12890–12897. [PubMed]

127. Arrizabalaga, O.; Moreno-Cugnon, L.; Auzmendi-Iriarte, J.; Aldaz, P.; Ibanez de Caceres, I.; Garros-Regulez, L.; Moncho-Amor, V.; Torres-Bayona, S.; Pernia, O.; Pintado-Berninches, L.; et al. High expression of MKP1/DUSP1 counteracts glioma stem cell activity and mediates HDAC inhibitor response. *Oncogenesis* **2017**, *6*, 401. [CrossRef] [PubMed]

128. Lin, Y.M.; Jan, H.J.; Lee, C.C.; Tao, H.Y.; Shih, Y.L.; Wei, H.W.; Lee, H.M. Dexamethasone reduced invasiveness of human malignant glioblastoma cells through a MAPK phosphatase-1 (MKP-1) dependent mechanism. *Eur. J. Pharmacol.* **2008**, *593*, 1–9. [CrossRef] [PubMed]

129. Jan, H.J.; Lee, C.C.; Lin, Y.M.; Lai, J.H.; Wei, H.W.; Lee, H.M. Rosiglitazone reduces cell invasiveness by inducing MKP-1 in human U87MG glioma cells. *Cancer Lett.* **2009**, *277*, 141–148. [CrossRef]

130. Ahmad, M.K.; Abdollah, N.A.; Shafie, N.H.; Yusof, N.M.; Razak, S.R.A. Dual-specificity phosphatase 6 (DUSP6): A review of its molecular characteristics and clinical relevance in cancer. *Cancer Biol. Med.* **2018**, *15*, 14–28. [CrossRef] [PubMed]

131. Messina, S.; Frati, L.; Leonetti, C.; Zuchegna, C.; Di Zazzo, E.; Calogero, A.; Porcellini, A. Dual-specificity phosphatase DUSP6 has tumor-promoting properties in human glioblastomas. *Oncogene* **2011**, *30*, 3813–3820. [CrossRef] [PubMed]

132. Ortega, F.; Perez-Sen, R.; Morente, V.; Delicado, E.G.; Miras-Portugal, M.T. P2X7, NMDA and BDNF receptors converge on GSK3 phosphorylation and cooperate to promote survival in cerebellar granule neurons. *Cell. Mol. Life Sci.* **2010**, *67*, 1723–1733. [CrossRef] [PubMed]

MDPI

St. Alban-Anlage 66

4052 Basel

Switzerland

Tel. +41 61 683 77 34

Fax +41 61 302 89 18

www.mdpi.com

International Journal of Molecular Sciences Editorial Office

E-mail: ijms@mdpi.com

www.mdpi.com/journal/ijms

www.ingramcontent.com/pod-product-compliance
Lightning Source LLC
Chambersburg PA
CBHW051730210326
41597CB00032B/5671